STATE OF THE WORLD

2005

Other Norton/Worldwatch Books

STATE OF THE WORLD

2005

A Worldwatch Institute Report on Progress Toward a Sustainable Society

Michael Renner, Hilary French, and Erik Assadourian
Project Directors

Lori Brown
Alexander Carius
Richard Cincotta
Ken Conca
Geoffrey Dabelko
Christopher Flavin
Gary Gardner
Brian Halweil
Annika Kramer
Lisa Mastny
Danielle Nierenberg
Dennis Pirages
Thomas Prugh
Janet Sawin
Aaron Wolf

Linda Starke, *Editor*

W·W·NORTON & COMPANY
NEW YORK   LONDON

The text of this book is composed in Galliard, with the display set in Gill Sans. Book design by Elizabeth Doherty; cover design by Lyle Rosbotham; composition by Worldwatch Institute; manufacturing by Phoenix Color Corp.

First Edition
ISBN 0-393-06020-9
ISBN 0-393-32666-7 (pbk)

W.W. Norton & Company, Inc., 500 Fifth Avenue, New York, N.Y. 10110
www.wwnorton.com

W.W. Norton & Company Ltd., Castle House, 75/76 Wells Street, London W1T 3QT

1 2 3 4 5 6 7 8 9 0

⊕ This book is printed on recycled paper.

Worldwatch Institute Board of Directors

Worldwatch Institute Staff

Erik Assadourian
Staff Researcher

Ed Ayres
Editorial Director
Editor, World Watch

Courtney Berner
Administrative Assistant

Lori A. Brown
Research Librarian

Zoë Chafe
Staff Researcher

Steve Conklin
Webmaster

Barbara Fallin
Director of Finance and
Administration

Susan Finkelpearl
Communications Manager

Christopher Flavin
President

Hilary French
Director, Globalization and
Governance Project

Gary Gardner
Director of Research

Joseph Gravely
Mail & Publications
Fulfillment

Brian Halweil
Senior Researcher

Mairead Hartmann
Development Associate

John Holman
Director of Development

Lisa Mastny
Research Associate

Anne Platt McGinn
Senior Researcher

Leanne Mitchell
Director of
Communications

Danielle Nierenberg
Research Associate

Tom Prugh
Senior Editor

Mary Redfern
Foundations Manager

Michael Renner
Senior Researcher

Lyle Rosbotham
Art Director

Janet Sawin
Research Associate

Molly O'Meara Sheehan
Senior Researcher

Patricia Shyne
Director of Publications
and Marketing

Acknowledgments

Each year a small group of researchers comes together to report on the challenges that face human society and the environment as well as the progress the world has made in responding to them. While these challenges have evolved significantly over the past 22 years, one conclusion has been constant throughout all editions of *State of the World*: we could never write this report without the assistance of countless individuals both inside and outside the Institute. Whatever success we have achieved in this undertaking is in large measure a tribute to the support and insights of a large number of people, many of whose names do not appear on the cover. These friends of Worldwatch all deserve our sincere thanks for their contributions to this year's special report on global security.

For the 2005 edition of *State of the World*, the Institute drew on the talents of a record number of outside authors, including leading experts on human and environmental security. Geoffrey D. Dabelko, Director of the Environmental Change and Security Project at the Woodrow Wilson International Center for Scholars, and Alexander Carius, Director of Adelphi Research in Berlin, Germany, contributed to the chapters on environmental peacemaking and water cooperation. They were joined by Aaron T. Wolf of Oregon State University, Annika Kramer of Adelphi Research, and Ken Conca of the University of Maryland. Dennis Pirages

of the University of Maryland wrote the chapter on the connections between health and security. Richard Cincotta of Population Action International worked with Lisa Mastny on the chapter on population. We are also pleased to include *Security Links* from nonproliferation experts Joseph Cirincione of the Carnegie Endowment for International Peace and Paul Walker of Global Green USA; from Pekka Haavisto of the U.N. Environment Programme's Post-Conflict Assessment Unit; from Rhoda Margesson of the Congressional Research Service; and from Jason Switzer of the International Institute for Sustainable Development.

In addition, chapters include Boxes contributed by Peter Croll of the Bonn International Center for Conversion; Moira Feil of Adelphi Research and Gianluca Rampolla of the Organization for Security and Co-operation in Europe; Chris Huggins and Herman Musahara of the African Centre for Technology Studies; Anders Jägerskog of the Expert Group on Development Issues at the Swedish Ministry for Foreign Affairs; Barbara Rose Johnston from the Center for Political Ecology in Santa Cruz, California; and Manuela Mesa and Mabel González Bustelo from the Peace Research Center in Madrid, Spain.

Chapter authors are grateful too for the enthusiasm and dedication of the 2004 team of staff researchers, fellows, and interns, who pursued elusive facts and produced a number of graphs and tables. Research fellow Renate Duckat, on loan from Worldwatch's German counterpart, Germanwatch, graciously juggled endless requests for Chapters 1, 6, 8, and 9; Molly Norton spent her summer digging up facts for Chapter 3; Molly Aeck and Corinna Kester tenaciously gathered information for Chapter 6; and staff researcher Zoë Chafe, along with her many other responsibilities, tirelessly helped in assembling Chapter 9. Chapter 9 also drew on the thought-provoking analysis of Robinne Gray and the diligent assistance of interns Kyoko Okamoto, Kotoko Ueno, Lauren Kritzer, and Roman Ginzberg. We would also like to thank Kun Qian for helping us with the early development of our China program. Joining Worldwatch throughout the year, all these individuals not only provided indispensable support but kept the Institute energized and in good spirits.

The immense job of tracking down articles, journals, and books from around the world fell to Research Librarian Lori Brown. In addition, Lori once again assembled a list of significant global events for the *Year in Review* timeline, drawing on her remarkable knack for gathering and organizing information.

Reviews from outside experts, who graciously gave us their time, were also indispensable to this year's final product. For their thoughtful comments and suggestions, as well as for the information many people provided, we are particularly indebted to: Daniele Anastasion, Chuck Bassett, Bidisha Biswas, Chris Bright, Amy Brisson, David Brubaker, Grant Cope, John Dimento, Paul Ehrlich, Robert Engelman, José Esquinas Álcazar, Moira Feil, Johanna Mendelson Forman, Cary Fowler, Uwe Fritsche, Benjamin Goldstein, Mary Kaldor, Anja Köhne, Bill Moomaw, Pat Roy Mooney, Patrick Mulvany, Leif Ohlsson, Meaghan Parker, Jim Riccio, Hope Shand, Robert Sprinkle, and Jacob Wanyama.

Further refinement of each chapter took place under the careful eye of independent editor Linda Starke. Linda's energy and long experience with Worldwatch publications ensured that we were able to convert our unpolished first drafts to the sculpted chapters they turned out to be—and within the deadlines she set.

After the edits and rewrites were complete, Art Director Lyle Rosbotham skillfully crafted the design of each chapter, the timeline, and the *Security Links*. His creative vision helped establish several innovations, including the photographs that complement each *Link*. From Dexter, Oregon, Ritch Pope once again assisted in the final production phase by preparing the index.

Writing is only the beginning of getting *State of the World* to readers. The task then passes to our committed communications department, which works on multiple fronts to ensure that the *State of the World* message circulates widely beyond our Washington offices. Communications Manager Susan Finkelpearl leads this effort—using her boundless enthusiasm to craft our messages for the press, public, and decisionmakers around the world. This year she was aided by Administrative Assistant Heather Wilson, who left the Institute in September to start a promising new adventure on Capitol Hill. Thanks also goes to Courtney Berner, who joined Worldwatch just in time to help organize the *State of the World* outreach efforts. And to Editorial Director Ed Ayres, who quietly plots out future issues of our monthly *World Watch* magazine while the rest of us are buried in book preparations.

With the Internet becoming increasingly central to our outreach efforts, we also owe a great deal of thanks to the industriousness of Webmaster Steve Conklin. He has used his technical expertise and creativity to develop a vibrant Web site, including several new innovations, such as the "Global Security" online feature. Our Information Technology Management Team from the company All Covered, under the direction of Raj Maini, ensured that the lines of communication ran smoothly both within and outside the office—even during the unnerving process of making the transition to a new server as our old one disintegrated.

This edition of *State of the World* launches a broader Global Security project, in which the Institute will work with an expanding network of partners to generate a deeper understanding of global security challenges and policy opportunities. A special thanks is due to those we are working with in this effort, including Adelphi Research, the Heinrich Böll Foundation, the Carnegie Endowment for International Peace, FUHEM, Germanwatch, GLOBE International, Green Cross International, the International Institute for Environment and Development, Population Action International, the Royal Institute of International Affairs, the U.N. Environment Programme, the Woodrow Wilson International Center for Scholars, and many others that plan to collaborate with us over the coming years. (For a complete listing, see the "Global Security Network" pages at www.worldwatch.org.)

These new relationships come on top of the many long-standing partnerships that have strengthened Worldwatch over the years. It is only through the assistance of our global publishing network that we are able to release the *State of the World* in 21 languages and 26 countries. These publishers, civil society organizations, and individuals provide invaluable advice as well as translation, outreach, and distribution assistance for our research. We offer our gratitude to them and would particularly like to acknowledge the help we receive from Øystein Dahle, Magnar Norderhaug, and Helen Eie in Norway; Anja Köhne, Brigitte Kunze, Christoph Bals, Klaus Milke, Bernd Rheinberg, Gerhard Fischer, and Günter Thien in Germany; Soki Oda in Japan; Gianfranco Bologna and Anna Bruno Ventre in Italy; Lluis Garcia Petit and Marisa Mercado in Spain; Benoit Lambert in Switzerland; Jung Yu Jin in South Korea; George Cheng in Taiwan; Yesim Erkan in Turkey; Viktor Vovk in Ukraine; Tuomas Seppa in Finland; Marcin Gerwin in Poland; Ioana Vasilescu in Romania; Eduardo Athayde in Brazil; and Jonathan Sinclair Wilson in the United Kingdom.

In the United States, for the past 22 years W.W. Norton & Company has published *State of the World*. We want to express our appreciation to Norton and its staff—especially Amy Cherry, Leo Wiegman, Nancy Palmquist, Lucinda Bartley, and Anna Oler. Through their dedication, *State of the World*, *Vital Signs*, and other Worldwatch books are available in bookstores and on university campuses across the country.

Thanks go to our friends at Sovereign Homestead, especially Mark Hintz, Bonnie Ford, Sherrie Reed, Terry Schwanke, and Ken Fornwalt, who help serve our customers and readers, answer their questions, fill orders, and spread the word about our new publications.

We also offer a special recognition to our new Director of Publications and Marketing, Patricia Skopal Shyne. Patricia's energy and experience has breathed new life into the daunting task of distributing Worldwatch's work as far and wide as possible. And the quiet and persistent efforts of Director of Finance and Administration Barbara Fallin and of Joseph Gravely, responsible for mail and in-house publication fulfillment, allow Worldwatch to function day in and day out. Without them, the gears of the institute would have long ground to a halt.

ACKNOWLEDGMENTS

And of course, without our many supporters none of our work would be possible. Our sincerest appreciation goes to the Institute's individual donors, including the 3,500+ Friends of Worldwatch, who with their enthusiasm have demonstrated their strong commitment to Worldwatch and its efforts to create a vision for a sustainable world. We are particularly indebted to the Worldwatch Council of Sponsors—Adam and Rachel Albright, Tom and Cathy Crain, John and Laurie McBride, and Wren and Tim Wirth— who have consistently shown their confidence and support of our work with especially generous annual contributions.

We give another special thanks to the generous support of the foundation community. Support has been provided by the Aria Foundation, the Blue Moon Fund, the Richard & Rhoda Goldman Fund, The William and Flora Hewlett Foundation, The Frances Lear Foundation, the Steven C. Leuthold Family Foundation, the Merck Family Fund, The Overbrook Foundation, the V. Kann Rasmussen Foundation, the Rockefeller Brothers Fund, the A. Frank and Dorothy B. Rothschild Fund, The Shared Earth Foundation, The Shenandoah Foundation, The Summit Fund of Washington, the Turner Foundation, Inc., the U.N. Population Fund, the Wallace Genetic Foundation, Inc., the Wallace Global Fund, the Johanette Wallerstein Institute, and The Winslow Foundation. Additional thanks goes to the assistance provided by government agencies, including the Norwegian Royal Ministry of Foreign Affairs and the German Society for Technical Co-operation.

Engaging these many donors falls on the backs of our dedicated development staff, John Holman, Mary Redfern, and Mairead Hartmann. Their behind-the-scenes efforts keep the lights on at the Institute (all compact fluorescents, of course).

Finally, we are particularly grateful for the hard work and loyal support of the members of the Institute's Board of Directors, who have provided key input on strategic planning, organizational development, and fundraising over the last year.

It is the support of the individuals mentioned as well as many more who remain unnamed that has allowed Worldwatch to devote itself for 30 years to creating a vision for a sustainable world. Their support gives us great hope that humankind will one day come together to lay the foundations for a more secure, peaceful, and sustainable world.

Michael Renner
Hilary French
Erik Assadourian
Project Directors

Contents

List of Boxes, Tables, and Figures

Boxes

Figures

Foreword

Five years ago, all 191 United Nations member states pledged to meet eight Millennium Development Goals by 2015, including eradicating extreme poverty and hunger and ensuring environmental sustainability. These critical challenges were reaffirmed by health officials from across the globe in October 2004 at the tenth anniversary of the landmark International Conference on Population and Development held in Cairo.

The overarching conclusion from this 2004 meeting was that while considerable, albeit erratic, progress was indeed being made in many areas, any optimism must be tempered with the realization that gains in overall global socioeconomic development, security, and sustainability do not reflect the reality on the ground in many parts of the world. Poverty continues to undermine progress in many areas. Diseases such as HIV/AIDS are on the rise, creating public health time bombs in numerous countries. In the last five years, some 20 million children have died of preventable waterborne diseases, and hundreds of millions of people continue to live with the daily misery and squalor associated with the lack of clean drinking water and adequate sanitation.

We must recognize these shameful global disparities and begin to address them seriously. I am delighted that the 2004 Nobel Peace Prize was awarded to Wangari Maathai, a woman whose personal efforts, leadership, and practical community work in Kenya and Africa inspire us all by demonstrating the real progress that can be made in addressing environmental security and sustainable development challenges where people have the courage to make a difference.

Humankind has a unique opportunity to make the twenty-first century one of peace and security. Yet the many possibilities opened up to us by the end of the cold war appear to have been partially squandered already. Where has the "peace dividend" gone that we worked so hard for? Why have regional conflict and terrorism become so dominant in today's world? And why have we not made more progress on the Millennium Development Goals?

The terrible tragedies of September 11, 2001, the 2004 terrorist attacks in Beslan in Russia, and the many other terrorist incidents over the past decade in Japan, Indonesia, the Middle East, Europe, and elsewhere have all driven home the fact that we are not adequately prepared to deal with new threats. But better preparation means thinking more holistically, not just in traditional cold war terms.

I believe that today the world faces three interrelated challenges: the challenge of security, including the risks associated with weapons of mass destruction and terrorism; the challenge of poverty and underdevelopment; and the challenge of environmental sustainability.

The challenge of security must be addressed by first securing and destroying the world's arsenals of weapons of mass destruction. Both Russia and the United States have taken numerous positive steps in this direction. But we must accelerate these nonproliferation and demilitarization efforts and establish threat-reduction programs around the world if we are to be truly successful.

The world's industrial nations must also commit greater resources to the poorest countries and regions of the globe. Official development assistance from the top industrial countries still represents but a tiny percentage of their gross national products and does not come close to the pledges made over a decade ago at the Rio Earth Summit. The growing disparity between the rich and the poor on our planet and the gross misallocation of limited resources to consumerism and war cannot be allowed to continue. If they do, we can expect even greater challenges and threats ahead.

Regarding the environment, we need to recognize that Earth's resources are finite. To waste our limited resources is to lose them in the foreseeable future, with potentially dire consequences for all regions and the world. Forests, for example, are increasingly being destroyed in the poorest countries. Even in Kenya, where Wangari Maathai has helped plant over 30 million trees, forested acreage has decreased. The global water crisis is also one of the single biggest threats facing humankind. Four out of 10 people in the world live in river basins shared by two or more countries, and the lack of cooperation between those sharing these precious water resources is reducing living standards, causing devastating environmental problems, and even contributing to violent conflict. Most important of all, we must wake up to the dangers of climate change and devote more resources to the crucial search for energy alternatives.

It is for reasons such as these that I founded Green Cross International 12 years ago and continue to advocate for a global value shift on how we handle Earth, a new sense of global interdependence, and a shared responsibility in humanity's relationship with nature. It is also for these reasons that I helped draft the Earth Charter, a code of ethical principles now endorsed by over 8,000 organizations representing more than 100 million people around the world. And it is for these reasons that Maurice Strong, Chair of the Earth Council, and I have initiated the Earth Dialogues, a series of public forums on ethics and sustainable development.

We need a Global Glasnost—openness, transparency, and public dialogue—on the part of nations, governments, and citizens today to build consensus around these challenges. And we need a policy of "preventive engagement": international and individual solidarity and action to meet the challenges of poverty, disease, environmental degradation, and conflict in a sustainable and nonviolent way.

We are the guests, not the masters, of nature and must develop a new paradigm for development and conflict resolution, based on the costs and benefits to all peoples and bound by the limits of nature herself rather than by the limits of technology and consumerism. I am delighted that the World-watch Institute continues to address these important challenges and goals in its annual *State of the World* report. I urge all readers to seriously consider their personal commitments to action after finishing this volume. Only with the active and dedicated participation of civil society will we be successful in building a sustainable, just, and peaceful world for the twenty-first century and beyond.

Mikhail S. Gorbachev
Chairman, Green Cross International

Preface

When the Nobel Peace Prize was awarded to Kenyan environmental activist Wangari Maathai in October 2004, the Nobel Committee's decision was met with dismay in some circles. To many traditional security specialists, it seemed frivolous at a time of military conflict, civil wars, terrorism, and proliferating nuclear materials to give this most prestigious of international awards to a person known for planting trees rather than signing treaties. Indeed, a leading politician in Norway, which sponsors the prize, commented, "It is odd that the committee has completely overlooked the unrest that the world is living with daily, and given the prize to an environmental activist." [1]

In our view, the award could not have been more fitting. The life history of Wangari Maathai is testimony to the fact that the insecurity the world struggles with today is inextricably linked to the ecological and social problems she has devoted her life to addressing. In 1977, she founded the Green Belt Movement to organize poor women to plant millions of trees—the group's goals included replenishing Kenya's dwindling forests, providing desperately needed cooking fuel, and making women active participants in improving their lives and those of their families.

Maathai's success and her subsequent challenge to government conservation policies put her in direct conflict with the country's autocratic president. She and her followers were beaten and jailed—but in the process they spurred thousands of followers to action in Kenya and around the globe. The civil society movement that Wangari Maathai leads helped pave the way for Kenya's peaceful transition from virtual dictatorship to elected government in 2003. Capping the historic transition, she is now a member of the Kenyan parliament and assistant environment minister in the current government.

By coincidence, the Nobel Peace Price was announced just as we were putting the finishing touches on *State of the World 2005*—the twenty-second edition of our annual book and the first to focus on global security, the topic that has so dominated private and political discourse over the past few years. As longtime admirers of the Green Belt Movement, my colleagues and I were heartened by the news of Wangari Maathai's award, and inspired by the hope that this latest Nobel Prize will help convince millions of people around the world to stop viewing global security as something that can be safeguarded solely through diplomatic skills or military power.

Our focus in the pages that follow is on the deeper roots of insecurity—many of them found in the destabilization of human societies and the natural world that has accompanied the explosive growth in human numbers and resource demands over the past several decades. Drawing on the varied expertise and insights of our own staff, as well as on a record number of collaborators from around the world, we have sought to unravel the often hidden links between such disparate phenomena as falling water tables, the spread of AIDS, transnational crime, environmental refugees, terrorism, and climate change. In doing so, we have found ample reason to fear that the profound insecurity that has gripped the world for the past three years may grow even deeper in the years ahead.

Demographic imbalances are one destabilizing force. As Lisa Mastny and Richard Cincotta describe in the second chapter, in roughly one third of the world's countries—most of them in Africa, the Middle East, and South and Central Asia—a large generation of teenagers is faced with limited economic prospects and often little in the way of education. Most of the world's civil wars, emigration, and terrorism emerge from those countries—exacerbated in many cases by ethnic and religious differences and by the breakdown of the social and ecological systems people depend on.

In many of these same countries, the spread of infectious diseases, particularly AIDS, is also tearing societies apart, killing many of the young people who are best equipped to lead their nations forward economically and politically. Growing human pressures on natural resources—triggering the collapse of fisheries and the drying up of rivers, for example—are further undermining some societies. The latest humanitarian crisis to hit the world's headlines in 2004 was in Darfur in Sudan, where the immediate clashes between Arab nomads and African villagers was preceded by years of desertification that led herders to encroach on farmland to their south, heightening tensions and eventually leading to open conflict, forced eviction of villagers, and genocide.

Access to oil is another cause of instability that has commanded recent attention.

The dramatic run-up in prices to over $50 per barrel in the fall of 2004 coincided with growing instability in the Persian Gulf, where the world's richest oil resources are located. The dominance of the oil industry in the Middle East has undermined the economic and political development of the region while flooding it with petrodollars that have increased economic disparities and financed the rise of terrorism. The dependence of the United States and Europe on Middle Eastern oil has led to highly skewed economic flows and heavy military investments that have created deep resentments on both sides. The prospect of world oil production beginning a long decline within the next decade, just when large countries like China and India stake their claims to remaining reserves, would be reason enough for concern even without the crisis caused by the U.S. invasion of Iraq. Together, they have created a global powder keg.

The possibility of disruptive climate change may be an even greater threat to the security of societies. Amid new signs of accelerated global warming—from the rapid melting of Arctic ice to the spread of diseases and pests into new territories—scientists are focusing on the potential for the sudden collapse of economically essential ecosystems such as forests, underground water resources, and coastal wetlands. The unprecedented four hurricanes that devastated Florida in 2004, combined with the record number of typhoons that hit Japan, left weather forecasters studying the possibility that catastrophic weather events could soon become the norm—with immense human consequences, particularly in the world's poorest countries. An October 2004 report by a coalition of aid and environmental agencies warned that climate change is likely to worsen poverty. By flooding valuable coastal areas and undermining forests and watersheds, a changing climate will exacerbate competition for resources.[2]

One tragic consequence of the September 11th terrorist attacks is that they substantially reduced world attention to many of the underlying causes of insecurity. Aid to the world's poorest countries has barely risen, and international commitments to combat problems such as AIDS and global warming are seriously underfunded. Moreover, with even traditional allies such as the United States and several European nations at loggerheads on many issues, we may be not only losing the struggle against terrorism in a narrow sense but setting in motion a range of additional instabilities that could lead the world into a dangerous downward spiral.

We devote this book to reversing that spiral and to building the international cooperation that is essential for achieving a secure world. Just as Wangari Maathai planted trees to improve the economic security of her people, it is time now to plant hope by working together to reach essential goals: a less oil-dependent energy system, a more equal society in which women's roles are strengthened, and a natural world that is stable and productive. Our authors demonstrate the need for a robust security policy—one that links traditional strategies such as disarmament, peace-keeping, and conflict prevention with underlying efforts to meet health and education needs and to restore ecosystems.

It is fitting that the Foreword of *State of the World 2005* is by another Nobel Peace Prize winner: former Soviet president Mikhail Gorbachev, who is now Chairman of Green Cross International. Gorbachev, who played a starring role in the conclusion of the late twentieth century's biggest security challenge, the cold war, has devoted much of his energy over the last decade to one of the great challenges of the twenty-first century—creating an environmentally sustainable world.

Wangari Maathai and Mikhail Gorbachev represent living bridges between the environment and security. Our futures will be shaped in large measure by how quickly the world follows their lead.

Christopher Flavin

President
Worldwatch Institute

1776 Massachusetts Ave., N.W.
Washington, DC 20036
worldwatch@worldwatch.org
www.worldwatch.org

November 2004

State of the World: A Year in Review

Compiled by Lori Brown

This timeline covers significant announcements and reports from October 2003 through September 2004. It is a mix of progress, setbacks, and missed steps around the world that are affecting society's environmental and social goals.

There is no attempt to be comprehensive.

But we hope to highlight global and local events that increase your awareness of the connections between people and the world's environment. An online version of the timeline with links to Internet resources is available at www.worldwatch.org/features/timeline.

STATE OF THE WORLD: A YEAR IN REVIEW

WOMEN
Shirin Ebadi, a human rights activist and one of Iran's first female judges, becomes the first Muslim woman to receive the Nobel Peace Prize.

OZONE
Report reveals elaborate CFC smuggling operations span three continents and threaten the Montreal Protocol.

MINING
Illegal mining of coltan, a mineral used in cell phones, is linked to deaths of hundreds of rare lowland gorillas in Democratic Republic of the Congo.

TRANSPORATION
A study of 75 metropolitan areas shows that traffic delays cost US motorists about $8 billion a year in wasted fuel and 3.5 billion hours in lost time.

POLLUTION
30,000 Ecuadorian indigenous peoples bring a class action suit demanding that ChevronTexaco clean up its pollution and provide compensation.

CLIMATE
Scientists report that atmospheric concentrations of methane, the second most potent greenhouse gas, have leveled off after two centuries of growth.

HEALTH
First known case of "mad cow disease" in the US is discovered in a Washington state cow imported from Canada.

OCTOBER   NOVEMBER   DECEMBER

2003 STATE OF THE WORLD: A YEAR IN REVIEW

2 4 6 8 10 12 14 16 18 20 22 24 26 28 30   2 4 6 8 10 12 14 16 18 20 22 24 26 28 30   2 4 6 8 10 12 14 16 18 20 22 24 26 28

CHILDREN
UN report details plight of children in war zones where thousands of children currently serve in more than 50 armed groups in 15 countries.

WILDLIFE
Officials report a surge in demand for skins of tigers, leopards, and other endangered wildlife as the fashion industry once again embraces fur.

URBANIZATION
UN-Habitat reports that 1 billion people worldwide—32 percent of the people in cities—live in slums and that the number may double in 30 years.

COMMUNICATIONS
Survey of global digital access indicates that 10 countries, including South Korea, have better access to information and communication technology than US does.

CLIMATE
World Meteorological Organization announces that 2003 was the third hottest year in nearly 150 years and part of a global warming trend.

ENERGY
Royal Dutch/Shell Group causes widespread shock to energy markets by downgrading the size of its proven oil and gas reserves by 20 percent.

GOVERNANCE
European Commission launches a Web site giving the public access to information on pollution by 10,000 of Europe's largest industrial facilities.

HUMAN RIGHTS
UN reports that China risks losing 40–60 million girls to abortion and murder in the next decade.

BIODIVERSITY
Study notes that if global temperature rises 2–6 degrees as now predicted, 18–35 percent of the world's species could be gone by 2050.

FISHERIES
Researchers report that farmed salmon contain far more toxic chemicals than wild salmon do and they recommend limiting consumption.

HEALTH
US agency reports that obesity is nearing smoking as leading cause of US deaths, with more than 30 percent of adult Americans now obese.

MARINE SYSTEMS
UN says the number of oceans and bays with "dead zones" of water, so devoid of oxygen that little life survives, has doubled to 146 since 1990.

JANUARY | **FEBRUARY** | **MARCH**

2004

4 6 8 10 12 14 16 18 20 22 24 26 28 30 | 2 4 6 8 10 12 14 16 18 20 22 24 26 28 | 2 4 6 8 10 12 14 16 18 20 22 24 26 28 30

POLLUTION
Report says principal water source for Buenos Aires is Argentina's most polluted river—a toxic soup of dioxins, heavy metals, pesticides, and sewage.

HEALTH
WHO announces that polio has reappeared in seven African countries after the disease spreads from Nigeria, where immunization campaigns were suspended.

NATURAL DISASTERS
Munich Re reports that five major natural catastrophes in 2003 killed 75,000 people, seven times as many as in 2002, and that economic losses rose 18 percent.

DESERTIFICATION
Erosion and desertification in northwestern China contribute to region's worst dust storms in many years, reducing visibility to 10 meters in some areas.

HEALTH
Emergence of a lethal bird flu in East Asia leads to mass slaughter of more than 100 million birds; by September, 31 people die from this avian flu.

BIODIVERSITY
Scientists report that oceanic white tip sharks have declined by 99 percent in the Gulf of Mexico due to over-fishing and demand for sharkfin soup.

CLIMATE
Concentration of carbon dioxide, the main global warming gas in Earth's atmosphere, posts largest two-year increase ever recorded.

ENERGY
US agency projects that world energy demand will grow 54 percent by 2025, with oil use rising from 81 million to 121 million barrels a day.

HEALTH
India, second only to China in tobacco use, bans smoking in public places, all tobacco advertising, and the sale of tobacco to minors.

ENERGY
More than 150 countries attend *Renewables 2004*, the largest-ever meeting of government and private-sector leaders focused on achievable renewable energy goals.

WILDLIFE
Study shows that Africa's black rhino population has increased to more than 3,600 due to improved law enforcement and expansion of protected habitat.

TOXICS
Report warns that toxic metals from discarded cell phones threaten both groundwater and the health of recyclers in Pakistan, India, China, and elsewhere.

NATURAL DISASTERS
Swarms of locusts, up to 10,000 hoppers per square meter, begin to spread across West Africa, causing significant damage to crops and pastures.

WATER
UN study finds that up to 90 percent of the water in two rivers destined for the Aral Sea is diverted for hydropower, irrigation, and other uses.

APRIL MAY JUNE

2004 STATE OF THE WORLD: A YEAR IN REVIEW

2 4 6 8 10 12 14 16 18 20 22 24 26 28 30 2 4 6 8 10 12 14 16 18 20 22 24 26 28 30 2 4 6 8 10 12 14 16 18 20 22 24 26 28

FORESTS
Researchers report that rising international demand for Brazilian beef is encouraging high rates of Amazon deforestation.

CLIMATE
British scientists warn that Chile's San Rafael glacier is melting at an alarming rate due to climate change.

BIOINVASION
Australian researchers report $720 million a year in damage from feral animals, with foxes and cats beating rabbits as the most expensive pests.

CLIMATE
New, more accurate supercomputer modeling system reveals that global temperatures could rise more rapidly than previously projected.

WATER
Report finds that World Bank is boosting its funding of large dam projects to the detriment of the environment and local peoples.

TOXICS
The Stockholm Convention on Persistent Organic Pollutants enters into force to rid world of 12 hazardous chemicals, including PCBs, dioxins, and DDT.

HEALTH
WHO report details severe economic burdens arising from interpersonal violence, which some countries spend at least 4 percent of GDP combating.

BIODIVERSITY
Treaty on plant genetic resources ensures continued availability of crops and their genes, so that countries will be able to feed their people.

MARINE SYSTEMS
Australia's Great Barrier Reef becomes the largest protected marine network as fishing and shipping are banned on one third of the reef.

WILDLIFE
Thousands of tern chicks die in the first documented wildlife fatalities during construction of a shipping canal from the Danube delta to the Black Sea.

POLLUTION
New evidence shows that Arctic polar bears are harmed by industrial chemicals swept in by winds and currents from the southern hemisphere.

ENERGY
China orders emergency shipments of coal by road and waterways to ease what some call the country's worst energy shortage in two decades.

NATURAL DISASTERS
Death toll across South Asia from six weeks of monsoon storms reaches 1,823, with nearly two thirds of Bangladesh submerged.

MARINE SYSTEMS
Report says that two thirds of Caribbean coral reefs are threatened by human activities, including overfishing and pollution runoff from agriculture.

JULY · AUGUST · SEPTEMBER

See page 181 for sources.

4 6 8 10 12 14 16 18 20 22 24 26 28 30 · 2 4 6 8 10 12 14 16 18 20 22 24 26 28 30 · 2 4 6 8 10 12 14 16 18 20 22 24 26 28

WASTE
Japan requires manufacturers to charge drivers for recycling of vehicles, aiming to raise the recycling rate to 95 percent by 2015.

POLLUTION
Radioactive plutonium particles from US nuclear weapons tests in the Pacific 50 years ago are detected for the first time in Japanese waters.

URBANIZATION
UN reports the world will soon become predominantly urban, with 60 percent of people living in cities by 2030.

CLIMATE
Russia's Cabinet approves the Kyoto Protocol on climate change, virtually ensuring it will soon come into effect worldwide.

MARINE SYSTEMS
After major confrontations with Galapagos fishers over the right to fish, Ecuadorian court upholds park authority's right to limit harvests of sea cucumbers.

ENERGY
Western oil companies agree to spend $50 million drilling oil in Peru's northern jungle, despite fierce resistance from indigenous groups and environmentalists.

NATURAL DISASTERS
Due to rapid urbanization and deforestation, Haiti suffers more damage in major storms, with torrential rains and hurricanes killing at least 4,000 people.

STATE OF THE WORLD

2005

CHAPTER 1

Security Redefined

Michael Renner

A little more than a decade after the end of the cold war seemed to herald a new era of peace, security concerns are once more at the top of the world's agenda. A heightened sense of insecurity, reflected as much in headlines as in opinion polls worldwide, is palpable. The September 11th terror attacks in the United States were no doubt a pivotal event. Subsequent bombings in countries from Spain to Kenya, Saudi Arabia to Russia, and Pakistan to Indonesia reinforced a widespread feeling of vulnerability. And the growing chaos in Iraq following the U.S.-led occupation feeds unease about the repercussions of a destabilized Middle East.

But terrorism is only symptomatic of a far broader set of deep concerns that have produced a new age of anxiety. Acts of terror and the dangerous reactions to them are like exclamation marks in a toxic brew of profound socioeconomic, environmental, and political pressures—forces that together cre-

ate a tumultuous and less stable world. Among them are endemic poverty, convulsive economic transitions that cause growing inequality and high unemployment, international crime, the spread of deadly armaments, large-scale population movements, recurring natural disasters, ecosystem breakdown, new and resurgent communicable diseases, and rising competition over land and other natural resources, particularly oil. These "problems without passports" are likely to worsen in the years ahead. Unlike traditional threats emanating from an adversary, however, they are better understood as shared risks and vulnerabilities. They cannot be resolved by raising military expenditures or dispatching troops. Nor can they be contained by sealing borders or maintaining the status quo in a highly unequal world.[1]

In a late 2003 Gallup International poll of some 43,000 individuals in 51 countries, twice as many respondents rated international security as "poor" as those who answered "good." Almost half of those interviewed think the next generation will live in a less safe

Units of measure throughout this book are metric unless common usage dictates otherwise.

world, while only 25 percent said they expected an improvement. Similarly, a June 2003 poll of 2,600 "opinion leaders" in 48 countries found a broad sense of pessimism, with at least two thirds in every region of the world describing themselves as "dissatisfied" with the current world situation. And in a series of World Bank–facilitated consultations involving some 20,000 poor people in 23 developing countries, a large majority said they were worse off than before, had fewer economic opportunities, and lived with greater insecurity than in the past.[2]

The need for international cooperation has grown stronger in this new century, even as rifts and divides have opened up.

In sharp contrast to the cold war's bipolar standoff involving nuclear arsenals and competing core ideologies, today's security challenges tend to be more diffuse, less predictable, and more multidimensional. Fears of a violent showdown between two superpowers have given way to concerns about local and regional wars fought predominantly with small arms, post-conflict volatility, instability emanating from weak and failed states, and the rise of international criminal and terror networks. Yet some of the old perils still exist. Progress toward nuclear disarmament has ground to a halt, for instance, while the danger of nuclear and other highly lethal weapons spreading to a growing number of countries—or falling into the hands of extremist groups—looms.

The challenges the world faces are compounded by weak and corrupt public institutions, the lack of recourse to justice, and unconstitutional or irregular means of political change, such as coups d'état and revolts. And they are heightened by an uneven process of globalization that draws nations and communities together in often unpredictable ways that entail real risks for many and that allow extremist groups to operate more easily than in the past.[3]

The East-West confrontation that used to stand in the way of enhanced cooperation has given way to a more vexing North-South relationship marred by enormous imbalances of livelihood, wealth, and power. The sole remaining superpower has an increasingly uneasy and contentious relationship with the rest of the world. And the critical structural changes and innovations needed to generate effective global governance—proposals to reform the U.N. Security Council or create a much stronger U.N. environmental body—have fallen victim to political paralysis.

The need for international cooperation has thus grown stronger in this new century, even as new rifts and divides have opened up, provoked in part by the Iraq crisis. Yet Fred Halliday, professor of international relations at the London School of Economics and Political Science, warns that "the world appears further away than ever from addressing the fundamental issues confronting it, and to be moving ever more deeply into a phase of confrontation, violence, and exaggerated cultural difference."[4]

Policies that seek security primarily by military means but fail to address underlying factors of instability will likely trigger a downward spiral of violence and instability, and quite possibly a collapse of international rules and norms. Policies derived from a new understanding of global security can avoid these dangers and promote constructive alternatives. A robust and comprehensive approach to creating a more stable world entails measures designed to stop environmental decline, break the stranglehold of poverty, and reverse the trend toward growing inequity and social insecurity that breeds despair and extremism. A fundamental shift in priorities is essential to accomplish these tasks. Ultimately, security must be universal.

The Roots of Insecurity

Awareness of the threats and challenges that cannot be resolved within the traditional framework of national security led a wide range of nongovernmental organizations (NGOs), scholars, and others to refine and redefine our understanding of security over the past two decades. What is the object of security? What is the nature of the threats? Who is to provide security? And by what means? These questions and discussions gathered momentum after the end of the cold war. The core insights they led to are even more relevant today:

- Weapons do not necessarily provide security. This is true for adversarial states armed with weapons of such destructive power that no defense is possible. It is true in civil wars, where the easy availability of weapons empowers the ruthless but offers little defense for civilians. And it was true on September 11th, when a determined group of terrorists struck with impunity against the world's most militarily powerful country.
- Real security in a globalizing world cannot be provided on a purely national basis. A multilateral and even global approach is needed to deal effectively with a multitude of transboundary challenges.
- The traditional focus on state (or regime) security is inadequate and needs to encompass safety and well-being for those living there. If individuals and communities are insecure, state security itself can be extremely fragile. Democratic governance and a vibrant civil society may ultimately be more imperative for security than an army.
- Nonmilitary dimensions have an impor-

tant influence on security and stability. Nations around the world, but particularly the weakest countries and communities, confront a multitude of pressures. They face a debilitating combination of rising competition for resources, severe environmental breakdown, the resurgence of infectious diseases, poverty and growing wealth disparities, demographic pressures, and joblessness and livelihood insecurity.[5]

The pressures facing societies and people everywhere do not automatically or necessarily trigger violence. But they can translate into political dynamics that lead to rising polarization and radicalization. Worst-case outcomes are more likely where grievances are left to fester, where people are struggling with mass unemployment or chronic poverty, where state institutions are weak or corrupt, where arms are easily available, and where political humiliation or despair over the lack of hope for a better future may drive people into the arms of extremist movements.

Insecurity can manifest itself in ways other than violent conflict. The litmus test is whether the well-being and integrity of society are so compromised that they lead to possibly prolonged periods of instability and mass suffering. Measured by the number of victims and mass dislocations caused, the repercussions of intense poverty and other societal failures tend to loom far larger than outbreaks of armed conflict. Whereas about 300,000 people were killed in armed conflicts in 2000, for example, as many people die each and every month because of contaminated water or lack of adequate sanitation.[6]

In abstract terms, issues such as infectious disease, unemployment, or climate change may or may not constitute security challenges. But do they cross thresholds of magnitude or trigger dynamics that render them something more momentous? Alone or in combination with other factors, they may well create conditions that call into question the basic fabric of communities and nations. As Alyson Bailes, director of the Stockholm International Peace Research Institute, asks, "Which 'hits' can a society bounce back from relatively easily, and which are the ones that risk undermining its whole viability?" The task, then, is to enhance our understanding of the interactions and dynamics among these factors and the combinations that are likely to bring about destabilizing results.[7]

Natural resources are at the core of a number of conflicts. Throughout human history, big powers have repeatedly intervened in resource-rich countries, militarily and by other means, in order to control lucrative resources. The result has often been enduring political instability. Against the backdrop of surging demand for oil, geopolitical rivalries for preferential access are again intensifying among major importers. (See Chapter 6.)

The benefits and burdens of oil extraction, mining, and logging projects are often distributed quite unequally, triggering disputes with indigenous peoples across the planet. Resource wealth has also fueled a series of civil wars, with governments, rebels, and warlords in Latin America, Africa, and Asia clamoring over resources such as oil, metals and minerals, gemstones, and timber. The revenues derived from such commodities help pay for weapons and sustain wars that have had devastating consequences for civilians caught in the crossfire; fighting and looting shred civilian infrastructures, disrupt harvests, and prevent delivery of vital services.[8]

Disputes also arise over access to renewable natural resources such as water, arable land, forests, and fisheries. This is particularly the case among groups—such as farmers, nomadic pastoralists, ranchers, and resource extractors—who depend directly on the health and productivity of the resource base but have incompatible needs. Such tensions intensify with the growing depletion of natural resources and rising demand owing to population pressures and growing per capita consumption. Local violence in countries like Brazil, Côte d'Ivoire, Haiti, Mexico, Nigeria, Pakistan, the Philippines, and Rwanda is in part driven by these factors.[9]

Water is the most precious resource. Both the quantity and quality are crucial for such fundamental human needs as food and health. Given population growth, nearly 3 billion people—40 percent of the projected world population—will live in water-stressed countries by 2015. Although there may not be any interstate water wars, as some have predicted, local disputes and clashes are likely to proliferate. (See Chapter 5.)[10]

Climate change is certain to sharpen a broad range of environmental challenges, thus intensifying many of these struggles. More frequent and intense droughts, floods, and storms will play havoc with harvests, undermine the habitability of some areas, escalate involuntary population movements, and severely test national and international institutions.

Different social groups and communities experience the effects of resource depletion and environmental degradation unevenly. These divergences can reinforce social and economic inequities or deepen ethnic and political fault lines. It is not a given that competition over scarce resources or the repercussions of environmental degradation will lead to armed conflict. But they often do sharpen hardships and burdens, heighten the desperation of those most affected, and reinforce the perception that disputes have a "zero-sum" nature. The challenge is to avoid such polarization and instead turn shared environmental problems into opportunities for conflict prevention and peacemaking. (See Chapter 8.)

A reliable supply of food is one of the

most basic determinants of how secure or insecure people are. Food security is at the intersection of poverty, water availability, land distribution, and environmental degradation. But war and social disruptions also play an important role in some cases. And the proliferation of factory farming and the promotion of monocultures have triggered growing worries about the safety and quality of food supplies. (See Chapter 4.)

About 1.4 billion people, almost all of them in developing countries, confront environmental fragility. Of these, more than 500 million people live in arid regions, more than 400 million people eke out a meager living on soils of very poor quality, some 200 million small-scale and landless farmers are compelled to cultivate steep slopes, and 130 million people live in areas cleared from rainforests and other fragile forest ecosystems. The soil productivity of these areas tends to be exhausted relatively swiftly, forcing people to move on to seek opportunity elsewhere, sometimes in distant cities or in competition with other rural dwellers.[11]

The U.N. Food and Agriculture Organization found that hunger—after falling steadily during the first half of the 1990s—grew in the latter part of the last decade, now afflicting some 800 million people worldwide. Inadequate food supplies make people more susceptible to disease. But there is also a reverse impact. The AIDS epidemic has a particularly devastating impact on farm production and food security because it incapacitates and kills primarily young adults during their peak productive years. AIDS is projected to claim a fifth or more of the agricultural labor force in most southern African countries by 2020, heightening the risk of famine.[12]

Disease burdens can in some cases be sufficiently severe to undermine economies and threaten social stability. Although the poor are most vulnerable, societies across the planet are now confronting a resurgence of infectious diseases. (See Chapter 3.) Pathogens are crossing borders with increasing ease, facilitated by growing international travel and trade, migration, and the social upheaval inherent in war and refugee movements. Logging, road-building, dam construction, and climate change enable diseases like malaria, dengue fever, and schistosomiasis to spread to previously unaffected areas or bring people into closer proximity with new disease vectors.[13]

> **Throughout history, big powers have repeatedly intervened in resource-rich countries in order to control lucrative resources.**

In the poorest developing countries, infectious diseases are weakening and impoverishing families and communities, deepening poverty and widening inequality, drastically reducing life expectancy, and severely taxing overall economic health. AIDS not only decimates farmers, it strikes many others in the prime years of life—including soldiers, teachers, health practitioners, and other professionals—and is turning an alarming number of children into orphans. Zambian school teachers, for example, are dying faster than the country can train replacements. The disease cripples societies at all levels, undermining a state's overall resilience and its ability to govern and provide for basic human needs. It is hard not to conclude that the impact on political stability will be profound in years to come.[14]

A combination of resource depletion, ecosystem destruction, population growth, and economic marginalization of poor people has set the stage for more frequent and more devastating "unnatural" disasters—natural disturbances made worse by human actions. Three times as many people—250 million—were affected by such events in 2003 as in 1990. Deforestation left Haiti extremely vulnerable to devastating hurricanes that in late 2004 caused massive mudslides and flash floods. The pace is likely to accelerate as climate change translates into more intense storms, flooding, heat waves, and droughts. In addition to sudden disasters, there is also the "slow-onset" degradation of ecosystems, in some cases sufficiently extreme to undermine the habitability of a given area. This is most calamitous for the poor because they tend to be far more directly exposed, have inadequate protection, and have little in the way of resources and wherewithal to cope with the consequences.[15]

Most worrisome in some ways is the vast reservoir of unemployed young people in many developing countries.

They may have scant choice other than to search for new homes. Although there are no reliable data for the numbers of such "environmental refugees," it is clear that many millions are affected and that their ranks are likely to skyrocket in the years ahead. Desertification, for example, puts an estimated 135 million people worldwide at risk of being driven from their lands. In February 2004, Canadian Environment Minister David Anderson claimed that "global warming poses a greater long-term threat to humanity than terrorism because it could force hundreds of millions from their homes and trigger an economic catastrophe." The displaced may not be welcome elsewhere, causing tensions over access to land, jobs, and social services.[16]

Lack of employment, uncertain economic prospects, and rapid population growth make for a potentially volatile mix even in the absence of displaced populations. (See Chapter 2.) A 2004 report from the International Labour Organization found that three quarters of the world's workers live in circumstances of economic insecurity. Most worrisome in some ways is the vast reservoir of unemployed young people in many developing countries, particularly where young adults aged 15–29 account for 40 percent or more of the total population. The United Nations projects that by 2005, some 138 countries will confront such a "youth bulge." Youth unemployment is skyrocketing to record levels, with the highest rates found in the Middle East and North Africa (26 percent) and in sub-Saharan Africa (21 percent). At least 60 million people aged 15–24 cannot find work, and twice as many—some 130 million—are among the world's 550 million working poor who cannot lift their families out of poverty.[17]

When large numbers of young men feel frustrated in their search for status and livelihood, they can be a destabilizing force. Their uncertain prospects may cause criminal behavior, feed discontent that could burst open in street riots, or foment political extremism. Whether these things happen depends on a number of factors—among others, the extent to which political systems are open to dissent and capable of change, people's sense of identity and civic engagement, and the role of education. U.N. Habitat Executive Director Anna Tibaijuka has warned that urban slums may well be incubators of extremism if governments fail to tackle the poverty and desperation engulfing them.[18]

Particularly if political grievances linger, the malcontented may be easy to recruit into insurgent groups, militias, or organized crime—as experiences in places like Rwanda, Kosovo, and East Timor have shown in recent years. Among Palestinians, support for political violence has flourished under the combination of a harsh occupation, a breakdown in political leadership, and joblessness that averaged about 35 percent in 2003. An educated and once relatively affluent society saw its poverty rate shoot up from 20 percent to about 50 percent between 1999 and 2003. Similar dynamics are now at work in Iraq, where the official unemployment rate is 28 percent and underemployment is at 22 percent, although some estimates peg the numbers as even higher.[19]

Bad Neighborhoods and Shared Vulnerabilities

Severe social, economic, and environmental problems—particularly if mixed with festering political grievances—can radicalize societies and may even bring about state failure. Dysfunctional, fragile, and violence-prone, so-called failed states are breeding grounds of despair and chronic instability, where warlords, criminal networks, or extremist groups are able to exploit a vacuum of governance and legitimacy.

Prior to September 11th, poverty, instability, and warfare in poor countries were widely regarded as marginal to the interests and welfare of the rich. But after the attacks it became clear that conditions of political turmoil and social misery cannot forever be confined to the periphery. "If we've learned anything from Sept. 11," wrote *New York Times* columnist Thomas Friedman, "it is that if you don't visit a bad neighborhood, it will visit you." Afghanistan, torn to pieces by geopolitical power struggles and then largely forgotten after the cold war ended, became an ideal sanctuary for al Qaeda, harbored by the Taliban regime. There is also evidence that al Qaeda operatives were able to use Liberia as a sanctuary from 1998 until 2002; along with warlord-turned president Charles Tay-

lor, the organization was apparently involved in trafficking diamonds from neighboring Sierra Leone.[20]

Why do states fail? Clearly, there are many internal reasons for this, and it happens in many parts of the world, from Haiti to Liberia and from Rwanda to Afghanistan. Corruption and patronage are rife. Coups d'état and dictatorial regimes trump democratic rule and trigger cycles of repression and upheaval. "Shadow state" structures deliberately weaken public institutions, while revenues and services are diverted to parallel networks that benefit only a small elite. Ethnic, tribal, and class divisions are exploited by opportunistic leaders. And population and resource pressures continue unabated.[21]

Such failures give a boost to extremist forces. In Iraq, for instance, repeated warfare and harsh international sanctions between 1990 and 2003 caused the virtual disappearance of the middle class and the collapse of the secular education system, resulting in widespread illiteracy and despair, which facilitated the growth of fundamentalist religious forces.[22]

But the term "failed state" hides an inconvenient truth: external factors are equally important. More appropriately, Thomas Friedman might have written: "If you help create a bad neighborhood, it will eventually come to haunt you." The current global trading and investment regime serves primarily the interests of the 20 percent or so of humanity who claim 80 percent of the planet's resources. It tends to marginalize the poor, sharpen social and economic inequalities, and weaken the state's ability to provide much-needed services and cope with challenges.[23]

Another crucial factor is outside interventions that sow the seeds of turmoil. In Afghanistan, for instance, the United States, Pakistan, and Saudi Arabia recruited Mujahedeen fighters in the 1980s to force Soviet occupation troops out. This struggle, and the subsequent ferocious civil war among the victorious resistance groups, devastated the country. The unraveling of Afghan society permitted the most ruthless elements to emerge victorious. The Taliban were the product of this long descent into impunity and societal breakdown, and Osama bin Laden's al Qaeda network was born out of the anti-Soviet recruiting drive. Supporting the Mujahedeen, which included some of the most violent and extremist leaders, in a great power "game" seemed to make good sense in the 1980s from a narrow geopolitical point of view. But the September 11th attacks were a fateful boomerang effect of the Afghan proxy war.[24]

Somalia, also often cited as a failed state, disintegrated in part because militarization sponsored first by the Soviets and then by the United States led to a disastrous war with Ethiopia in the late 1970s that left the country awash in weapons. The extreme neglect of civilian needs paved the way for a popular revolt, the overthrow of the Siad Barre dictatorship, and civil war. Some 500,000 weapons ended up in the hands of competing warlords who ravaged the country.[25]

Some commentators have urged fresh military interventions to ward off trouble emanating from "disorderly societies." Max Boot of the U.S. Council on Foreign Relations has written that "Afghanistan and other troubled foreign lands cry out for the sort of enlightened foreign administration once provided by self-confident Englishmen in jodhpurs and pith helmets." But the unfortunate history of "blowback"—unintended negative consequences of actions undertaken by intervening powers—suggests that the likely outcome is ongoing cycles of violence rather than any durable stability.[26]

The discovery that failed states may represent a broader security threat falls short of a complex reality. Long before such cases showed up on northern radar screens, they had already failed their own people. More to the point, even if particular failed states never make it onto the northern agenda—if they are never labeled as such—they still fail their own people. Pakistan, for example, may fit that description, given its entrenched poverty, endemic corruption, religious schools that indoctrinate more than they impart skills, and scarce budget resources going to the military and to nuclear weapons development instead of toward meeting basic needs. The misery accompanying state breakdown needs to be addressed in its own right—not only because the rich and powerful have identified such a condition as a threat to themselves.

North and South, and rich and poor, tend to view security challenges in very different ways. But U.N. Secretary-General Kofi Annan has warned: "We now see, with chilling clarity, that a world where many millions of people endure brutal oppression and extreme misery will never be fully secure, even for its most privileged inhabitants." Annan urged the world in March 2004 to move away from the idea that some threats, such as terrorism and weapons of mass destruction, are of interest to only northern countries, while threats such as poverty and the struggle to secure the basic necessities of human existence only concern the South. "I think we need a clear global understanding of the threats and challenges that we all have to face, because to neglect any one of them might fatally undermine our efforts to confront others."[27]

Overcoming the divides that increasingly separate disparate communities, cultures, and nations and dramatically improving international cooperation is clearly a Herculean task. Individual countries have vastly divergent powers, varying capacities to confront challenges, and different perspectives on the proper course of action. A shared conception of security can only be developed if the connections among different challenges are acknowledged and if there is a much better understanding that many of them are in fact shared risks and vulnerabilities that require joint solutions.

Controlling Weapons, Defusing Conflicts

Achieving shared security will depend in part on meeting the traditional security challenge of limiting the spread of weapons and resolving conflicts before they become violent. Unfortunately, the recent record on this is mixed. Worldwide holdings of tanks, artillery, jet fighters, warships, and other so-called heavy conventional weapons were reduced by one quarter between 1985 and 2002. Nuclear warhead stockpiles fell by 68 percent, military expenditures were cut 30 percent, and arms exports fell by 58 percent. The number of soldiers shrank by 27 percent, and the ranks of workers in arms industries by 54 percent. (See Figure 1–1.)[28]

Small arms control became an accepted part of the international agenda. (See Chapter 7.) Considerable progress was made against one type of weapon in this category, anti-personnel landmines. These indiscriminate weapons impose debilitating burdens on public health systems, make fertile land unusable, cause economic activity to grind to a halt, and hinder reconstruction

efforts after wars end. A landmark 1997 treaty outlawing anti-personnel landmines led to reduced use of mines, a dramatic drop in production and a nearly complete halt to exports, the destruction of more than 50 million mines in stockpiles, and a significant reduction in the number of mine victims.[29]

In another key normative achievement, many of the world's governments signed on to a new International Criminal Court, intended as a tool to bring to justice the perpetrators of genocide, war crimes, and other acts of impunity. The statute for the Court was adopted in 1998 and entered into force in 2002; by October 2004, 139 countries had signed and 97 countries had ratified the statute.[30]

These accomplishments would have been unthinkable without the rise of what some have called a "second superpower"—world public opinion. The 1990s saw "soft power"—a combination of diplomacy, persuasion, and marshaling of public opinion—wielded by NGOs frequently acting in concert with "like-minded" governments. NGO

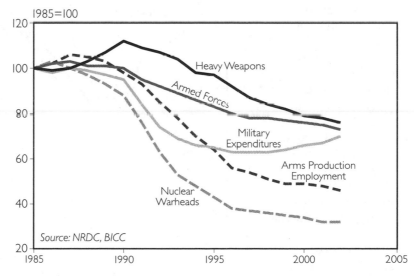

Figure 1–1. Progress in Global Disarmament, 1985–2002

involvement helped widen the scope of the security discussion and promoted new concepts of security. Nonmilitary issues were also elevated at a series of U.N. conferences on environment, social development, population, and women.[31]

Nonetheless, the 1990s were a decade with highly contradictory results—an era of missed opportunities as much as of notable accomplishments. Disarmament had its limits. Although NATO and Warsaw Pact states trimmed their armaments substantially, a considerable portion of the surplus was not destroyed but instead transferred to developing countries—which now for the first time possess more heavy weapons than the industrial states of the North do.[32]

The number of deployed nuclear weapons decreased, but since 1995 the pace of disarmament has slowed considerably. Russia in particular needs large-scale assistance to safeguard its warheads against theft and to dismantle its surplus stocks. Meanwhile, there is also a danger that fissile materials from the civilian nuclear power industry might be lost or diverted for weapons purposes. The quantities of plutonium and highly enriched uranium derived from both military and civilian reactors continue to expand. Estimated at more than 3,700 tons at the end of 2003 in some 60 countries, this is enough to produce hundreds of thousands of nuclear weapons.[33]

While South Africa and more recently Libya forswore

nuclear weapons, power politics and regional rivalries pushed India, Israel, North Korea, and Pakistan toward acquiring a nuclear capability and may yet persuade others, such as Iran, to follow suit. The existing nuclear weapons powers have not given any indication that they will fulfill their commitment under the Non-Proliferation Treaty to move toward disarmament. To the contrary, the United States is developing more usable warhead concepts and designs, including earth-penetrating warheads and low-yield nuclear weapons, and its 2001 Nuclear Posture Review asserted that nuclear weapons "provide credible military options to deter a wide range of threats" and help "achieve strategic and political objectives."[34]

The overall number of wars declined during the 1990s. (See Figure 1–2.) Although this is clearly good news, there are some questions about whether available statistics capture the full extent of armed violence in the world. Besides methodological limitations, the distinction between war and peace has become blurred in a variety of ways. Vio-

Number of Conflicts

Source: Gleditsch et al., PRIO

Figure 1–2. Armed Conflicts, 1955–2002

lence is often sporadic rather than continuous, and instability continues to fester in many societies even after the formal end of warfare. Regular armies have shrunk in size, yet warlords, crime networks, and private military companies signal a growing privatization of violence and forms of insecurity that are not necessarily picked up in war statistics.[35]

Because of chronic instability, refugee flows, and other spillover effects, the world has an obvious interest in preventing the outbreak of violent conflicts and in terminating ongoing wars as quickly as possible. The number of peacekeeping missions has grown considerably since the early 1990s. Yet most of these efforts are handicapped by inadequate resources, erratic political support, and the lack of a permanent structure to ensure the deployment of well-trained peacekeepers in a timely manner and in sufficient numbers.[36]

The end of the cold war did open up peacemaking opportunities previously unavailable, allowing the Security Council to work far more productively. The five permanent members cast only 18 vetoes between 1990 and late 2004—far fewer per year than the 199 vetoes between 1946 and 1989. Still, the permanent members have increasingly resorted to a "hidden" veto—threatening to use the veto in order to keep unwanted items off the Council agenda. The Council never takes action on conflicts that permanent members regard as their own sphere of influence, such as Chechnya, Tibet, or Northern Ireland. Both actual and hidden vetoes by the United States have prevented Council action on the Israeli-Palestinian conflict.[37]

In addition, lack of big-power interest has repeatedly prevented Council involvement where warfare and humanitarian disasters would have warranted action. The result is an unpalatable choice between paralysis (as occurred during the Rwandan genocide in 1994) and intervention by self-appointed "willing nations" (as in Kosovo in 1999, when Russia blocked Council action and NATO initiated an air war against Serbia). Undoubtedly, the authority of the Security Council has suffered tremendously.[38]

In retrospect, the 1990s provided a brief window of opportunity after the cold war to build institutions and mechanisms that could address new challenges and act on a broader understanding of security. The opportunity was largely squandered: on the whole, the international community's investment in conflict prevention, peacekeeping, and post-conflict reconstruction was inadequate. The failure to move forward more decisively during the post–cold war "honeymoon" is now coming back to haunt us in the post–September 11th era.[39]

The Impact of the War on Terrorism

The fear set in motion by the September 11th attacks triggered a dangerous reaction: a war on terrorism of essentially unlimited scope and duration that has caused government policies and media coverage in many countries to once again focus on an overly narrow slice of security challenges and to revert to a stronger reliance on military tools. Whether terrorism can be "defeated" by military means is questionable since extremist groups do not represent easily identifiable targets. Terrorism is a path chosen by protagonists who tend to be politically desperate and militarily weak. Acts of terror are not going to disappear as long as the roots of extremist violence are not tackled.[40]

A number of measures have been undertaken in the name of anti-terrorism that may well perpetuate a cycle of violence. These actions have undermined international cooperation, weakened human rights laws and other international norms, and played into the hands of extremists who thrive on a "clash of civilizations." And this militarized response is draining resources and political attention away from the underlying socioeconomic and environmental issues that generate growing tension and instability.

To be sure, governments have been far from unanimous in their reaction to September 11th and other terrorist acts. In fact, the transatlantic and inter-European rifts over Iraq revealed fundamental policy differences and caused deep fissures in the western alliance. Whereas the United States embraced the use of force with little hesitation, Europe has been more ambiguous. In December 2003, the European Council adopted a European Security Strategy declaration. Arguing that "in an era of globalization, distant threats may be as much of a concern as those that are near at hand," the document concludes that "the first line of defense will often be abroad." It acknowledges that "none of the new threats is purely military; nor can any be tackled by purely military means." But then the document endorses more resources for defense and the transformation of European armies "into more flexible, mobile forces." In the final analysis, it gives priority to military intervention, but relatively short shrift to non-military ways of dealing with security challenges.[41]

The war on terror threatens to sideline the struggle against poverty, health epidemics, and environmental degradation.

A number of governments—China, Colombia, India, Indonesia, Israel, the Philippines, and Russia among them—have seen the war on terrorism as an opportunity to strike against insurgents, separatists, or other political opponents with greater impunity, branding them terrorists. Military campaigns have been accompanied by a heavy-handed law-and-order approach eroding human rights, curbing civil liberties, intimidating domestic political dissent, and adopting punitive measures against refugees and asylum-seekers. And in the name of fighting terrorists, supplier nations have shown little reluctance providing weapons and military aid to states that have committed grave human rights violations. Amnesty International worries that "human rights and humanitarian laws are under greater threat worldwide than at any time since the UN was founded more than half a century ago."[42]

The unprecedented wave of global empathy with the United States following the events of September 11th held out the promise that humanity could rally around a common purpose. Yet the Bush administration rejected a multilateral approach. It reversed earlier U.S. support for or stiffened its opposition to several treaties, such as the International Criminal Court statute, the nuclear test ban treaty, a proposed verification regime for the Biological and Toxin Weapons Treaty, and the inspection and verification provisions for a still-to-be-negotiated treaty banning production of fissile materials for nuclear weapons.[43]

Most critically, the administration asserted a blanket right to conduct preemptive wars in contravention of the U.N. Charter. The National Security Strategy of September 2002 warns that "to forestall or prevent…hostile acts by our adversaries, the United States will, if necessary, act preemptively." This is a dangerous precedent that some other governments may be inclined to follow as well. Russia, embroiled in a brutal fight with Chechen separatists, has announced that it, too, will resort to preemptive raids. There has also been speculation that Israel may launch a strike against Iranian facilities suspected of producing nuclear-weapons materials. But the result could be an anarchic future of dueling preemptive strikes and wars. Even if such a bleak scenario does not come to pass, there is a danger that the international rule of law, already often observed in the breach, will be weakened further.[44]

According to the Bush administration, Iraq is the "central front in the war on terror." But Jeffrey Record, an analyst at the U.S. Army War College, argues that the administration has fused rogue states, weapons of mass destruction proliferators, and terrorist organizations into a monolithic threat, "and in so doing…may have set the United States on a course of open-ended and gratuitous conflict with states and nonstate entities that pose no serious threat to the United States."[45]

Indeed, the Iraq occupation opened a Pandora's box of violence and chaos. In the face of mounting violence, "security" measures absorb much of the money ostensibly allocated to reconstruction. Even though little has been spent on actual rebuilding, the United States decided to divert almost $3.5 billion from water, sewage, and electricity projects toward a range of security measures. A July 2004 U.S. National Intelligence Estimate paints a gloomy picture, including the possibility of a descent into civil war. Kurdish separatism, Sunni-Shi'a animosities, and power struggles that pit Islamic and secular-minded forces against each other are among the triggers that could fragment the country. If this happens, instability may spread to Iraq's neighbors as well.[46]

Rather than striking a blow against terrorism, the occupation of Iraq has accelerated the radicalization of an Islamic world already seething over events in the occupied Palestinian territories, Kashmir, and Chechnya. Iraq has become a potent new recruiting ground for extremists. The London-based Institute for International Security Studies reported in May 2004 that al Qaeda has been galvanized by the war in Iraq; it believes that the organization is present in more than 60 countries and has "18,000 potential terrorists at large." Indeed, a U.S. State Department report shows an increase in the number of "significant" terrorism incidents and victims in 2003 over 2002.[47]

The Iraq war has drained away badly needed funds from the momentous task of disarmament, demobilization, and reconstruction in Afghanistan—the country that hosted al Qaeda and is now once more in danger of falling prey to warlords and the burgeoning drug trade. According to A. Yusuf Nuristani, Minister for Irrigation, Water Resources, and Environment, Afghanistan receives only $1 for every $30 going to Iraq. The nation has received far less international donor support than other countries undergoing post-conflict reconstruction: just $67 annually per capita, compared with $74 for Haiti, $114 for Rwanda, $249 for Bosnia, and $814 for Kosovo.[48]

From the perspective of a broader conception of security, the war on terror threatens to sideline the struggle against poverty, health epidemics, and environmental degradation, drawing scarce financial resources and political capital away from the root causes of insecurity. Driven primarily by surging U.S. spending, world military expenditures are now close to $1 trillion a year.[49]

Surprisingly modest investments in health, education, and environmental protection could tap the vast human potential now shackled by poverty and break the vicious circles that are destabilizing large areas of our planet. Estimates suggest that programs to provide clean water and sewage systems would cost roughly $37 billion annually; to cut world hunger in half, $24 billion; to prevent soil erosion, another $24 billion; to provide reproductive health care for all women, $12 billion; to eradicate illiteracy, $5 billion; and to provide immunization for every child in the developing world, $3 billion. Spending just $10 billion a year on a global HIV/AIDS program and $3 billion or so to control malaria in sub-Saharan Africa would save millions of lives. All this adds up to a little more than half the $211 billion likely to be appropriated for the Iraq war by the end of 2004.[50]

At the same time, aid flows to the developing world declined during the 1990s, from about $73 billion in 1992 to $57 billion in 2002. Tallying all financial flows, the United Nations reports that in 1994–2002 developing countries suffered a cumulative outflow of $560 billion. And the budget allocations of many poor countries themselves heavily favor their armed forces. For some—Burundi, Eritrea, and Pakistan among them—military spending equals or surpasses combined public expenditures for health and education.[51]

There is a distinct danger that the critical health, education, and anti-poverty gains envisioned in the U.N. Millennium Development Goals adopted by the world community in September 2000 will not be achieved because international attention and resources have been diverted to military budgets and the war on terror. (See Chapter 9.) Yet it is precisely these underlying factors— and the way they translate into political dynamics and tensions—that are among the key drivers of much of the world's instability.[52]

It is the dimming of hope for a better future that helps fuel extremism and makes it easier for agitators to recruit. Poverty is actually on the rise in some parts of the world, including sub-Saharan Africa, where it grew from 42 to 47 percent of the population between 1981 and 2001. "A world not advancing towards the Millennium Development Goals," warned Kofi Annan in September 2004, "will not be at peace. And a world awash in violence and conflict will have little chance of achieving the Goals."[53]

Because of the preeminence of the United States in the world, future directions in its policies will be crucial in determining which path humanity will choose. Writing prior to the November 2004 U.S. elections, Anatol Lieven of the Carnegie Endowment for International Peace anticipated that "the U.S. war on terrorism will be conducted much more cautiously from now on, whether Bush or Kerry wins in November. A cautious policy is, however, not the same thing as a new policy." A rededication to multilateralism and to finding common approaches to the world's challenges will be crucial. Still, the Iraq invasion cannot be undone, and its destabilizing consequences in the Middle East and the Islamic world cannot be wished away. There is no putting this genie back in a bottle.[54]

Principles for a More Secure World

The effort to reconceptualize security is not an academic exercise. The point is to persuade policymakers to adopt a different outlook on the world—to interpret trends, developments, and news events in a new light and ultimately to promote different agendas and policies. At least three core principles derive from a redefinition of security.

First, a new security policy needs to be transformative in nature, strengthening the civilian institutions that can address the roots of insecurity. In linking environment, health, poverty, migration, and other issues with security, there is a definite risk of "securitizing" these issues—that is, applying the language and rationality of traditional security institutions and thus promoting adversarial rather than cooperative thinking. Merely relabeling certain challenges as security threats may accord them a more prominent spot on the political agenda, but it would accomplish little more than broadening the purview and power of traditional security institutions. To avoid militarizing policy, it is important to apply the language of human rights, equity, and livelihood to this new outlook on the world. In effect, this means reclaiming the term security.

The second principle flows directly from this insight: a new security policy must above all be preventive in nature. Conflict prevention is too often seen as a narrow, last-ditch effort where the outbreak of violence seems imminent. But understanding the root causes of conflict and insecurity implies a far broader, earlier applicability, not merely an effort to address symptoms. Donor countries tend to be relatively generous when it comes to Band-Aid measures. Too much (although at the same time, ironically, not enough) is being spent on humanitarian measures such as disaster relief and other emergency aid, on refugee support and resettlement, and even on too little–too late peacekeeping efforts.[55]

Robert Picciotto, a former World Bank official and now head of the London-based Global Policy Project, argues that "the economics of international security resemble the economics of public health. Just as public health policy goes beyond curative measures, security policy extends to conflict prevention." Conflict prevention needs to be built into a broad range of social and economic policies. In effect, there is a need to conduct security impact assessments analogous to the environmental impact assessments carried out in some countries.[56]

The third principle is that a new security policy needs to be cross-cutting and integrative. Understanding complex security challenges, allowing a sophisticated assessment of the dynamics leading to instability, and undertaking a more effective diagnosis of policies needed to prevent conflict and provide mean-ingful security will require bringing together insights from a broad range of disciplines—political science, economics, sociology, geography, history, public health, and many others.

A new security policy must above all be preventive in nature.

The international conferences of the past decade or so have given credence to the need to connect environment, development, and security. Development and peace are closely correlated and symbiotic; their absence is what often causes state failure. While poverty does not necessarily lead to violence, there is no doubt that the absence of beneficial development breeds insecurity and permits at best a fragile peace. For development to take place in turn requires peace and political stability. And development needs to be infused with sustainability and equity; the simple-minded maximization of economic growth may end up imperiling environmental integrity, destroying poor communities' livelihoods, and producing highly unequal outcomes.[57]

But translating this last principle into actual policy remains a challenge. It requires transcending academic and bureaucratic boundaries and overcoming the constraints of narrow specialization in an expert-dominated world—whether in government, international organizations, academia, or NGOs. And it requires fusing these sources of expertise by promoting inter- and transdisciplinary thinking and encouraging the development of a shared "language." Given conflicting operating cultures, agendas, and time horizons, this is an uphill struggle.[58]

There are also important imbalances among different government institutions. The political staying power and the resources at the command of defense establishments are vast in comparison with those of development and environment ministries. Foreign and security policymakers typically can ensure political attention and bureaucratic muscle, but they may simply sweep human security concerns under the rug of the traditional security agenda. In fact, foreign aid has long been subordinated to narrow "national security" concerns. Institutions tasked with environmental protection or development aid are able to bring strong expertise to bear, yet they suffer from limited political influence and meager financial means.

There is a danger that a global form of apartheid—highly unequal power relations—will only be further cemented and ratified.

These principles are being put to the test in the growing debate over the notion of "humanitarian intervention" in failed states where governments are unable to protect their own citizens against mass killings or expulsions or have even actually targeted them. The horrors in Bosnia, Rwanda, Kosovo, East Timor, and most recently Darfur in Sudan have given voice to a growing chorus demanding new instruments to avert large-scale humanitarian disasters. The argument that a state's sovereignty entails responsibilities toward its citizens has been embraced by the U.N. Secretary-General, among others.[59]

But it has triggered an intense normative discussion on how to balance the competing values of sovereignty (and nonintervention) and human rights. No consensus has emerged on such questions as, Who has the right to intervene? Under what conditions? And by what means? The International Commission on Intervention and State Sovereignty (ICISS), sponsored by the Canadian government, addressed these questions in careful detail in its December 2001 report, *The Responsibility to Protect*. ICISS makes it clear that such intervention must be a last resort, needs broad international support, and must adhere strictly to international law. The scale, duration, and intensity of operations must be geared not to defeating a state but rather to protecting a population. By implication, the use of certain types of weapons is not acceptable.[60]

These are good principles, and ideally they would be enshrined in an international convention. But critics argue that humanitarian interventions will invariably be carried out by the strong against the weak and that states capable of intervening will do so only if it advances their own national interests. Humanitarianism can easily become a convenient excuse for other objectives, opening the door, in British journalist George Monbiot's words, "to any number of acts of conquest masquerading as humanitarian action."[61]

Indeed, there is a danger that a global form of apartheid—highly unequal power relations—will only be further cemented and ratified. Already, new forms of interventionism are being proposed. Writing in *Foreign Affairs*, Lee Feinstein and Anne-Marie Slaughter suggest "a corollary principle in the field of global security: a collective 'duty to prevent' nations run by rulers without internal checks on their power from acquiring or using weapons of mass destruction." The selective nature of this proposal is quite clear. The authors write: "To be practical, the duty has to be limited and applied to cases when it can produce beneficial results. It applies to Kim Jong Il's North Korea, but not to Hu Jintao's (or even Mao's) China." Presumably those doing the intervening would be the very states that already possess nuclear weapons and thus have a parochial interest in denying such arsenals to all other governments.[62]

Instead of pursuing universally binding

disarmament measures, western countries are now focusing much more on nonproliferation—in other words, on the disarmament of others. The chosen tools are export controls, sanctions, and measures like the Proliferation Security Initiative, under which the United States and several of its key allies are intercepting planes and ships suspected of carrying chemical, biological, or nuclear weapons or missile components.[63]

On a more fundamental level, there is the question of whether military intervention can ever be a cure for violence and its underlying conditions. Perhaps the most basic problem with humanitarian intervention is that it ultimately fails the test of prevention. It addresses symptoms, but not the underlying reasons for humanitarian calamities. It is driven by a passion to stop headline-making violence, but overlooks the death and misery caused by poverty and environmental breakdown.[64]

If conflict prevention that addresses the core dynamics and structural reasons for insecurity is not forthcoming, then the world will always be confronted by the stark choice of military intervention or doing nothing. In such a situation, any action taken elevates the role of the military and ends up cementing the power of traditional security thinking and institutions.

But there is no need to limit ourselves to dead-end choices. As this book demonstrates, there are many social, economic, and environmental policies that can help create a more just and sustainable world and that can turn shared vulnerabilities into opportunities for joint action. Such policies make sense in their own right, but they offer the added bonus of creating real security in a way that the force of arms never can.

Transnational Crime

With a globalizing world comes the freer movement of people, products, and money as well as an increased demand for goods—including illicit ones like drugs, weapons, environmentally sensitive resources, even human beings. Transnational crime syndicates, the main carriers of these goods, pose a considerable threat to global security. They distribute harmful materials, weapons, and drugs, they exploit local communities and disrupt fragile ecosystems, and they control significant economic resources. In 2003, transnational crime syndicates may have grossed up to $2 trillion—more than all national economies except the United States, Japan, and Germany.[1]

The bulk of crime syndicates' revenue comes from drug trafficking. Sales of illegal drugs raked in anywhere from $300 billion to $500 billion in 2001. Illicit drugs take a considerable toll on society. Not only do they contribute directly to more than 200,000 deaths each year, but drug addiction disrupts tens of millions of lives, and the use of intravenous drugs can spread diseases like HIV and hepatitis. Illicit drug sales are also a major revenue stream for insurgent groups and terrorist organizations.[2]

Another major source of revenue for transnational crime syndicates is environmental products—everything from protected plants, animals, and natural resources to hazardous waste and banned chemicals. Annually, the trafficking of these goods earns transnational crime organizations $22–31 billion. Hazardous waste is often shipped secretly with legally exported trash and recyclables, for example, earning $10–12 billion while creating toxic waste sites around the world.[3]

The sale of banned chemicals also poses a significant environmental threat. For example, the annual smuggling of 20,000–30,000 tons of ozone-depleting substances has weakened the ability of the Montreal Protocol to protect the ozone layer effectively. While the smuggling of these chemicals is declining in industrial countries as they switch to less harmful alternatives, new markets are forming in developing countries, where the ban on chlorofluorocarbons is now tightening.[4]

The trafficking of endangered species pulls in another $6–10 billion from the sale of more than 350 million specimens of protected species. Along with jeopardizing the survival of these species, smuggling these plants and animals can compromise global security by spreading diseases and introducing nonnative species into sensitive, new habitats. Fortunately, the 1973 Convention on International Trade in Endangered Species of Wild Fauna and Flora (CITES) has helped reduce wildlife trafficking by banning trade in 900 endangered species and restricting trade in another 32,000 threatened species.[5]

Perhaps one of the most tragic businesses that transnational criminal organizations are involved in is the trafficking of humans. While the number is difficult to know for sure, the U.S. State Department estimates that at least 600,000–800,000 people are sold internationally each year—exploited for labor, sexual services, or even the removal of their kidneys or other organs for transplant purposes. This lucrative trade brings in an estimated $10 billion each year, but at the expense of millions of individuals, their families, and their communities. Along with disrupting countless lives, human trafficking helps fuel the illicit sex industry—another major means of transmission of HIV/AIDS and other diseases.[6]

Weapons trafficking earns comparatively little—one estimate suggests an annual revenue of less than $1 billion—but it takes a tremen-

U.S Customs & Border Protection

U.S. Customs officer removing drugs from a van on the Mexican border

dous toll on security and human well-being. Trafficked weapons, primarily small arms such as automatic rifles, pistols, and shoulder-fired missiles, are used regularly in civil conflicts and by local criminal groups. Small arms contributed to a half-million deaths in 2002— 300,000 through violent conflicts and another 200,000 through homicides (40 percent of all murders). While authorities are constantly battling gun smugglers, there are few international laws on arms trafficking, and U.N. arms embargoes often go unenforced—none have led to a conviction since the mid-1990s. Due to the lack of global political commitment, arms trafficking continues to be a significant threat.[7]

Indeed, only a global commitment to combat transnational crime will succeed in helping reduce this security threat—and for the first time such a commitment may be evident. In September 2003, the United Nations Convention Against Transnational Organized Crime was ratified. This binds ratifying states to adopt broad new legal frameworks that will foster international cooperation in investigating, prosecuting, and punishing crimes committed by transnational groups, which in turn will help prevent offenders from taking advantage of discrepancies in national laws.[8]

The new treaty will also help create a stronger set of laws around money laundering, increasing the difficulty and risk for transnational criminal organizations when investing in new ventures. Several protocols will strengthen this convention. Two, ratified in December 2003 and January 2004, will help coordinate laws on human trafficking and migration. A third, if ratified, will help counter the lack of national legislation on arms trafficking. Currently, only 25 countries have such laws.[9]

Without stronger financial commitments, however, the achievements of CITES, the U.N. crime convention, or other treaties on transnational criminal operations will remain limited. Most secretariats set up to oversee environmental laws are underfunded and understaffed, significantly constraining their effectiveness. The CITES Secretariat, for example, had an annual budget of just $5 million in 2002—less than a thousandth of what animal and plant smugglers bring in each year.[10]

It will also be important to reduce the demand for illicit goods. Until demand shrinks—whether for opium or ivory—there will always be groups willing to take risks to make significant profits. Recognizing this, some countries are putting a larger share of their funds into demand-reduction programs. Sweden, for instance, now directs two thirds of its support for the U.N. Office of Drug Control to reducing demand for drugs. Over time, lowering demand and raising the risks of providing supply will help diminish the profitability of transnational crime and, in the process, build a more secure world.[11]

—*Erik Assadourian*

CHAPTER 2

Examining the Connections Between Population and Security

Lisa Mastny and Richard P. Cincotta

In the early 1990s, the U.S. Central Intelligence Agency assembled a team of prominent researchers, academics, and analysts to tackle a timeless but nagging question: why are some countries more prone to violence and armed conflict than others? The group, known as the State Failure Task Force, sifted through hundreds of social, political, economic, and environmental variables from the 1950s to the 1990s, looking for factors that could predict "state failure"—a collapse of national order caused by mass political or ethnic killings, coups d'état, or civil wars. They hoped to get at the root of the widespread instabilities that have continued to impede economic and human development in regions from sub-Saharan Africa to South Asia.[1]

One particularly strong but surprising finding stood out: a high rate of infant mortality—the proportion of newborns dying before they reach their first birthday—was the best single predictor of instabilities worldwide. It was an even better predictor than such factors as low levels of democracy or a lack of openness to trade. Infant mortality, it turns out, serves as a proxy for a whole range of other indicators—including economic performance, education levels, health, environmental quality, even the presence of democratic institutions—thus providing a good indication of a country's overall quality of life.[2]

It is not surprising that demographic factors have strong statistical links to instability. The dynamics of human population—in particular, fluctuating birth and death rates—can be a powerful force. Not only do they influence the overall growth rate, size, and makeup of a population, they also shape its age structure—the proportion of people in each age group relative to the population as a whole. This in turn determines important

Richard P. Cincotta is Senior Research Associate at Population Action International in Washington, D.C. This chapter draws extensively on a 2003 report he wrote with his colleagues Robert Engelman and Daniele Anastasion entitled *The Security Demographic: Population and Civil Conflict After the Cold War.*

economic variables, such as the number of people entering and leaving the workforce and the ratio of young or elderly dependents to working people. The ebb and flow of people also influences important trends like urbanization as well as the demand for and availability of such critical resources as food, water, and energy. All these forces can exert strong political, social, economic, or environmental pressures on a society and its institutions and can have important implications for domestic stability and even international security. In this case, security is not just the absence of conflict, but a reasonable confidence among people that conflict is not imminent or likely.[3]

The world is going through interesting demographic times. Countries from every major political and religious background and in virtually every region of the world have experienced momentous change in their numbers and the structure of their populations over the past few decades. Due in part to international efforts, average family size is just over half what it was in the early 1960s—at around three children per couple—and infant mortality has declined by two thirds. Global population growth is decelerating more dramatically than was anticipated even a decade ago and is now growing at 1.2 percent, nearly half as fast as it was 35 years ago. And even with the alarming rebound of malaria and emergence of HIV/AIDS, life expectancy in the developing world has climbed from 41 years in the early 1950s to 63 years today, largely because of greater child survival.[4]

Yet these statistics provide an incomplete overview, masking diverse trends both among countries and regions and within them. The human population remains large, at 6.4 billion people, and is still growing by more than 70 million people every year, most of whom are in developing countries. Meanwhile, the global demographic transition—the trans-formation of national populations from short lives and large families to the longer lives and small families seen in much of the industrial world—remains woefully incomplete. Roughly one third of all countries, including many in sub-Saharan Africa, the Middle East, and South and Central Asia, are still in the early stages of the transition, with fertility rates above four children per woman.[5]

Recent studies suggest that these countries bear the highest risks of becoming embroiled in an armed civil conflict—warfare within countries that ranges from political and ethnic insurgencies to state-sanctioned violence and domestic terrorism. Most are bogged down by a debilitating demographic situation. They are home to large and growing proportions of young people, many of whom are entering the ranks of the jobless or the underemployed. Many of these countries are also experiencing rapid urban population growth—often beyond what they can accommodate—as well as exceedingly low levels of available cropland or fresh water per person. Meanwhile, the rising pandemic of HIV/AIDS is striking lethal blows to the basic services and government operations of several countries. These conditions act as "demographic risk factors" that can contribute greatly to the cycle of recurrent conflict and political deterioration that is inhibiting economic and social progress in the world's weakest and most unstable countries.[6]

The Clash of the Ages

Just before dawn on April 28, 2004, a band of attackers wielding machetes and knives launched a surprise attack on a police post in Thailand's southern province of Pattani. Failing to overtake the structure, the militants fled to the nearby Krue Se mosque, where they soon came under siege from heavily armed government security forces. For three hours, troops riddled the sixteenth-century building with grenades and automatic weapon fire, gunning down more than 30 attackers.[7]

As news of the massacre spread, analysts attributed the tensions to rising ethnic unrest among the south's largely Muslim population, which has long complained of cultural, religious, and economic repression by the central government in Bangkok. But in an address to the nation soon after the attacks, Prime Minister Thaksin Shinawatra pointed to an additional variable: the age and prospects of the combatants, most of whom were under the age of 20. "They are poor and have little education and no jobs," he noted. "They don't have enough income and have a lot of time, so it creates a void...to fill."[8]

Thailand is not the only country feeling the effects of a demographic imbalance. According to the United Nations, more than 100 countries worldwide had "youth bulges" in 2000—a situation where people aged 15 to 29 account for more than 40 percent of all adults. All these extremely youthful countries are in the developing world, where fertility rates are highest, and most are in sub-Saharan Africa and the Middle East. (In North America and Europe, in contrast, young adults account for only about 20–25 percent of all adults.) (See Table 2–1.)[9]

In most cases, a youth bulge is the result of several decades of rapid population growth. It typically occurs in countries at the earlier stages of the demographic transition: though birth rates remain high, infant and child mortality have begun to fall due to advances in health care and nutrition, resulting in higher proportions of children surviving overall. Birth rates tend to decline less readily than death rates not only because of cultural preferences for many children and long lives but also because techniques for unwanted pregnancy have tended to be more complicated, less diverse, and much more controversial

Table 2–1. Share of Young People in Selected Countries, 2005 Projections

Country	Share of Adult Population Aged 15–29	Total Fertility Rate
	(percent)	(children per woman)
Zimbabwe	59	3.9
Zambia	57	5.6
Burundi	56	6.8
Uganda	55	7.1
Mali	55	7.0
Rwanda	54	5.7
India	40	3.0
China	30	1.8
United States	27	2.1
Norway	23	1.8
Japan	21	1.3
Italy	19	1.2

SOURCE: See endnote 9.

than those available for extending life. Disproportionately high young populations can also be present in countries where a baby boom occurs, where large numbers of adults emigrate, or where AIDS is a major cause of premature adult death.[10]

An abundance of youth is not necessarily a bad thing. In the United States and other industrial countries, where most young adults have been educated or technically trained, employers view young people as an asset, and companies actively seek out their energy and ingenuity. Economists have long recognized that a large proportion of young workers can provide a "demographic bonus" to economic growth when the productivity, savings, and taxes of young people support smaller populations of children and elderly. In Thailand's more prosperous regions, for instance, young, educated, and industrious workers—including a large proportion of young women working in the manufacturing and financial sectors—have contributed significantly to the country's dynamic growth.[11]

Where economic opportunities are scarce, however, the predominance of young adults can constitute a social challenge and a political hazard. Over the past decade, youth unemployment rates worldwide have jumped from 11.7 percent to a record 14.4 percent in 2003, more than double the overall global unemployment rate. According to the International Labour Organization, an estimated 88 million young people aged 15 to 24 were without work in 2003, accounting for nearly half the world's jobless. In the developing world—home to 85 percent of all young people—unemployment in this group is particularly high, with rates 3.8 times higher than among adults overall.[12]

Leif Ohlsson, a researcher at the University of Göteborg in Sweden, notes that young men in rural areas who are unable to inherit land are often the hardest hit. In some cases, their fathers and grandfathers have long since divided up the family property into tiny parcels that would be unworkable if they were divided again. In other cases, the land has degenerated as a result of unsustainable practices, or larger commercial agricultural enterprises have swallowed up any remaining cropland. In the absence of a secure livelihood, these men may find themselves unable to marry or earn the respect of their peers. British researcher Chris Dolan has coined the expression "the proliferation of small men" to describe the growing number of disenfranchised young men in northern Uganda who cannot fulfill their culture's expectations of being a "full man." Dolan has found that such men disproportionately become alcoholic, engage in violence, commit suicide—or join a militia.[13]

In countries that are economically downtrodden and politically repressive, insurgent organizations can offer youngsters social mobility and self-esteem. During the recent civil war in Sierra Leone, young people constituted roughly 95 percent of the fighting forces, in part because they had few other options in life. Sierra Leone ranked as the world's least developed country on the Human Development Index prepared by the U.N. Development Programme in 2003, and the gross national income per capita in 2002 totaled only $150 (compared with $36,006 in the United States). An official with the Christian Children's Fund in the capital city, Freetown, said of the large body of young soldiers: "They are a long-neglected cohort; they lack jobs and training, and it is easy to convince them to join the fight."[14]

According to the United Nations, more than 100 countries worldwide had "youth bulges" in 2000.

But it is not just the poor or uneducated who are discontented. "We have a large number of youth between 18 and 35 who are properly educated, but have nothing to do," lamented William Ochieng, a former government official, in Kenya's *The Daily Nation* in January 2002. Studies show that the risks of instability among youngsters may increase when skilled members of the elite classes are marginalized by a lack of opportunity. Sociologist Jack Goldstone has noted that the rebellions and religious movements of the sixteenth and seventeenth centuries were led by young men of the ruling class who, when they reached adulthood in an overly large group the same age, found that their state's patronage system could not afford to reward them with the salary, land, or bureaucratic position that matched their class and educational achievements.[15]

It is not difficult to find contemporary parallels. Goldstone attributes the collapse of the Communist regime in the Soviet Union in the early 1990s in part to the mobilization of large numbers of discontented young men who were unable to use their technical educations due to party restrictions on entering the elite. And Samuel P. Huntington, Harvard professor and author of the controversial treatise on the "clash of civilizations," has pointed to connections between the unmet expectations of skilled youth and ongoing tensions in the Middle East, where 65 percent of the population is under the age of 25. Many Islamic countries, he argues, used their oil earnings to train and educate large numbers of young people. But with little economic growth, few in this rapidly growing workforce have the opportunity to use their skills. Young, educated men in this region, Huntington concludes, often face one of three paths: migrate to the West, join fundamentalist organizations and political parties, or enlist in guerilla groups and terrorist networks.[16]

Death rates have actually reversed their decline in more than 30 countries worldwide.

Discontented elites may in turn mobilize less-educated groups to their cause. Investigations into Thailand's recent upsurge in violence point to the possible involvement of Muslim extremist groups, who may be actively targeting young men of strong religious faith and little formal education to further their broader Islamist goals. In the town of Suso, which lost 18 men under the age of 30 in the April 2004 uprising, most of the dead had graduated from the country's privately run Islamic schools, which are often a last resort for families who cannot afford mainstream college educations. In Pakistan, meanwhile, studies estimate that as many as 10–15 percent of the country's 45,000 religious schools

have direct ties to militant groups.[17]

How strong is the link between young people and conflict? Political scientists Christian Mesquida and Neil Wiener of Canada's University of York reviewed conflicts over the latter half of the twentieth century and found that in conflict-torn regions such as the Balkans and Central Asia, countries with younger populations experienced more battle-related deaths per thousand people than other countries. More recently, researchers with the Washington-based group Population Action International (PAI) reported that countries with high youth bulges—where young adults accounted for more than 40 percent of all adults—were roughly two-and-a-half times more likely to experience an outbreak of civil conflict during the 1990s than countries below this benchmark.[18]

The good news is that large youth bulges will eventually dissipate if fertility rates continue their projected worldwide decline. Already, during the 1990s, the number of countries where young adults account for 40 percent or more of all adults decreased by about one sixth, primarily because of falling birth rates in East Asia, the Caribbean, and Latin America. At the same time, however, a smaller subset of countries in the earlier stages of their demographic transitions—most of them in sub-Saharan Africa and the Middle East—have experienced rapid growth in their populations aged 15 to 29, who sometimes account for more than half of all adults. Until these youthful populations decline and employment prospects improve, these countries will likely continue to pose a challenge to regional development and international security.[19]

In general, completing the demographic transition is viewed as a welcome accomplishment. Children in smaller families typically grow up healthier and more skilled than others in their same income class. But reaching the end of the transition bears its own challenges. In some post-transition countries, including Russia, Japan, and most of Europe, what was once a postwar youth bulge has now matured into an alarming bulge in senior citizens. While aging and sustained population decline are unlikely to be as threatening to global security as large numbers of unemployed young men are, policymakers and economists alike are increasingly concerned about the implications of this development for economic growth and military preparedness. (See Box 2–1.)[20]

The Emerging Threat of HIV/AIDS

In 2003, nearly 3 million people died from HIV-related infections, bringing to more than 20 million the total number of AIDS deaths since the first cases were identified in 1981.[21] Largely because of this rising pandemic, death rates have actually reversed their decline in more than 30 countries worldwide. The global spread of HIV/AIDS threatens to create a lethally imbalanced age structure—but in a way never before seen in history.

BOX 2–1. ARE AGING AND DECLINING POPULATIONS A PROBLEM?

In recent years, population and security analysts have begun to question the implications of demographic change in countries that have fully completed their demographic transitions, where populations are increasingly aging. In Japan, the United States, and many European countries, birth rates are now slowing to the point where the share of working-age adults has shrunk and elderly populations represent one fifth or even one third of the total. Russia, Italy, much of Eastern Europe, and a dozen or so other countries have experienced such low fertility in recent years that the population is actually in decline—a situation that could only be offset in the short run by very high levels of immigration. With a total fertility rate of barely more than one child per woman, Russia's population is now shrinking by 0.7 percent annually—roughly a million people every year.

So far, no country has shown overt signs of economic or political instability because of population aging. (In fact, both Japan and Russia have recently experienced economic upsurges.) Economists, however, are alarmed by demographic projections indicating that the ratio of workers to retirees in Europe will likely drop by half, from four to only two—putting stress on retirement systems and upward pressure on

wages. Labor shortages and wage increases in many sectors could in turn affect enlistment rates in the military forces of these countries, leading to a scarcity of career soldiers.

Meanwhile, health care costs for senior citizens are rising at double-digit rates. By 2040, the costs of health care and other public benefits to the elderly are projected to exceed 27 percent of the gross domestic product in Italy, Spain, Japan, and France. And in the United States, funding for seniors has failed to keep pace with government promises: the deficit between what has been pledged and actual funding now tops $44 trillion.

Unlike the many developing nations grappling with the consequences of rapid population growth, however, most industrial countries have considerable capacity to adjust to the challenges of population aging. Several European governments, for example, have accepted additional immigrants, extended the retirement age, and attracted more women into the workforce while enhancing child care benefits. And Japan is returning some of the responsibility for elderly care to families and relying more heavily on technology and outsourcing for low-skill or labor-intensive jobs.

SOURCE: See endnote 20.

No disease in human experience debilitates and kills exactly as AIDS does, laying low by the tens of millions not the young and the old but people in the most productive years of their lives. Nearly 90 percent of fatalities associated with the disease occur among people of working age. Nine countries in sub-Saharan Africa are now losing more than 10 percent of their working-age adult populations every five years, largely as a result of high HIV prevalence. (See Table 2–2.) (By comparison, industrial countries typically lose about 1 percent of this age group to death every five years, while even in war-torn countries with relatively low HIV prevalence, such as Afghanistan and Sudan, the figure was about 4–6 percent in the late 1990s.)[22]

The International Labour Organization predicts that, in the absence of treatment, an estimated 74 million workers worldwide could die from AIDS-related causes by 2015—the equivalent of the loss of an entire country the size of South Africa or Thailand. As Peter Piot, the executive director of UNAIDS, has noted, "AIDS is devastating the ranks of the most productive members of society with an efficacy [that] history has reserved for great armed conflicts."[23]

Rather than provoking direct confrontations over, say, access to AIDS drugs or treatments, HIV/AIDS will likely make its mark more insidiously—affecting industrial development, reducing agricultural production, weakening military and political integrity, and eroding the capacity to respond to chronic domestic discontent and sudden crises in the world's weakest countries. Where the epidemic is most advanced, as in sub-Saharan Africa, the disease is remarkably widespread, disturbing the operations of government, schools, factories, farms, health care facilities, and the armed forces. The impact of AIDS on the military

Table 2–2. Countries with the Highest Rates of Adult Death, 2000–05

Country	Death Rate Among Adults Aged 15–64, 2000–05
	(percent dying)
Zimbabwe	18.1
Swaziland	15.9
Zambia	15.2
Lesotho	14.7
Botswana	14.0
Malawi	11.2
Namibia	10.7
Central African Republic	10.5
Mozambique	10.2

SOURCE: See endnote 22.

forces of some 20 countries worldwide is particularly alarming, posing threats to operational readiness and peacekeeping commitments—and representing a dangerous reservoir for further spread of the disease. (See also Chapter 3.)[24]

AIDS-affected countries could also become vulnerable to political instability as the growing number of children orphaned by the disease increases the proportion of dependent people, exacerbates poverty, and widens inequalities. According to a recent joint report by UNICEF, UNAIDS, and the U.S. Agency for International Development, between 2001 and 2003 the number of AIDS orphans worldwide increased from 11.5 million to 15 million—the vast majority of them in Africa. In the absence of government support for placing homeless orphans in families and in schools, these children could become a source of future urban discontent, criminal activity, and recruits for insurgencies.[25]

So far there are no outward signs of AIDS-inspired mass violence or rebellion. But this could change as the pandemic continues its global spread. In 2003 the number of people living with HIV grew by nearly 5 million, to a total of 38 million—the greatest single-year increase since the start of the epidemic. While births still more than offset AIDS deaths in most of the 53 countries now considered AIDS-affected, in a handful of countries this situation could be reversed within the next few years, causing population declines. The Central African Republic, Lesotho, Malawi, Mozambique, Swaziland, Zambia, and Zimbabwe now have life expectancies under 40 years, while at least 13 countries have suffered measurable reversals in human development since 1990. Tragically, some of the most dramatic slowing in population growth worldwide is now occurring in countries experiencing both high rates of HIV prevalence and significant declines in fertility. But unlike the birth rate–driven slowdown that proved economically beneficial to many countries over the last half century, slowed population growth due to AIDS is likely to hinder economic growth.[26]

Overall, the AIDS epidemic is creating a pernicious combination of deepening poverty and a loss of trust in governments that are less and less capable of delivering basic services, let alone promoting economic development. U.S. Secretary of State Colin Powell has portrayed the disease as a looming "destroyer of nations...[with] the potential to destabilize regions, perhaps even entire continents." And former U.S. President Bill Clinton called the pandemic a prelude to "more terror, more mercenaries, more war...and the failure of fragile democracies." Without the capacity to plan for and resolve long-term problems and to respond to acute crises, the countries most heavily affected by AIDS risk stalling in their paths toward industrialization, democracy, and the final phases of the demographic transition.[27]

Rapid Urban Growth

For several years now, relentless cycles of drought and flooding have wreaked havoc on the tiny country of Malawi, in the heart of southern Africa. In 2002 and 2003, torrential rains caused massive mudslides, washing away bridges and homes and devastating harvests of maize, the main food staple. Unable to eke a living from the ravaged countryside, rural residents have flocked in droves to the country's burgeoning cities—giving Malawi the dubious distinction of being the world's fastest-urbanizing nation today. But like many rapidly growing developing countries, this impoverished nation is ill prepared to deal with the side effects of this urban onslaught, including rising homelessness, crime, and unemployment.[28]

Since 1950, the world's urban population has more than quadrupled, from 733 million to just over 3 billion, and it is now growing faster than world population as a whole. Roughly 60 percent of this growth is the result of natural increase—urban births minus deaths. But an unrelenting wave of migration accounts for nearly all the rest, as people are simultaneously "pushed" to cities to get away from stagnating or war-torn rural areas and "pulled" by the allure of more promising jobs, education, and the attractions of modernization. As this trend continues, humanity is approaching a historic milestone: by 2007, for the first time ever, more people will live in cities than in rural areas.[29]

Urbanization is by and large a positive demographic trend. Traditionally, the movement to cities has contributed to economic growth and global integration, as more people find homes close to schools, health clinics, workplaces, and communication networks. Because cities are centers of industry and education, people in them are almost always ahead of those in rural areas in gaining access to new technologies, information, and goods. Urban dwellers are also generally the first to experience declines in infant mortality and fertility, as governments give priority to urban clinics that can provide more cost-effective health care delivery. In Africa, for instance, recent changes in childbearing behavior and fertility declines are mostly an urban phenomenon.[30]

Countries with rapid rates of urban population growth were roughly twice as likely as other countries to experience civil conflict during the 1990s.

But urbanization is a double-edged sword. The very features that have made cities in the industrial world prosperous—a youthful population, ethnic and religious diversity, a middle class, and proximity to political power—are potential sources of volatility for many surging and economically depressed cities in the developing world. The remarkable rates of growth that many developing-country cities sustained during the 1970s and 1980s—Jakarta and Delhi, for instance, roughly doubled in size during this period—could not help but deplete city budgets, flood job markets, and challenge the adequacy of existing services and infrastructure. Still today, fewer than half the residents of most urban centers in Africa, Asia, and Latin America have fresh water piped to their homes and fewer than a third have good-quality sanitation. Municipal governments in the poorest countries are typically least capable of mustering the human and financial resources to contend with these problems, especially when the poorest, non-taxable segment of the population continues to grow rapidly.[31]

When local governments and community leaders have the will and resources to overcome ethnic, religious, and regional differences, close interaction among disparate groups can have positive social implications. "Cities force people to mix and become familiar with members of groups whose paths might never cross in rural areas," notes Marc Sommers of CARE International. But when group interests conflict, cities can harbor intense economic and political competition, particularly if housing and job markets illuminate disparities in access to education, capital, and political power.[32]

When historic grievances and cultural misunderstandings resurface, cities can become a locus for ethnic and religious conflict. In 1992, outside the normally sleepy, rural northern Indian town of Ayodhya, some 150,000 Hindu militants descended on a virtually abandoned sixteenth-century mosque, attacking security forces and destroying the building. Rather than spreading through the nearby countryside, where Muslim and Hindu communities coexisted peaceably, the hatred exploded hundreds of kilometers away, in Mumbai, Calcutta, Ahmedabad, and Delhi. Three days of intense violence in Ahmedabad and nearby Vadodara left more than 850 people dead and thousands homeless. In total, nearly 95 percent of the 1,500 people killed in the communal riots that ensued were urban dwellers. The incidents, some of which were reportedly condoned by local government leaders, not only unraveled efforts by moderates to broker cooperation between Muslim and Hindu politicians, they also aggravated already delicate relations between India and neighboring Pakistan.[33]

Overall, according to Population Action International, countries with rapid rates of urban population growth—4 percent a year or more—were roughly twice as likely as other countries to experience civil conflict during the 1990s. Disenchanted urban youth—whether politicized students or the angry unemployed—are often among the first recruits. For decades now, the landless sons of Pakistani farmers have crowded into dismal urban slums around Karachi and Islamabad, where many have found an outlet for their frustrations through political discontent and sectarian violence.[34]

Urban unrest will likely only increase as the largest cities in the developing world extend further into the countryside and as crossroads and market towns rapidly transform into population centers. Today, 16 of the world's 20 "megacities"—those with more than 10 million residents—are in developing countries, and the urban share of the developing world's population is projected to reach 60 percent by 2030, compared with 42 percent in 2003 and just 18 percent in 1950. Asia already has more people living in cities than live in all industrial countries combined, and about half of its urban growth still lies ahead.[35]

As the associations between urban growth and conflict become more apparent, analysts will need to pay more attention to the role of urbanization in their assessments. Over the short term, policymakers should consider programs that improve the quality and capacity of municipal governance, stimulate job creation, and strengthen ethnic-community relations in rapidly urbanizing regions. Over the longer term, however, only the slowing of population growth in most of the world's countries—particularly those in earlier stages of the demographic transition—offers hope that cities, too, will grow at a more manageable and stable pace.[36]

Competing for Water and Cropland

In the dry and contentious region of the world once known as the Fertile Crescent, three countries—Iraq, Syria, and Turkey—have long pondered schemes to reap additional water from the Tigris and Euphrates Rivers. But there is a problem. In 2002, the combined population of the three countries grew by more than 2 million, to 110 million people. Neither the Euphrates, which flows from central Turkey through Syria and Iraq, nor the Tigris to the east can provide enough water to satisfy the rising annual demands of these users, particularly during drier years. "Each country has acknowledged the impossibility of marrying their schemes," wrote Douglas Jehl in the *New York Times* in 2002. "But none has shown any willingness to scale back."[37]

Many regions of the world are experiencing rapid declines in both the quality and the availability of critical natural resources. More than 30 countries—most of them in Africa and the Middle East—have now fallen below even the most conservative benchmarks for scarcity of either cropland (0.07 hectares per person) or renewable fresh water (1,000 cubic meters per person). (See Table 2–3.) Some countries have reached this situation due to a combination of harsh climate or terrain and a rapidly growing population; others, however, are experiencing these scarcities almost exclusively as a result of population growth. (Four countries—Egypt, Israel, Kuwait, and Oman—have reached critical levels in both areas, with population growth rates above the developing-world average of 1.5 percent.)[38]

Faced with this reality, analysts have expressed growing concern about the inevitability of "resource wars" in the coming decades, particularly over fresh water. Nearly half the land on Earth lies within some 263 international river basins that span the borders of two or more countries—accounting for 60 percent of the world's freshwater supply. (See Chapter 5.) Egypt's 71 million inhabitants, for example, depend

Table 2–3. Top Countries Facing Per Capita Scarcity of Cropland or Fresh Water, 2005

Country	Available Cropland Per Capita	Population Growth Rate, 2000–05	Country	Available Renewable Fresh Water Per Capita	Population Growth Rate, 2000–05
	(hectares per person)	(percent)		(cubic meters per person)	(percent)
Kuwait	<0.01	3.46	Kuwait	7	3.46
Singapore	<0.01	1.69	Saudi Arabia	78	2.92
Maldives	0.01	2.98	Libya	173	1.93
Bahrain	0.01	2.17	Jordan	174	2.66
Brunei	0.02	2.27	Yemen	186	3.52
Iceland	0.02	0.79	Israel	299	2.02
Malta	0.02	0.42	Oman	331	2.98
Bahamas	0.03	1.13	Algeria	426	1.67
Oman	0.03	2.98	Tunisia	498	1.07
Qatar	0.03	1.54	Burundi	547	3.10
Egypt	0.04	1.99	Rwanda	581	2.16
Japan	0.04	0.14	Egypt	775	1.99

SOURCE: See endnote 38.

on the Nile for more than 97 percent of their water needs, but they must share the river with Ethiopia and eight other upstream countries, all militarily weaker and struggling to boost agricultural production and urban services in the face of rapid population growth, cyclical long-term drought, and seasonal rainfall.[39]

Yet so far, history has shown us that nations picking fights over fresh water—and other renewable natural resources, for that matter—have mostly brandished words rather than guns. International resource-related hostilities have typically ended in nonviolent outcomes, such as negotiated agreements or the formation of regulatory institutions to resolve disputes. For the near future, at least, the greatest risk will likely be population-influenced resource disputes not between countries but within them. While theorists still hotly debate whether such disputes ever lead directly to large-scale civil conflicts or collapse of a state, there is gen-

eral agreement that scarcities of water, cropland, and other resources can increase the risks of smaller, localized friction and even sporadic violence. These disputes tend to escalate into serious armed conflicts only when the institutions that should help manage a country's resources and resolve scarcities are too weak, poorly funded, or corrupt to do so. In the future, however, continued rapid population growth in the developing world—and the unprecedented demands on supplies this will bring—could challenge even the most capable institutions.[40]

One source of rising tension is the allocation of fresh water among diverse local users—particularly among farmers and the more politically influential and growing set of urban and industrial users. The International Food Policy Research Institute estimates that at current rates of population growth and water consumption, global household water use will increase by more than half by 2025. At least some of this growth will occur at the expense of farmers—particularly in the developing world, where crops depend more heavily on irrigation than in North America or Europe. Not only will this increasingly pit the economic integrity of rural communities against the growth of urban and industrial centers, it could also pose threats to national or regional food security by hampering food production efforts.[41]

Competing claims over fresh water may complicate efforts to resolve long-standing conflicts in the Middle East. For more than three decades, Israel has restricted Arabs in the occupied West Bank from drilling new wells for agriculture, while Israeli settlers continue to drill deeper—in some cases causing water tables to fall far below the reach of the Palestinian wells. Since 1967, the proportion of their cropland that Palestinian farmers irrigate has dropped from 27 percent to around 5 percent, contributing to unemployment and productivity loss as well as to a list of grievances against Israeli rule.[42]

Despite the importance of fresh water to a country's economic and human development, studies indicate that civil disputes over water have tended to be less volatile than those over cropland. This may reflect differences in how the two resources are owned and priced as well as the access people have to them. Water is traditionally viewed as a community resource, and disputes over water rights are frequently defused by water agencies or in local or provincial courts. Land, in contrast, lends itself to long-standing private and often inequitable ownership. When cropland becomes locally scarce, disputes over farmland distribution may surface between peasants who recognize age-old ethnic communal or squatters' rights and landlords or ethnically different landholders who secured their holdings through deeds or ancestral conquest.[43]

In some cases, local disputes over land have mushroomed into larger threats. The Zapatista rebellion in the southern Mexican state of Chiapas, for instance, grew out of centuries-old tensions between locally powerful landowners and land-poor Mayan peasants who had been hemmed in by other poor settlers and excluded from government-protected forest reserves. The scale and organization of the rebellion unnerved foreign investors and, according to some analysts, may have contributed to the nationwide monetary crisis that eventually undermined the power of the ruling regime.[44]

Competing claims over fresh water may complicate efforts to resolve long-standing conflicts in the Middle East.

Notwithstanding the many historical examples of resource-related violence, research suggests that the links between resource scarcity and conflict may not be as strong as those between conflict and other demographic factors such as urbanization or the youth bulge. Most theorists agree that environmental changes—whether cropland or water scarcity or increased deforestation and soil erosion—are only a small part of a complex mix of stresses that promote mass civil instability. Sociologist Jack Goldstone, for instance, argues that discontent among the elite classes is a more critical element in the evolution of civil conflict. Environmental scarcities and "ecological marginalization," Goldstone contends, are not particularly powerful determinants of vulnerability because these factors rarely adversely affect the livelihoods or power of the elite.[45]

Moreover, a wide range of opportunities exist to mediate potentially explosive resource scarcities through legislation and sound economic policies, trade, and technical cooperation. Rwanda, for instance, is now finalizing a new national land policy that, if effectively implemented, could address long-standing concerns over farmland distribution. (See Box 2–2.) And many resource-scarce European countries, as well as industrialized Asian nations such as South Korea, Taiwan, and Japan, have boosted imports of food and animal feed to ease burgeoning agricultural demands on land and fresh water. Others have begun to import fresh water via pipelines and other direct means, have become more water-efficient, or are turning to desalinization to supplement sparse drinking water.[46]

For most developing countries faced with dwindling resources and rapidly growing populations, there is little immediate promise of attracting the capital needed to industrialize rapidly.

Of course, most wealthy industrial countries can afford to invest in resource-efficient technology and possess ample foreign currency to import grain—making them much less vulnerable to a conflict involving natural resources. For most developing countries faced with dwindling resources and rapidly growing populations, there is little immediate promise of attracting the capital needed to industrialize rapidly or to transform land and water use practices radically. Whatever the historical record, the outcomes of increasing competition for these critical resources are uncertain and less than reassuring. For this reason, the need to address the underlying demographic forces that drive resource scarcities—and to invest in programs that will help to slow population growth in these affected regions—is increasingly urgent.

Minimizing Risks, Moving Forward

In most cases, the four "demographic risk factors"—growing proportions of young people, the HIV/AIDS crisis, rapid urbanization, and reduced availability of cropland or fresh water—do not occur in isolation. Rather, they interact with each other and with nondemographic variables, including historic ethnic tensions, unresponsive governance, and weak institutions, to produce stresses that challenge government leadership and the capacity of countries to function effectively. While unlikely to lead directly to political chaos or warfare, they can greatly compound a country's vulnerability to conflict. According to Population Action International, countries displaying two or more of either a high proportion of young people, high rates of urban growth, or shortages in the per capita availability of cropland or fresh water accounted for 23 of the 36 countries that experienced new outbreaks of civil conflicts during the 1990s.[47]

Fortunately, demographics is not destiny. But the likelihood of future conflict may ultimately reflect how societies choose to deal with their demographic challenges. For instance, PAI found that roughly half the countries that should have been at very high demographic risk during the 1990s in fact navigated the post–cold war period peacefully. Why? In at least some of these cases, countries were able to offset these risks through strong governance, conflict resolution, ethnic mediation, or successful economic policies—including creating jobs in cities, importing critical resources, distributing farmland, and encouraging emigration. The West African country of Cape Verde, for instance, may have offset its vulnerability during the 1980s and 1990s by making it easier for its citizens to migrate to Europe for work

BOX 2–2. LAND POLICY REFORM IN RWANDA

Access to land is a critical issue in Rwanda, which is experiencing severe land scarcity. Over the last 40 years, the population density on agricultural land has increased from 121 people to about 321 people per square kilometer. Some 60 percent of households have less than a half-hectare of land. Land distribution is also highly unequal: the quarter of the population who own more than one hectare per household account for nearly 60 percent of the country's holdings. Because of population pressures, the fallow periods are minimal, which leads to decreased soil fertility, and many households now cultivate soils ill suited for agriculture.

In recent years, successive waves of population displacement and resettlement have had profound implications for land settlement and ownership. Rwanda has been affected by internal conflict since independence, culminating in the 1994 genocide that resulted in the deaths of some 800,000 people. Arguments over land are now among the most numerous as well as the most intractable legal disputes in the country. Though post-conflict governance has generally been more inclusive, Rwanda has had little experience with participatory governance, and the ability of civil society to influence policy remains questionable.

In response to these challenges, the Rwandan government is now finalizing a new National Land Policy and National Land Law. The policy aims to establish appropriate plans and guidelines for land allocation and use, promote land consolidation, establish local land commissions, and undertake registration of all plots. It also aims to encourage community participation and to ensure that women benefit from land through inheritance.

Land consolidation is one of the most difficult issues. The process of combining smaller plots into more productive and economically viable holdings will likely involve some degree of compulsion, raising the danger that poorer farmers will lose control over their land-based

livelihoods. The composition, technical capacity, and accountability of the new land commissions charged with implementing land resettlement will be important.

Registering Rwanda's land plots, meanwhile, will require massive financial and technical resources. The new policy calls for landowners to shoulder the costs of registration—leading to fears that the wealthy will be able to register land at the expense of the poor, though it seems that a two-tier system may be established, with the system subsidized for the majority. The policy calls for the creation of a land reserve to allocate land to those who have none, but its current narrow definition of landlessness excludes a high proportion of this population.

The establishment of district land commissions could provide effective means for resolving local disputes over land access—many of which arose from the "land-sharing" that occurred after 1994 to accommodate the competing claims of refugees. To be successful, however, the new land commissions must be accountable to the local communities rather than simply to higher levels of government.

Overall, the new land policy and law provide several potentially positive opportunities for improved land management and environmental governance. But they will need to be implemented cautiously and transparently. Already, confusion has arisen at the community level over the possible implications for household landownership. The involvement and coordination of all sectors of government will need to improve, and community-based organizations and nongovernmental groups will need to play a key role in supporting the new land commissions and monitoring implementation.

—*Chris Huggins and Herman Musahara,*
African Centre for Technology Studies

SOURCE: See endnote 46.

and by encouraging them to send a share of their income back home.[48]

East and Southeast Asian countries like South Korea and Malaysia, meanwhile, turned a growing youth bulge in the 1960s and early 1970s into a positive economic force by making critical investments in education and job training. And many oil-rich countries created urban employment, large armies, and bloated bureaucracies to absorb their surging populations—in addition to ruthlessly repressing any dissent that might lead to conflict. In North Korea, China, and Turkmenistan, the expansion of military and internal security forces probably helped repressive regimes maintain political stability during the post–cold war era despite large proportions of young adults.[49]

Over the long term, however, the only way to ease potentially volatile demographic pressures will be by tackling population growth head-on. The significant drop in fertility that has occurred in some 20 developing countries in East Asia, the Caribbean, and Latin America in recent decades is an encouraging trend. Much of the credit for this transformation goes to countries that have invested in vital reproductive health services, including improved access to family planning and maternal and child health care. Tunisia and Thailand, for example, were able to move more quickly into the last stage of the demographic transition through proactive social policies, cost-effective and widely accessible reproductive health programs, and technical assistance from abroad. Many countries also promoted policies that helped girls stay in school and increased women's opportunities for employment outside the home—bolstering women's status and income, improving child nutrition and child survival, and expanding demand for modern contraception.[50]

Unfortunately, most countries at high demographic risk today lack the institutional capacity to respond to these challenges. They need the stable financial systems and markets, adequate law enforcement, clearly delineated property rights, and functioning educational and health care systems that are the foundations of stronger countries. And in most cases, the level and quality of services that might keep the demographic situations of these countries from worsening—including family planning, girls' education, maternal and child health, and HIV/AIDS prevention—are woefully inadequate.

Extending these services to the world's weakest and poorest countries will require far greater international collaboration and assistance than is evident today. Wealthier countries will need to step in to contribute technical expertise, funding, and supplies. The world is now facing critical shortages in supplies needed for contraception, HIV/AIDS prevention, and other sexual and reproductive health care services. According to one recent study, in 1999 fewer than five condoms per reproductive-age man were available from international donors and governments for all of sub-Saharan Africa. And the annual cost of supplying enough free and affordable condoms worldwide is projected to more than double within the next decade or so, from $239 million in 2000 to $557 million in 2015.[51]

Unfortunately, just when the need is most urgent, international support for family planning and related services continues to wane. In 2000, it came only to half the $17 billion goal that the United Nations set in 1994 at the International Conference on Population and Development in Cairo. The U.S. share of that U.N. goal works out to $1.9 billion, but in 2000, U.S. funding for reproductive health—which includes programs in family planning, HIV/AIDS, and maternal and child care, and a contribution to the U.N. Population Fund (UNFPA)—amounted to only about one third of that pledge. In 2004, for the third consecutive year, the U.S. government withheld the $34 million it owes UNFPA—roughly 10 percent of that key agency's budget. Continued failure to live up to international commitments will hamper significantly the progress of the demographic transition and make it even more difficult to halt the spread of HIV.[52]

While policies and programs that influence population trends have traditionally been the sphere of international donors and of health and social service providers, the international security community, too, has begun to take notice. In April 2002, in a written response to congressional questioning, the U.S. Central Intelligence Agency noted that "several troublesome global trends—especially the growing demographic youth bulge in developing nations whose economic systems and political ideologies are under enormous stress—will fuel the rise of more disaffected groups willing to use violence to address their perceived grievances." The agency warned that unless U.S. counterterrorist operations seek to address many of the underlying causes that drive terrorists—including demographics—they may ultimately fail to eliminate the threat of future attacks.[53]

Despite such warnings, security strategists and policymakers have been slow to take action. For the most part they have chosen to focus their attention on the promotion of democracy and market reforms that, in the absence of parallel changes in the social sphere, ironically may destabilize countries. Helping countries approach the final phase of the demographic transition—a phase in which people live long lives and families are typically small, healthy, and educated and where population has nearly stopped growing—promises to help reduce the frequency of conflicts and

bring about a more peaceful world. There is now ample evidence that by addressing key factors related to demographic change, governments could strengthen the security of strategic countries, pivotal regions, and the world as a whole.[54]

Just when the need is most urgent, international support for family planning and related services continues to wane.

Without the backing of the national security community, international reproductive health programs risk being ignored—or, much worse, being sacrificed amid domestic struggles for political advantage. National security and defense analysts as well as military officials often have influence that transcends changes in government leadership and shifts in the political climate—making them important allies in the push for demographic change. By including population data and projections in area studies, operational environment forecasts, and other security and threat assessments, they can provide accurate information and guidance to policymakers, the media, and opinion leaders on the global benefits that can be expected from a completed demographic transition. Their support can help to secure funding for programs in family planning, girls' education, maternal and child health, and HIV/AIDS prevention and treatment.[55]

The security community can take more direct action as well, by helping to facilitate access to reproductive health services for refugees, civilians in post-conflict situations, and all military personnel via peacekeeping and other operations. Senior military and diplomatic officers are often the only ones in direct authority in restricted areas and are thus uniquely positioned to help boost the availability of reproductive health services to these people. Military commands can also lend logistical and organizational support to outside groups charged with offering reproductive health care in post-conflict environments—including UNFPA, the U.N. High Commissioner for Refugees, government health agencies, and various nongovernmental organizations.[56]

Military and diplomatic officers and international aid organizations can help ensure that women, in particular, have the opportunity to represent their own interests and those of their children—not just in refugee situations, but also during the peace process and the rebuilding of conflict-affected countries. The U.K. Department for International Development, for instance, has strengthened its inclusion of women in mediation and peace processes as well as in legal and political affairs in post-conflict environments. By supporting social and political changes that encourage girls to stay in school and offer women wider opportunities, these groups can help boost women's status, lower fertility rates, and possibly even shift national priorities away from strife and toward human development. "The issue of equal participation by women is not simply an issue of gender equality and human rights but could represent the decisive factor in maintaining peaceful development in a troubled region," noted Lul Seyoum, an Eritrean women's rights activist, at a 1999 conference on women and conflict.[57]

In the area of HIV/AIDS prevention and treatment, governments with exemplary programs in their own armed forces can share these activities more widely with military and civilian populations elsewhere. Donor governments, meanwhile, can expand their reproductive health programs to ensure that high-quality care—including comprehensive contraceptive information and ser-

vices, prevention of sexually transmitted diseases, and maternal and child health care—is available to military personnel and their families. Such initiatives could help armed forces in Africa and Asia reduce the high rates of HIV transmission now threatening military personnel, their families, and neighboring communities.

Clearly, fostering demographic change can bring great opportunity. But it is not a security cure-all. Obviously, demographic progress alone cannot guarantee that a country will oust oppressive leadership, join the global family of democratic nations, or resist collusion with insurgents and terrorists. Demographic changes cannot be relied on to reduce risks when armed conflict is already persistent or recurrent, when a culture of violence and retribution is firmly entrenched, or when troubles spill over from unstable neighboring countries. Colombia, Northern Ireland, and Sri Lanka are all areas where demographic advances in the 1990s should have helped alleviate risk. Yet in each case costly civil conflicts that had started in previous years remained active throughout the decade.[58]

By that same token, moving into the later stages of the demographic transition is not the only way a country can reduce its vulnerability to instability or conflict. Demographic processes rarely—if ever—act alone in raising or reducing the chance of political instability. Civil conflict, in particular, is a complex, multistage process that builds upon a country's historic and current vulnerabilities and is driven over time by largely country-specific and unpredictable events. International conflict resolution and peacekeeping efforts have done much to hold down the frequency of minor and incipient conflicts. And there is strong evidence that countries can reduce their vulnerability to conflict through the spread of democratic institutions, economic development and the alleviation of poverty, and expanded international trade relations.[59]

Nevertheless, demography needs to be part of the analysis. If the relationships between the demographic transition and conflict hold in the coming decades, decisions made today that affect this momentous transition could have an enormous influence not only on demographic prospects but also on the future of global security. For countries in the early stages of their demographic transitions, it could take nearly two decades after fertility begins to fall to observe a significant reduction in population growth. The many risks of delaying this transition underscore the need for governments to put supportive policies into effect sooner rather than later.

Environmental Refugees

In 2003, roughly 1 in every 370 people on Earth—17.1 million in all—were classified as "persons of concern" by the United Nations High Commissioner for Refugees (UNHCR), the U.N. agency dedicated to the protection of refugees and other displaced populations. This figure included 9.7 million refugees (people fleeing persecution or fear of persecution), 1.1 million returned refugees, 4.2 million internally displaced persons (IDPs), 233,000 returned IDPs, 995,000 asylum seekers, and 912,000 people who fall into other categories, including "statelessness."[1]

This estimate does not include the growing number of environmental refugees—people "forced to leave their traditional habitat, temporarily or permanently, because of a marked environmental disruption...that jeopardized their existence and/or seriously affected the quality of their life." Among the natural or human-caused environmental factors that force people to move are resource scarcity and inequitable distribution of natural resources, deforestation and other environmental degradation, natural or industrial disasters, climate change, systematic destruction of the environment as an instrument of war and remnants of war, overpopulation, and development projects. In 2004, Essam El-Hinnawi of the Natural Resources and Environmental Institute in Cairo estimated there are now 30 million environmental refugees worldwide.[2]

In 2002, the Red Cross reported that the number of people killed by natural disasters, including floods, droughts, and earthquakes, fell by 40 percent between the 1970s and 1990s (mainly attributed to better disaster preparedness), while the number whose lives were adversely affected by these events increased by 65 percent. The toll is predicted to rise as the projected impacts of climate change intensify: according to the Intergov-

ernmental Panel on Climate Change, there could be as many as 150 million environmental refugees by 2050.[3]

Furthermore, mass population displacements may trigger instability or conflict in the country of origin, in host countries, or within the region. Scarce resources may be further degraded or depleted and give rise to competition over access; insufficient infrastructure or inequitable distribution may reinforce social divisions and tensions. Overcrowding, unsanitary conditions, and lack of potable water could lead to deadly epidemics. If the global population movements predicted are realized, these and other impacts could have severe implications for global security.

Despite the seriousness of these trends, the issue of environmental refugees has received scant attention at the highest levels. The focus has been more on the impact that mass displacement has on the environment rather than on the role the environment itself plays in creating refugees. More analysis is needed to help define the concept of environmental refugees, identify underlying causes of the problem, explore short-term and long-term consequences, and draw up effective responses.

As possible responses to the consequences of environmental displacement, two distinct approaches suggest themselves: a re-examination of the international effort to address refugee crises and analysis of the humanitarian-environmental link concerning operations on the ground.

Under international law, refugees are given special status that grants them certain rights and makes them eligible for legal and material assistance. All countries party to the 1951 Convention Relating to the Status of Refugees or its 1967 protocol are obliged to provide certain basic minimum resources and protection to refugees. UNHCR implements

Girls practice writing in Kakuma Refugee Camp, Kenya

the convention and protocol and provides legal protection, coordinates emergency relief, and helps find long-term solutions.

UNHCR has opposed expanding the convention to include the environment as a source of persecution—arguing that although forced movement is a common theme, people leaving for environmental reasons can still seek protection from their own government, whereas those fleeing traditional categories of persecution often cannot. Furthermore, UNHCR is already struggling to focus its limited resources on the wide range of humanitarian crises covered by its existing mandate. Critics argue, however, that excluding the environmentally displaced is narrow-minded and especially unfair to poor countries whose environmental condition may be the result of other polluters' behavior.[4]

Given the current limitations of legal protection under the refugee convention, the lack of consensus to change it, and the shortfalls on UNHCR resources, many observers maintain that the plight of environmental refugees needs a broader response from the U.N. system and the international community. While debate continues on whether to reform the convention or create a new one, the lack of a designated organization to focus on environmental refugees has led to an inconsistent and often incomplete response to displacement crises.[5]

In addition to the institutional and legal response, organizations that provide assistance in crises could take further action. However, incorporating environmental concerns into existing relief efforts can be tricky. Unlike the consequences of many conventional crises, environmental damage is often less obvious, neither improving with time nor remaining within national borders. The aid community, though often aware of the environmental dimension, has tended to postpone action unless it is clearly linked to a humanitarian or security matter.

While sustaining human life is paramount during a humanitarian emergency, environmental mitigation need not be seen as a luxury or a burden but rather as another tool that aid agencies can use to improve their response to the displaced. One study suggests that "an average $1 invested in mitigation can save $7 in disaster recovery costs." Although a number of aid agencies have already established programs and guidelines that focus on environmental mitigation, further research may be needed on how third parties might manage these environmental effects more efficiently and why they need to do so in the overall humanitarian response.[6]

The anticipated increase in environmental refugees threatens to undermine local stability, environmental security, and the quality of life for millions of people. This increases the pressure on the international community—both institutions and those groups who provide assistance in times of need—to develop a concerted, focused effort to better define the problem of environmental refugees and to find a solution that meets basic human needs and helps preserve the quality of the environment.

—*Rhoda Margesson,*
U.S. Congressional Research Service

CHAPTER 3

Containing Infectious Disease

Dennis Pirages

Moving slowly westward with merchants and other travelers along trade routes from China, the Black Death arrived in Kaffa in the Crimea in 1346. Carried from place to place by the black rat (*Rattus rattus*) and associated fleas, the disease moved slowly and inexorably across the European continent, reaching France, Italy, and Spain in 1348. The plague moved on and reached the eastern North Sea and the Baltic Sea in 1350. In much of Western Europe, nearly 40 percent of the population died during this epidemic.[1]

In a remote part of southern China in late 2002, a new respiratory disease that became known as SARS (severe acute respiratory syndrome) jumped from animals to people and quickly spread to other parts of the country. In a matter of only a few weeks the disease had swiftly moved with travelers on trains and planes across Asia and then on to much of the rest of the world. Within just six months SARS had been reported in

29 countries, killing nearly 800 people and sickening more than 8,000. Fortunately, the extent of the outbreak was limited by the fact that the virus was mainly transferred in respiratory droplets, making the disease susceptible to low-tech countermeasures such as surgical masks. Had the virus been more easily transmitted, the resulting pandemic (a global epidemic) could have killed millions around the world.[2]

Measured in numbers of premature deaths and associated physical suffering, the biggest source of human insecurity, past and present, is the dreaded Fourth Horseman of the Apocalypse—infectious disease. In 2002, for example, warfare accounted for 0.3 percent of deaths from all causes worldwide. Communicable diseases accounted for 26 percent. The acceleration of international travel and the growth of global commerce are making the rapid spread of infectious diseases a much more pressing security challenge.[3]

Dennis Pirages is Harrison Professor of International Environmental Politics in the Department of Government and Politics at the University of Maryland.

Over the centuries the number of deaths and injuries from military combat have paled in comparison to those from disease. It is estimated that all the wars of the twentieth century killed 111 million combatants and civilians, an average of about 1.1 million a year. Communicable diseases are now killing 14 times as many people annually. Even while the world was caught up in the military horrors of World War I, an influenza virus was spreading around the world. Estimates of the number of deaths from this pandemic vary, but it is thought that the virus took the lives of 20–40 million soldiers and civilians, many times the number of deaths directly attributable to the war.[4]

Given the very visible legacy of violent conflict among people, it is understandable why security has come to be defined mainly as a military matter. Warfare is vivid, violent, and destructive. But more important, people historically have been able to comprehend the nature of military threats and have come up with strategies to cope with them. In contrast, although pathogens have claimed far larger numbers of human casualties, relatively few resources from public treasuries have been devoted to dealing with them since the causes of and remedies for infectious diseases have been poorly understood.

Another reason disease has not been considered a serious security threat is that during much of the last century the campaign against infectious disease was presumably on the verge of victory. Buoyed by improved sanitary conditions, advances in medical research, new vaccines, and better antibiotics, health officials in the 1970s declared that industrial countries no longer need worry about the scourge of communicable diseases. Thirty-five years ago, Surgeon General W. H. Stewart told the U.S. Congress that the time had come to close the book on infectious disease. And 30 years ago, well-known biologist John Cairns wrote that the western world had virtually eliminated death by infectious disease.[5]

Unfortunately, their optimism was not well founded. Not only is the campaign to eradicate infectious disease an ongoing one, but pathogens have exhibited remarkable resilience and flexibility. Old maladies such as tuberculosis, malaria, and cholera have persisted and spread geographically. And previously unrecognized diseases such as Ebola, hepatitis C, hantavirus, and HIV have emerged as new threats. Why, after repeated declarations that the struggle against infectious disease was over, are experts now much more wary about the challenges presented by pathogenic microorganisms?

Reducing the worldwide death toll from infectious disease should receive the highest priority.

Research and surveillance are now yielding a much better understanding of the dynamics of disease outbreaks. And new remedies are increasingly available to deal with them. Reducing the worldwide death toll from infectious disease should receive the highest priority. If even a small percentage of the money that the United States and other military powers devote to defense were diverted to promoting better public health around the world, human well-being would improve dramatically, adding immensely to the personal sense of security. If health care spending in the world's 60 poorest countries could be steadily increased from the present $13 per capita to $38 by 2015, experts say, on average 8 million lives could be saved each year. This would require a total contribution from industrial countries of about $38 billion—a fraction of what the United

States recently spent to unseat Saddam Hussein in Iraq.[6]

There are hopeful signs that security thinking and priorities are beginning to change. Because of their impact on economic and military concerns, infectious diseases are increasingly being treated as a conventional security threat. Military forces are being weakened significantly and economies are being devastated in areas of the world being seriously affected by HIV/AIDS. In April 2000, the realization that governments could potentially be toppled because of the disease led former U.S. President Bill Clinton to declare HIV/AIDS to be a threat to U.S. national security. Thus, perceptions of the challenges of new and resurgent diseases are changing, understanding of their causes and consequences is increasing, and the determination to do something about this crucial aspect of security is growing.[7]

The Dynamics of Disease Outbreaks

Getting sick is a common experience, and most encounters with disease organisms cause relatively little long-term damage to victims. Human immune systems have coevolved with a variety of potential pathogens and have developed defenses over time for dealing with most of them. Significant outbreaks of disease can occur, however, when people encounter new pathogens or new serotypes of old ones. But even most of these unfamiliar pathogens do relatively little damage to victims, and symptoms disappear after a few days. Occasionally, however, virulent and debilitating pathogens emerge for which immune systems have few defenses, and fatal diseases can then move quickly through human populations.

Large-scale disease outbreaks, epidemics, and even pandemics occur when something happens to disturb an evolutionary equilibrium that normally exists between people and pathogens. Disequilibrium can result from changes in human behavior or circumstances, mutations in or movement of pathogens, or changes in shared environments. When people travel to new environs they risk encounters with pathogens to which they have little immunity. International travelers, for example, often become ill during their trips or shortly after they return. Most of the time these illnesses do little permanent damage and are simply a nuisance. But as the outbreak of SARS in 2003 illustrated, travel can sometimes have deadly results.

By the same token, reasonably benign viruses and bacteria can mutate into more destructive serotypes. And previously unknown pathogens can jump from animals to people. The current worldwide spread of avian flu is of particular concern since this disease, found thus far mainly in birds, can apparently mutate into a form that passes from person-to-person. Finally, environmental change can upset established equilibriums between people and pathogens, facilitating new disease outbreaks. Climate change, for instance, is expected to alter temperature and rainfall patterns, thus permitting tropical diseases to thrive in previously cooler areas where they could not survive before.[8]

In at least three periods in recent world history, significant changes in relationships between people and microbes have facilitated disease outbreaks or even epidemics. The first wave of change began to gather momentum about 10,000 years ago, when the domestication of wild animals during the early stages of the Agricultural Revolution brought people and animals much closer together, thus providing more opportunity for disease organisms to move between them. Settlement in agrarian communities, and eventually cities, also brought people into close contact with accumulated wastes that often contained disease organisms.[9]

The second wave began about 2,500 years ago as increasing contact among centers of civilization accelerated the spread of diseases among previously unexposed peoples. The Roman Empire in the West and the Han Dynasty in the East came into closer contact as trade expanded, and germs were swapped in both directions. Thus the expanding Roman Empire was repeatedly afflicted by unknown maladies that seemingly originated on the periphery.[10]

A third wave of significant proportion gathered momentum during the era of transoceanic exploration and trade expansion that began in the fourteenth and fifteenth centuries. The bubonic plague arrived in Europe from Asia early in this period, and European explorers and settlers arriving in the western hemisphere brought smallpox, measles, influenza, and other diseases with them that decimated indigenous peoples.[11]

There is considerable evidence that a fourth wave is now building due to the dynamics of industrialization, globalization, population growth, and urbanization. The 1957 Asian flu and the 1968 Hong Kong flu pandemics that together killed more than 4 million people worldwide may well be precursors of worse things to come. And the slow-moving HIV/AIDS pandemic already has killed more than 20 million people and sickened between 34 million and 46 million. Why, in the face of all of the recent advances in medical care, does "microsecurity"—freedom from various adverse health effects of microbial-scale agents—once again seem to be diminishing? There are several significant changes taking place in people, pathogens, and their shared environments that are now facilitating the spread of infectious diseases.[12]

Most obvious among these destabilizing factors are changes in human circumstances and behavior. Demographic shifts, including population growth, rapid urbanization, and increasing migration, are now major contributors to insecurity. (See Chapter 2.) The present world population of 6.4 billion is expected to grow to more than 7.9 billion over the next 20 years. Almost all this population growth is expected to take place in developing countries. In 1965, 36 percent of the world's population lived in cities. This figure is now approaching 50 percent. The number of people living in crowded slums in poor countries is exploding. In Asia, the population packed into cities will grow from the current 1.5 billion to 2.6 billion by 2030. In Africa, the number of city-dwellers will grow from 297 million to 766 million over the same period. Obviously, as more people live in unsanitary conditions in more densely packed cities, it is easier for pathogens to spread rapidly.[13]

Environmental change can upset established equilibriums between people and pathogens, facilitating new disease outbreaks.

This population growth, as well as periodic ethnic violence in developing countries, is forcing two kinds of migration that also contribute to the spread of disease. Population growth in many of these countries is pressing people to settle on previously unoccupied land, often cleared forests. This newly settled land is frequently shared with numerous potentially dangerous pathogens. These microbes have remained in animal hosts in forested areas until people have intruded, thus offering them new pathways out of the forest into larger human populations. Similarly, there are approximately 17 million refugees and internally displaced people worldwide due to various kinds of humanitarian emergencies. Crowded refugee camps are ideal incubators for diseases, and the people who do get to leave them often bring diseases along to their new locations. In the troubled Darfur region in Sudan, for example, from late May to late August 2004 there were 3,573 reported cases of hepatitis E, which resulted in 55 deaths.[14]

The persistence of poverty in the face of rapid population growth in the urbanizing developing world is another factor that is increasing the potential for disease outbreaks. In spite of the growth in the global economy over the last few decades, the gaps between rich and poor both among and within countries are actually increasing. Between 1970 and 1995, real per capita income for the richest one third of all countries increased at 1.9 percent annually, the middle one third saw an annual increase of only 0.7 percent, while the bottom one third showed no increase at all. These figures have changed very little over the last decade. An estimated 2.8 billion people now live on less than $2 per day. Current income inequalities are reflected in comparative health expenditures. In the United States, total per capita expenditures on health stood at $4,887 in 2001. In Niger, the equivalent figure, at the average U.S. dollar exchange rate, was $6, in Sierra Leone it was $7, and in Nigeria it was $15. People living in such impoverished conditions have little access to medical care.[15]

Thus we currently live in an epidemiologically divided world. Many people suffer from infectious diseases of the underfed at the same time that an increasing number are afflicted with chronic diseases of the overfed. While persistent hunger is the condition of at least 1.2 billion people, a similar number

eat much more than is needed. In Africa, infectious and parasitic diseases account for about half of all deaths, while in Europe they account for only 2 percent. Each year more than 2.3 million people, primarily in poor countries, die from eight diseases that could easily be prevented by vaccinations.[16]

Low- and middle-income countries carry more than 90 percent of the world's disease burden but account for only 11 percent of health spending. The diseases that most commonly affect the poor attract scant research and development spending. Pharmaceutical companies see little profit in developing drugs for diseases endemic to poor countries. Between 1975 and 1997, only 13 out of the 1,233 drugs that reached the global market were applicable to the tropical diseases that are responsible for the greatest number of deaths from infectious disease.[17]

Changes in the ways that people behave (or misbehave) are also among the factors upsetting the balance between people and pathogens. Ecologically unsound practices, such as grinding up dead animal parts to feed to living ones, can only improve the position of pathogens. Changing patterns of sexual behavior, including unprotected sex with multiple partners, have dramatically increased the incidence of sexually transmitted diseases such as herpes, syphilis, and gonorrhea. And it is difficult to conceive of a more effective way to spread disease than intravenous drug use accompanied by needle sharing, which has greatly accelerated the spread of hepatitis, HIV/AIDS, and other diseases.

Although technological innovation is usually thought of as a strong ally in curbing infectious disease, there are aspects of technology that have quite the opposite effect. On the bright side, innovations in biomedical technologies create new tools with which to control disease. The dark side, however, is that technology is transforming the nature of the environment in which people and pathogens interact. The growing speed and pervasiveness of international travel means that more people, produce, and pathogens are moving swiftly across borders. The number of passenger-kilometers flown internationally grew from 28 billion in 1950 to 2.6 trillion in 1998. The amount of air freight moving internationally grew from 730 million to 99 billion ton-kilometers over the same period. More than 2 million people now cross an international border each day.[18]

These traveling people, goods, and agricultural commodities can carry pathogens from one part of the world to distant others in a matter of hours. Limited disease outbreaks due to increased international travel and commerce occur regularly. Occasionally, more serious outbreaks occur. The SARS virus moved out of southern China to other countries in Asia and then to much of the northern hemisphere in a matter of weeks. And since an ever larger quantity of the fruits and vegetables consumed in the United States now comes from other countries, public health officials have been reporting sharp increases in disease outbreaks linked to imported agricultural commodities and the microbes they often carry.[19]

Just as the era of exploration brought smallpox and various other European diseases to the western hemisphere with devastating results for Native Americans, the travels of domesticated and other animals spread their own pathogens. It is likely that the 300 pigs that Hernando de Soto brought with him to Florida in 1539 were responsible for passing more maladies to Native Americans, deer, and turkeys than his soldiers did. In the contemporary world, numerous infectious diseases are spreading among animal populations. In central Africa, Ebola outbreaks

have killed thousands of apes. The West Nile virus, which killed 241 people in the United States in 2002, also sickened 14,000 horses that year and spread to nearly 200 species of birds, reptiles, and mammals, causing an unknown number of deaths. Increased trade and travel is spreading viral, bacterial, and fungal pathogens to plants around the world as well.[20]

Each year more than 2.3 million people, primarily in poor countries, die from eight diseases that could easily be prevented by vaccination.

Avian flu is an animal virus that periodically threatens to spread widely to human populations. In 1997, 18 people were sickened in Hong Kong and 6 died when avian flu jumped directly from chickens to people. As a result, 1.4 million chickens were slaughtered in an effort to prevent further spread of the disease. Avian flu reappeared in Europe in 2003, and 30 million chickens were killed there as a preventive measure. In 2004 avian flu swept through eight Asian countries, killing more than two dozen people and leading to the culling of more than 100 million fowl. The disease also appeared in Canada and the United States. (See also Chapter 4.)[21]

The current form of the (H5N1) avian flu virus has demonstrated the ability to jump from fowl to only a limited number of people. But in late September 2004, WHO reported on the first probable case of human-to-human transmission of the disease, when a woman in Thailand died. The concern is that the flu could continue to change in ways that would let it pass easily from person to person, possibly setting the stage for a deadly pandemic. Disease experts continue to monitor avian flu for changes that could indicate the emergence of more lethal variants. There are similar concerns that recent genetic changes in swine flu viruses could create a variant that would be lethal to humans.[22]

The widespread and often indiscriminate use of antibiotics and other anti-bacterial agents is creating families of drug-resistant microbes.

Technological innovation is also linked to many environmental changes that are helping new diseases emerge and old ones resurface and spread. Changes in water quality, for instance, are often the result of intense urbanization and the industrialization of agriculture. Waterborne diseases account for about 90 percent of infections in developing countries, where nearly 95 percent of urban sewage is dumped untreated into rivers and lakes. In India, 114 cities dump untreated sewage and partially cremated bodies into the Ganges River. And runoff from agriculture, including animal and chemical wastes, threatens aquatic ecosystems in industrial as well as developing countries.[23]

Atmospheric changes, primarily due to industrial activity, are increasing people's exposure to disease as well. The buildup of greenhouse gases is increasing the potential for a higher incidence of disease, as climate change is expected to increase the geographic range of diseases that thrive in warm climates. Serious diseases such as cholera, malaria, and yellow fever—now pretty much restricted to the tropics—could spread to presently temperate areas as warming takes place. And the West Nile virus, carried by the temperature-sensitive mosquito *Culex pipiens*, arrived in New York City in the summer of 1999 and has since afflicted animals and

people across North America. It may well be a harbinger of things to come.[24]

Finally, technology also is playing an unintended role in transforming the microbial world. Paradoxically, antibiotics and other pharmaceuticals that are intended to control pathogens often rebound with detrimental effects. Some 50 million pounds of antibiotics are produced in the United States each year. More than a third of this total is fed to farm animals. The widespread and often indiscriminate use of antibiotics and other anti-bacterial agents is creating families of drug-resistant microbes. This is accelerated by patients who do not finish prescribed treatments, and by the large-scale use of antibacterial agents in soaps, lotions, and detergents.[25]

Antimicrobial resistance is a serious and growing problem. About 14,000 people die annually in the United States from drug-resistant microbes that infect them during hospital stays. Half the people infected with HIV harbor a strain that is now resistant to at least one drug used to fight the disease. Pneumonia remains a serious threat, killing 3–4 million people a year. In some areas of the world as many as 70 percent of chest infections are resistant to available antimicrobials. The bacterium that causes cholera is also becoming resistant to the principal antibiotics used against it. In Rwanda, for example, there is nearly 100 percent resistance to tetracycline and chloramphenicol, two major antibiotics used to treat cholera.[26]

Tuberculosis is similarly becoming multidrug resistant. A simple treatment course of six months for non drug-resistant tuberculosis can cost as little as $20. But treatment with second- and third-line antibiotics can cost between $7,000 and $8,500. In some of the worst cases, treatment for multiple drug–resistant tuberculosis can cost $250,000 and take up to two years.[27]

The Current State of "Microsecurity"

Given these and many other factors that are significantly changing relationships among people and pathogens, it is clear why the campaign against infectious diseases is far from over. In fact, 20 previously well-known diseases have reemerged or spread geographically and at least 30 diseases not previously known to be infectious have been identified over the last three decades. Yet, accurately assessing the current state of biosecurity in various regions of the world is a complex undertaking. Collecting and aggregating data in a timely fashion is difficult since many of the most serious infectious diseases are endemic in poor countries, where physicians are scarce and recordkeeping is sporadic. Only about one third of all deaths that occur in the world each year are actually captured by national vital registry systems; the rest are estimated.[28]

Keeping in mind these caveats about the data, an estimated 57 million people died of all causes worldwide in 2002; 5.2 million of these deaths were from various kinds of accidents and injuries, so theoretically most of them were preventable. The bulk of the world's deaths, 33.5 million, were due to noncommunicable and chronic diseases. Of these, 16.7 million people died of cardiovascular disease, 7.3 million died from various forms of cancer, 3.7 million died from noninfectious respiratory diseases, and 2 million died from digestive diseases. While steady progress is being made in treating these noncommunicable diseases, many of these victims were elderly, and other people died at least partially due to lifestyle choices.[29]

The remaining 18.3 million deaths were caused by maternal and perinatal conditions, nutritional deficiencies, and communicable diseases. Most of these deaths were avoidable and thus are appropriate security concerns. Communicable diseases were responsible for 14.9 million deaths in this category. (See Table 3–1.) Respiratory infections—largely influenza and pneumonia—were the big killers. They accounted for 4 million deaths, showing little change from two years earlier. HIV/AIDS was in second place, with 2.8 million deaths. (Although this was a slight decline from the 2000 figure, the number of people infected with HIV grew rapidly, and the latest reports indicate that deaths rose in 2003.) Deaths from diarrhea and tuberculosis declined slightly, while malaria took 200,000 more lives than in 2000. In addition to these major killers, measles accounted for 611,000 deaths, mainly among children.[30]

Life expectancy at birth is a good summary indicator of a country's current state of

Table 3–1. Deaths from Major Communicable Diseases, 2000 and 2002

Disease	2000	2002	Share of all Deaths in 2002
	(million)	(million)	(percent)
Respiratory infections	3.9	4.0	6.9
HIV/AIDS	2.9	2.8	4.9
Diarrhea	2.1	1.8	3.2
Tuberculosis	1.7	1.6	2.7
Malaria	1.1	1.3	2.2

SOURCE: See endnote 30.

health. Japan now has the world's longest life expectancy. A child born there today can, in theory, expect to live 81.9 years. A more nuanced health indicator, however, is healthy life expectancy (HALE). This is a measure of the number of healthy years that a newborn can now expect to live based on current rates of ill health and mortality. In Japan, the HALE now is 75, meaning that the average Japanese child born today can expect to experience 75 healthy years and 6.9 years of disability due to infectious or chronic diseases. In Sierra Leone, in stark contrast, a child born today has a healthy life expectancy of only 28.6 years.[31]

The most worrisome near-term threat from traditional diseases is posed by influenza.

There is a huge difference in healthy life expectancies in rich and poor countries. (See Table 3–2.) People in Japan, Sweden, and Switzerland can count on at least 73 years of healthy life. The United States, despite having by far the highest per capita medical expenditures of any country in the world,

only ranks twenty-eighth, with a healthy life expectancy of 69.3 years. At the lower end of the distribution are seven sub-Saharan African countries with HALEs of less than 35 years, a result of both poverty and the rampage of HIV/AIDS.[32]

Looking to the near future, the most pressing issue will remain HIV/AIDS. HIV is thought to have jumped to humans from chimpanzees and was first identified in the early 1980s. The virus is especially dangerous because it has a long incubation period, attacks the immune system directly, and mutates frequently, and because there is no vaccine or cure for it yet. The long incubation period means that it can be years before victims exhibit symptoms. Thus they can unknowingly pass the disease on to others over an extended period of time.

The latest estimates indicate that between 34 million and 46 million people are now living with HIV/AIDS worldwide. In 2003 there were 4.2–5.8 million new infections, and 2.5–3.5 million people—mainly in sub-Saharan Africa—died of the disease. To date there have been more than 20 million HIV/AIDS fatalities. Since the virus was first

Table 3–2. Healthy Life Expectancy in Selected Countries, 2002

Highest Life Expectancy		Lowest Life Expectancy	
Country	Years of Healthy Life	Country	Years of Healthy Life
Japan	75.0	Sierra Leone	28.6
Sweden	73.3	Lesotho	31.4
Switzerland	73.2	Angola	33.4
Italy	72.7	Zimbabwe	33.6
Australia	72.6	Swaziland	34.2
Spain	72.6	Malawi	34.9
Canada	72.0	Zambia	34.9
France	72.0	Burundi	35.1
Norway	72.0	Liberia	35.3
Germany	71.8	Afghanistan	35.5

SOURCE: See endnote 32.

identified, 4 million children have contracted HIV from their mothers during pregnancy, delivery, or breastfeeding. The disease is so pervasive in sub-Saharan Africa because for the most part the victims are have little access to medical care, live in poverty, and have an incomplete understanding of how the disease is transmitted.[33]

The number of HIV/AIDS victims is expected to grow considerably as the disease spreads to other countries. India, China, Nigeria, Ethiopia, and Russia are likely to see a rapid growth in HIV/AIDS over the next few years. The number of people infected in these five countries is projected to grow from today's 14–23 million to 50–75 million by 2010. India is expected to have 20–25 million cases and an adult prevalence rate of 3–4 percent. China is expected to have 10–15 million victims, with an adult prevalence rate of under 2 percent.[34]

The most worrisome near-term threat from traditional diseases is posed by influenza. A common disease, the flu is normally thought of as a nuisance rather than a serious threat. Outbreaks occur yearly as the virus continually undergoes small changes that help it evade some of the immunity from previous exposures or vaccines. Periodically, however, the virus shifts to new, more deadly forms that may spread rapidly from person to person. As noted earlier, the flu pandemics in 1957 and 1968 together killed more than 4 million people. More rapid and frequent worldwide travel, as well as urban crowding, mean that a future pandemic could spread much more rapidly and be much more deadly. The World Health Organization (WHO) has estimated that, even under today's conditions, such an outbreak could kill 650,000 and hospitalize 2.3 million people in industrial countries alone.[35]

Many of the factors responsible for new and reemerging diseases among people are having similar adverse impacts on animals. This is of concern both because of the effect on the animal kingdom and because diseases frequently jump from animals to people. Thus, West Nile virus passes from birds to people via mosquitos, Lyme disease from mice to people via deer ticks, and Hantavirus to people directly from field mice. (See also Chapter 4.) It is thought that the SARS virus jumped to people from palm civets—a tree-dwelling animal with a cat-like body. At the same time, human diseases often pass to animals. There is evidence that human diseases have infected primates as people have come in greater contact with them in their forest habitats. Some endangered mountain gorillas in Rwanda have died from human maladies such as measles, often passed on to them by tourists and scientists. Studies also document antibodies to human diseases such as influenza, measles, and tuberculosis in wild macaques and orangutans.[36]

Economic Consequences of Infectious Diseases

It is obvious that history's great epidemics and pandemics have had tremendous impacts on economic performance. The bubonic plague in fourteenth-century Europe, for example, devastated such a large portion of the population that it set the stage for the labor-saving innovations that shaped the Industrial Revolution. Contemporary disease outbreaks also have significant economic consequences, which is why governments often attempt to keep them under wraps. An outbreak of what was thought to be pneumonic plague in India in 1994 resulted in $1.7 billion in lost exports and tourism, and a cholera outbreak in Tanzania in 1998 meant an economic loss of $36 million, both significant jolts to developing economies.

But these pale in comparison with the economic damage done in Asia by the recent outbreak of SARS and in Africa by the ongoing HIV/AIDS pandemic.[37]

The 2003 SARS outbreak had a devastating impact on East Asia, a densely populated and economically dynamic area. It took only a few weeks for the first SARS-related death to put the economies of China, Taiwan, and Singapore into a tailspin. Trade and tourism are essential to the success of these economies, and SARS was a major restraint on travel and exports. Air traffic came to a near standstill in the region, with major carriers grounding up to 40 percent of their flights. Singapore's airport, which usually hosts 29 million passengers annually, saw its numbers slow to a trickle. In South China and Hong Kong, some hotels operated at only 10 percent of capacity. The Canton Trade Fair that year, which usually results in $17 billion in business deals, was an economic disaster. Few potential buyers were willing to attend.[38]

The ranks of the most productive people in some of the world's poorest countries are being systematically depleted by HIV/AIDS.

In China, the tourism industry lost an estimated $7.6 billion and 2.8 million jobs. The loss to China's overall travel economy in 2003 was thought to be around $20.4 billion. Singapore's tourism industry took a hit of some $1.1 billion and 17,500 jobs. In fact, economists shaved about 1.5 percentage points off the 2003 growth estimates for the economies of Hong Kong, Singapore, and Malaysia. It is extremely difficult to tally the direct and indirect worldwide economic costs of the SARS outbreak, but clearly it was

well in excess of $100 billion.[39]

The HIV/AIDS pandemic has exacted enormous direct and indirect economic costs since it began its slow worldwide spread a quarter of a century ago. It is hard to come up with an exact number because the harshest impact of the pandemic has been felt in some of the world's least-developed countries, where a great deal of economic activity takes place outside of the formal sector. In economically developed countries, where HIV prevalence among adults generally is much less than 1 percent, the main economic impact of the disease has been on escalating health expenditures. The indirect economic toll in lost productivity has been negligible. But in many of the least economically developed countries that are most affected by the disease, HIV prevalence among working-age adults is above 20 percent. (See Table 3–3.) Per capita health expenditures there are generally very low, and it is the indirect economic toll due to the impact of the disease on the labor force that has been most devastating.[40]

Sub-Saharan Africa is the area hit hardest by HIV/AIDS. In 2003, an estimated 26.6 million people in this region were HIV-positive and some 2.3 million people died from the disease. Overall, growth in the gross domestic product (GDP) of the 33 African countries that show a measurable economic impact from HIV/AIDS declined by an average of 1.1 percent a year between 1992 and 2002; by 2020, this will translate into a collective loss in economic growth of 18 percent—or roughly $144 billion. Anywhere from 9 to 18 percent of working-age adults die prematurely every five years in the most-affected sub-Saharan countries. Thus the ranks of the most productive people in some of the world's poorest countries are being systematically depleted by HIV/AIDS. The disease is retarding industrial development,

Table 3–3. Countries Most Affected by HIV/AIDS

Country	HIV Prevalence Among Adults Age 15–49 in 2001	Death Rate, Adults Age 15–64	HIV-Positive Population
	(percent)	(percent dying in 2000–05)	(thousand)
Botswana	39	14	330
Zimbabwe	34	18	2,300
Swaziland	33	16	170
Lesotho	31	15	360
Namibia	23	11	230
Zambia	22	15	1,200
South Africa	20	10	5,000
Kenya	15	9	2,500
Malawi	15	11	850
Mozambique	13	10	1,100

SOURCE: See note 40.

reducing agricultural production, devastating education, weakening the military (see Box 3–1), and ultimately could undermine political stability.[41]

With more than 5 million people HIV-positive, South Africa now has the largest number of victims in the world. More than 500,000 people there have already died from HIV/AIDS, and the number HIV-positive is projected to reach 6 million by 2010. The bulk of these deaths will occur in the young adult population. Between 1992 and 2002, South Africa lost an estimated $7 billion annually due to declines in its labor force. Efforts to build an educational system in a post-apartheid era are being stymied by the loss of personnel to the disease, and a severe shortage of teachers and professors over the next decade is anticipated. Human capital is being substantially depleted by illness, and foreign investors are hesitant to put more money into the country. As a result of the depletion of the labor force and a drop in foreign investment, GDP in South Africa in the period 2006–10 is projected to be 3.1 percent below what it would have been without HIV/AIDS.[42]

Managing Future Disease Outbreaks

A race is now under way between the accelerating pace of globalization, associated changes in the human condition, and related new challenges from the microbial world on the one hand and the growing capabilities of scientists, physicians, and health officials to locate, diagnose, and contain disease outbreaks on the other. Meeting the challenges of new and resurgent diseases in a more densely populated world requires the international community to take several important steps: increase surveillance to detect new disease outbreaks quickly, use anticipatory thinking and action to prepare for and avoid rapidly moving future pandemics, start a campaign to eradicate serious illness among the world's poor, encourage much greater transparency in countries where disease outbreaks are likely to occur, and shift security spending away from military pursuits and toward building effective public health systems.[43]

The recent experience with the rapid spread of SARS, the threat of a new influenza

BOX 3–1. HIV/AIDS IN THE MILITARY

Contagious diseases have had a disproportionately high impact on soldiers and combatants throughout history. In ancient times, typhus and cholera were the killers. Today the problem is AIDS.

The incidence of HIV/AIDS is considerably higher in many armed forces today than in the civilian population. The problem is most pronounced among soldiers in developing countries. In Zimbabwe and Malawi, for example, some estimates show infection rates of 70 and 75 percent respectively. In fact in Zimbabwe, an estimated three quarters of all soldiers leaving the forces will die of AIDS within a year.

The reasons for these disproportionately high infection rates are varied. The institutional ethos in the armed forces, for one, tends to encourage risk taking, which can carry over to soldiers' sexual relations, influencing crucial decisions such as whether to use condoms. And as someone infected with HIV may not show symptoms for years, HIV-positive soldiers remain on active duty without knowing they are ill. This lets HIV/AIDS spread much farther than would be the case otherwise. It has also allowed governments and military officials to ignore the gravity of the situation.

Military conflict itself can often help spread HIV, chiefly because of the high incidence of the disease among combatants and sexual violence during wars. In some war-afflicted countries, such as the Democratic Republic of the Congo and Angola, 40–60 percent of combatants are HIV-positive. HIV/AIDS is the major cause of disability and death in a number of African police and armed forces.

The danger of spreading HIV/AIDS through peacekeeping forces—which has long been ignored—is also now gradually being recognized. The fact that countries receive hefty payments for contributing peacekeepers to operations may have acted as a disincentive to collect any data that could undermine their troops' participation. Another aspect of the high incidence of HIV/AIDS among armed forces relates to demobilization after the end of conflict: by "reinserting" infected ex-combatants into less-affected home communities, programs that ignore HIV/AIDS can make matters even worse. In the absence of HIV-prevention education and affordable condoms, demobilization and reintegration of combatants can create a public health crisis across entire regions, as it has in sub-Saharan Africa.

A high HIV/AIDS prevalence rate in the military has numerous implications for a country's national security. The disease diminishes the operational efficiency of a country's armed forces in several ways. With great numbers of people sick or dying, demoralization among the troops is a problem; effectiveness and discipline are threatened. Preparedness and combat readiness can also deteriorate, as replacements of sick personnel have to be found and trained, all of which involves the expenditure of human and financial resources. Medical treatments for sick personnel are another cost. HIV/AIDS has a particularly draining effect on the skilled ranks: overall military capacity is weakened because of a loss of leadership capacities and professional standards, which are not easily or cheaply replaced. In certain countries with very high prevalence rates, some commanders fear that they will not be able to staff full contingents in the years ahead.

—*Peter Croll,*
Bonn International Center for Conversion

SOURCE: See endnote 41.

pandemic, the continuing spread of HIV/AIDS, and the specter of bioterrorism (see Box 3–2) have all fostered a new awareness of the need for improved surveillance on a global scale. Two Internet-based networks have been created to scan for and monitor disease outbreaks. ProMED-mail (the Program for Monitoring Infectious Diseases) was set up in 1994. This Internet-based alert system is dedicated to gathering and rapidly disseminating information about disease outbreaks worldwide. It currently links more than 30,000 subscribers in 150 countries. On any given day, correspondents from around the world post information on outbreaks affecting people, plants, or animals. WHO has also responded by using the Internet to link together disease experts and laboratories as needed.[44]

Because of the rapidity with which diseases can appear and spread around the world, health officials must now work in an anticipatory mode. A new influenza virus could easily spread worldwide well before an effective vaccine could be prepared. Thus WHO has set up a globe-spanning flu-monitoring network that is charged with scanning for viral mutations that could lead to a new influenza pandemic as well as helping to determine the composition of each year's flu vaccine. In the United States, each February the Food and Drug Administration (FDA) uses a panel of experts to select viral strains as candidates for vaccines that will be distributed the following fall. In February 2003, WHO identified a new flu strain (A/Fujian) that FDA advisers wanted to include in their annual flu vaccine. But an approved culture medium needed for vaccine production could not be made available in time; as a result, the vaccine lacked an A/Fujian antigen and was much less effective than it might have been. And in October 2004, the production facility of one of the only two flu vaccine suppliers for the United States was closed by British authorities because of bacterial contamination problems. This left the United States with only about half the vaccine it needed.[45]

Techniques are also being developed for

BOX 3–2. BIOWARFARE

Biological weapons link traditional military and biosecurity concerns. They do not necessarily require a great deal of expertise and money to produce. In fact, crude biological weapons were used in the 1346 siege of Kaffa, when Tatar victims of the plague were catapulted over the walls and into the besieged city.

In the contemporary world, the knowledge and equipment necessary to make more sophisticated biological weapons are widely available. But the major powers, while they have had programs on such weapons, have been hesitant to use them in warfare. This is due partially to moral prohibitions against their use, but also to the fact that they are not particularly effective in the battlefield and that better weapons have been available.

For small countries and terrorist organizations, however, biological weapons could be the equivalent of a poor man's atomic bomb. The anthrax-laced letters that killed five people in the United States in 2001 were a small demonstration of how terrorists could use biological weapons to create chaos in industrial countries. Terrorist groups could use pathogens to attack people, livestock, or crops. In the United States, in order to help prepare for this grim possibility, the Project Bioshield Act authorizes the Department of Homeland Security to spend up to $5.6 billion over 10 years to stockpile vaccines and medicines to counter bioterrorism.

SOURCE: See endnote 44.

---

using satellites to identify high-risk spots for deadly diseases before outbreaks occur. Satellites can monitor environmental conditions such as temperature, rainfall, and vegetation that are linked to population surges in disease-carrying animals. For example, the mosquitoes that carry malaria require pools of stagnant water in which to deposit their eggs and humidity between 55 and 75 percent so that they can survive. The parasite carried by the mosquito, the actual cause of the infection, requires temperatures warmer than 18 degrees Celsius. Relevant information can be gathered by satellites and fed into computers that can estimate outbreak risks for various geographical areas. India is now considering setting up a satellite-based malaria early warning system for the whole country.[46]

Campaigns to completely eradicate some infectious diseases represent another approach to enhancing biosecurity. The campaign to get rid of smallpox is one very visible success story. In 1967 the World Health Organization launched a worldwide effort to eliminate the disease. At that time there were 10–15 million cases of smallpox each year, resulting in about 2 million deaths annually. The campaign was able to eliminate the disease within 10 years; the last case of smallpox occurred in Somalia in 1977. More than $300 million was spent to achieve this goal, and more than 200,000 local workers were involved in the most affected countries.[47]

The smallpox eradication effort proceeded so smoothly for several reasons. Smallpox is an easily recognized acute illness. Scientists were able to create an inexpensive vaccine that conveyed long-term immunity. Large-scale vaccination campaigns were accompanied by close surveillance and case reporting. Finally, public education about smallpox was promoted to help ferret out hidden cases. Ironically, one of the less fortuitous consequences of smallpox eradication is that countries have now stopped vaccinating for the disease. In the United States, smallpox vaccinations ended in 1972, and it is unclear to what extent older Americans today have

residual immunity to the disease. Since both the United States and Russia retain frozen laboratory samples of the virus, there are fears that terrorists could get some of it and make supplies to be introduced into these extremely vulnerable populations.[48]

A similar campaign to eradicate polio, launched in 1988, is still under way. More than $3 billion already has been spent on this effort. The Global Polio Eradication Effort is currently spearheaded by WHO, the U.S. Centers for Disease Control and Prevention, Rotary International, and UNICEF. As a result of the campaign, which had aimed to eliminate polio by 2000, the Americas, the Western Pacific, and Europe have been certified as polio-free. The number of countries in which polio is endemic has declined from 125 to just 6—Afghanistan, Egypt, India, Niger, Nigeria, and Pakistan. The disease has been difficult to eradicate because, unlike smallpox, only about one case in 200 results in limb weakness or paralysis that lets physicians diagnose the disease. Thus widespread immunization (80 percent) is required to break the virus transmission train, a target that is difficult to reach in many low-income countries.[49]

Politics has played an ugly role in foiling efforts to eliminate polio, particularly in Nigeria. The predominantly Islamic northern state of Kano suspended vaccinations in late 2003 out of fear that the vaccine was deliberately contaminated by "western countries" to cause infertility or HIV/AIDS. Subsequently, hundreds of new cases of polio were confirmed in Nigeria, many in the Kano region, and the virus quickly spread to 10 other African countries. One region it spread to was Darfur in Sudan, where nearly 1.5 million people have been driven from their homes by a prolonged civil war. Most of them have settled in refugee camps and, because of the lack of clean water and adequate sanitation facilities, conditions are ripe for polio to spread. Given the transient nature of the population, it has been difficult to keep children in place to receive the required number of doses of vaccine.[50]

The political crisis in the Kano region was apparently resolved when a delegation of northern Nigerians traveled to Indonesia, an Islamic country, to test the vaccine and inspect the facility in which it is manufactured. Polio vaccine of exclusively Indonesian origin will be used in northern Nigeria in the future.[51]

A campaign to eradicate other major diseases is also now under way. In 2001, U.N. Secretary-General Kofi Annan called for the creation of the Global Fund to Fight AIDS, Tuberculosis and Malaria—three of the world's most devastating diseases. The fund is a partnership between governments, civil society, and the private sector. Its goal is to attract, manage, and disburse resources to fight AIDS, tuberculosis, and malaria but not to implement programs directly. So far, contributions have totaled about $1.6 billion a year—far short of the $8 billion the Secretary-General said would be needed. Although President George W. Bush promised in his 2003 State of the Union address to give $15 billion over five years to fight AIDS in Africa and the Caribbean, his 2004 funding request was for only $200 million.[52]

The Global HIV Prevention Group—50 health experts from 15 countries brought together by the Bill and Melinda Gates Foundation and the Henry J. Kaiser Family Foundation—is also active in combating the spread of HIV/AIDS. In 2004, the Gates Foundation launched a $200-million HIV prevention program in India, which now is second only to South Africa in the number of people infected by the disease. The program aims to contain the spread of HIV/AIDS in India through aggressive education and condom distribution efforts among prostitutes and truckers in the six states now most affected.[53]

One of the most important issues in the continuing campaign to contain HIV/AIDS and other infectious diseases in poor countries is the need to develop better mechanisms to distribute antiretroviral drugs and other pharmaceuticals to poverty-stricken disease victims. Advocates for victims who cannot afford treatments developed and priced in industrial countries disagree with intellectual property and profit concerns raised by pharmaceutical companies. At the same time, the politically powerful pharmaceutical industry in the United States has used its clout to oppose efforts to make cheaper generic versions of drugs available to fight diseases in poor countries. Their representatives argue that intellectual property rights (and high prices) must be protected if new drugs are to be developed. Others take the position that a humanitarian crisis exists and that much cheaper generics should be made available.[54]

The number of countries in which polio is endemic has declined from 125 to just 6—Afghanistan, Egypt, India, Niger, Nigeria, and Pakistan.

Progress in getting antiretroviral drugs to HIV/AIDS victims in sub-Saharan Africa has been slowed by this intellectual property dispute. And there is a related problem of training enough health care workers to administer these drugs and monitor the condition of patients, a problem made worse by the fact that health workers themselves are dying of AIDS in considerable numbers or emigrating. As a result, the WHO goal of putting 3 million HIV-positive people in poor countries on antiretroviral drugs by 2005 is unlikely to be met.[55]

The recent outbreak of the SARS virus illustrates both the successes of current disease prevention efforts and one of the challenges still to be met—greater transparency or openness. This limited tragedy could have been much worse had the newly established surveillance efforts not been in place. WHO's Global Outbreak Alert and Response Network, which includes a team in Geneva and 120 collaborating health organizations, quickly tracked the progress of the virus. Researchers were able to pinpoint its likely source in southern China. As the virus spread to other countries, victims were isolated and put in quarantine. As a result, one fifth of the SARS victims were health care workers. And the outbreak was brought to an end with slightly fewer than 800 deaths.[56]

There is a pressing need to create innovative mechanisms to provide affordable drugs to disease victims in poor countries.

At the same time, however, political and economic considerations impeded early progress in identifying the new disease, and lack of openness was a recurring problem. In the early days of the outbreak, Chinese bureaucratic red tape blocked the timely flow of information to physicians. In Guangdong Province, the health department initially received information about the disease in a "top secret" document from a central government health committee. Unfortunately, the document sat on a desk for three days because no one there at the time had a high enough security clearance to open it. Initially the Chinese Communist Party tried to cover up the outbreak in order to protect

commerce, especially the tourist trade. But citizen fear and anger led President Hu Jintao to reverse himself and open up communication on SARS. Several responsible party members were purged, and dormant "neighborhood watch" committees dusted off their armbands and started carefully monitoring people's health. In the end, President Hu probably increased his popularity considerably because of his efforts.[57]

The next stage in containing infectious diseases requires confronting difficult political and economic issues. Primary among these is redefining security funding priorities to reflect the serious nature of the challenges of new and resurgent diseases in an era of globalization. And as recent disease outbreaks have been accompanied by a reluctance on the part of government officials to supply timely information, greater transparency is clearly essential to cope with the threat of fast-moving viruses. There are also economic liability issues that must be addressed arising from the need for rapid development of vaccines in the case of swiftly moving diseases. Finally, there is a pressing need to create innovative mechanisms to provide affordable drugs to disease victims in poor countries as well as incentives for the development of new vaccines and medicines applicable to the serious diseases that are still endemic there.

The race thus continues between the growing ability of new and resurgent diseases to spread more rapidly and the ability of an increasingly sophisticated network of health officials and laboratories worldwide to respond quickly to new disease threats. The good news is that HIV/AIDS, SARS, and the threat of bioterrorism have alerted policymakers to the serious human security issues posed by infectious disease. The first step in responding to the increasing disease threat has been to use enhanced telecommunications capabilities to create more effective surveillance networks and to apply new medical expertise and technologies to the task of rapidly identifying potentially lethal diseases. But much more remains to be done. The public health infrastructure must be substantially improved in almost all countries, rich and poor. Most important, however, security priorities and expenditures should be revised drastically to reflect the seriousness of the threat posed by infectious disease in an increasingly interconnected world.

Bioinvasions

Vast increases in global trade and tourism are reshuffling ecosystems worldwide. The value of goods and services exported worldwide grew sixfold over the past three decades—from $1.5 trillion in 1970 to $9.1 trillion in 2003. Between 1990 and 2003 alone, international tourist arrivals rose 50 percent, to a peak in 2002 of 702.6 million arrivals. As people move around the world, especially by boat and plane, so do other species. And as invasive species—introduced organisms that have the potential to decrease biodiversity and harm economies and the environment—spread around the world, they are more likely to colonize ecosystems degraded by unsustainable resource use or simplified by intensified agriculture.[1]

Though some 50,000 species have reached the United States over the past 200 years, only about one in seven is considered invasive. In fact, intentionally introduced species—including corn, wheat, rice, cattle, and poultry—accounted for 98 percent of the U.S. food system in 1998. But many other types of organisms, especially invertebrates and microbes, are introduced unintentionally. These new plant and animal species often invade ecosystems by spreading rapidly and out-competing natives for local resources. They choke pastures, disrupt water systems, and even drive other species to extinction. In Mexico, 167 of the 500 native freshwater fish are now at risk of extinction—and 76 of these cases are tied directly to the spread of invasives.[2]

Invasive species have economic costs as well. In China, losses from invasive species now run to $14.5 billion annually, or 1.36 percent of the country's gross domestic product. In the United States, these losses surpass $138 billion each year. Invasive species also have local economic impacts. In Benin,

women are traditionally involved in transport and trading while men fish and farm. After a dense aquatic plant called the water hyacinth infested one of the country's rivers, men reported a 70-percent drop in their annual income, from $1,984 to $607, over seven years. And the women, who were earning $519 a year trading along the river before their aquatic routes became tangled with the hyacinth, took in only $137 annually during the infestation.[3]

Because of the potent ecological and economic effects of biological agents and the ease with which terrorists can obtain them, bioinvasion is now viewed as a possible terrorist threat. Robert Pratt, in the *US Army War College Quarterly*, suggests that terrorists could introduce invasive species in the United States to "confuse, disrupt, and demoralize the US government and its citizens over time." While not as dramatic as a release of smallpox, the covert introduction of exotic species in, for example, lakes and rivers, "might go undetected for years until the species are well implanted and impossible to counter." This could have long-term environmental and economic effects by weakening natural systems and exhausting financial resources in efforts to eliminate the invasives.[4]

All these concerns point to the urgent need to slow new bioinvasions and minimize the harm being caused by species that have already invaded other ecosystems. In early 2004, the International Maritime Organization (IMO) adopted a convention to combat the spread of waterborne species in ship ballast water—water taken in to stabilize empty vessels as they cross waterways. This water is then deposited at often-distant destination ports. As many as 4,000 invertebrate species are transported in ballast water each day. Once the convention is ratified by 30 IMO

USFWS

Zebra mussels covering a native mussel

member states, it will require all ships to be fitted, by 2016, with equipment for treating their ballast water. If properly implemented, the treaty could prevent the further spread of invasive species such as the European zebra mussel. These animals have infested freshwater bodies in nearly 40 percent of the United States, displacing other marine species and costing the country an estimated $1 billion in eradication costs during the 1990s alone.[5]

Several international environmental agreements, including the Convention on Biological Diversity (CBD) and the Ramsar Convention on Wetlands, highlight the threats to biodiversity posed by invasive species and propose actions to curb their introduction and spread. The CBD calls for other intergovernmental bodies, such as the World Trade Organization (WTO) and the Food and Agriculture Organization, to create relevant policies for addressing the invasive species problem. In particular, there is a need for better communication between the treaty secretariats so they can harmonize their invasive species policies. As a starting point, the Secretariat of the CBD has applied to become an observer to the WTO Committee on Sanitary and Phytosanitary Measures.[6]

In addition, governments must work to synchronize domestic efforts to combat the spread of invasive species. In the United States, customs and border agents have now taken over responsibility for searching for invasive species at ports of entry, as part of a decision to move sections of the U.S. Coast Guard and the Agricultural and Plant Health Inspection Service to the new Department of Homeland Security (DHS). The DHS also recently joined the National Invasive Species Council, an interagency panel formed in 1999 to address this problem in the United States. This effort to consolidate domestic security and invasive species control could yield important advances, but only if it receives sufficient funding within DHS and is not overshadowed by other security concerns.[7]

The European Union has, through its Financial Instrument for the Environment, allotted 27 million euros to 102 projects that address the control and eradication of invasive species in its member nations. But many other countries lack sufficient funding, monitoring technologies, and training to deal adequately with this growing problem. To fill this gap, the CBD calls on donor countries to help boost the capacity of island nations and other vulnerable countries to minimize the spread and impact of invasive species.[8]

Without carefully constructed international monitoring and eradication systems, each day planes and boats will continue to ferry invasive species around the planet, threatening economic and environmental security at their final destinations.

—*Zoë Chafe*

CHAPTER 4

Cultivating Food Security

Danielle Nierenberg and Brian Halweil

At the International Conference on AIDS in July 2004, participants from all over the world gathered in Bangkok to discuss the increasingly dire prospects for millions of people suffering from this disease. Media covering the event wrote scores of articles on how women are the fastest-growing segment of the AIDS population, on the explosion of AIDS in Asia, and on the lack of appropriate drugs in the developing world. One story most journalists missed, however, is how AIDS has become an accomplice of food insecurity. In fact, the disease is steadily stripping many countries in the developing world of their agricultural base.

In Africa, 7 million workers died between 1985 and 2000 in the 25 most affected countries. In Kenya, one study found that food consumption had dropped by 40 percent in homes where people have HIV/AIDS. Women, who make up to 80 percent of the farm workforce, now account for about 60 percent of people living with AIDS in sub-Saharan Africa. And many have had to abandon their fields to care for sick husbands and relatives. The region is also losing much of its knowledge about agriculture because parents are dying before they can pass their hard-earned skills to the next generation. As a result, orphaned children left to run farms are in some cases replacing traditional food crops such as beans, which are high in protein and nutrients, with root crops that are much easier to grow but less nutritious.[1]

The impact of AIDS on agricultural output may be new, but it is by no means the only threat to food security. Where people cannot afford to buy enough food, timeless problems like water shortages continue to be the main causes of hunger. Worldwide, 434 million people face water scarcity, and by 2025 between 2.6 billion and 3.1 billion people will be living in either water-stressed or water-scarce conditions. As water for agriculture becomes less available, nations become more dependent on expensive food imports. In addition, more than 80 percent of arable land worldwide has lost productivity because of soil degradation. Although global harvests increased during the second half of the twentieth century, experts estimate the harvests would have grown a further 10 percent were it not for this problem. Conflict, too, is threatening the ability of millions of people to get enough to eat: In Afghanistan, farmers were not able to get into their fields to plant crops in 2002; many were forced to kill their livestock in order to survive. And according to the U.N Food and Agriculture Organization (FAO), the violence in Greater Darfur, Sudan, in 2004 forced 1.2 million people from their homes and fields.[2]

The results of all this food insecurity are all too familiar and easy to see. We have become used to the images of Sudanese women so thin they can barely carry their children, Ethiopian men so malnourished they are no longer able to walk, and—perhaps most tragic of all—children crying for food, their bellies bloated. Indeed, the number of hungry people in developing countries increased by 18 million in the second half of the 1990s to some 800 million today. Worldwide, nearly 2 billion people suffer from hunger and chronic nutrient deficiencies. Behind the tragic photographs of these desperate individuals, however, are the less visible problems that threaten the global food supply. At both the local and the national level, the most important determinants of food security in the future may be quite different from those of the past.[3]

Among the major food security threats on the horizon are the loss of diversity of plant and animal species, the emergence of new diseases and foodborne illnesses, and food bioterror. The disturbing pictures of Asian chicken farmers who were forced to bury or incinerate millions of chickens because of avian flu may foreshadow a larger epidemic on the horizon. At the same time, uniformity in our livestock herds and the crowded, filthy conditions in which they live not only invite new diseases, they leave our farms wide open and vulnerable to the spread of foodborne pathogens and malicious biowarfare attacks. (See also Chapter 3.)

Perhaps the most important new threat will be the interplay between agriculture and climate change. Farming may be the human endeavor most dependent on a stable climate. The most serious threats will not be the occasional severe drought or heat wave but subtle temperature shifts during key periods in the crop's life cycle, as these are most disruptive to plants bred for optimal climatic conditions. Plant scientists from Asia have found that rising temperatures may reduce grain yields in the tropics by as much as 30 percent over the next 50 years.[4]

Uniformity in our livestock herds and the crowded, filthy conditions in which they live leave farms vulnerable to the spread of foodborne pathogens and biowarfare attacks.

Ironically, the technologies developed since the 1960s to revolutionize agriculture may actually be increasing vulnerability on our farms. For instance, chemical-based pesticides and insecticides initially allowed farmers to reduce their losses to bugs and disease. But they began to fail as pests developed resistance, and the chemicals left toxic residues in our water, soil, and food. Raising thousands of animals in factory farms lowered the price of meat, allowing more people to eat hamburgers, steaks, and chicken breasts on a daily basis. But society is paying the price for cheap meat in the form of a loss of domestic animal diversity and diseases that jump the species barrier and infect people.

Yet just as the threats—both new and old—to food security are numerous, so are the solutions. Our most important tool is not new chemicals or fertilizers or genetically engineered seeds but a new approach to farming that depends on the knowledge of farmers and a sophisticated use of the environment around them.

Losing Agricultural Diversity

In the late 1990s, French farmers began to notice something missing in their fields—the buzzing of bees. From apples to *haricot verts*, hundreds of crops in France depend on bees for pollination. The mystery of the bee's disappearance, however, was not hard to solve. The culprit was imidacloprid, an ingredient in the broad-spectrum insecticide Gaucho. This Bayer product is applied directly to the seeds of corn and sunflowers and absorbed throughout the entire plant. Bees pollinating those plants pick up the insecticide along with the pollen they use to make honey and carry it back to the hive, where it poisons other bees. Although Gaucho was banned in France in 1999, its replacement, Fipronil— an insecticide manufactured by another multinational agribusiness, BASF—is just as deadly.[5]

According to the National Union of French Apiculturists, the country's beekeepers, imidacloprid has killed hundreds of thousands of bees and pushed many of the country's small beekeepers out of business. Fipronil, too, has now been banned in France, but the government is letting farmers use up any remaining stores, infuriating many people. In February 2004 hundreds of farmers, led by activist Jose Bove, occupied the offices of France's national food agency, demanding that any use of the insecticide be banned because it is not only killing off the bees but also destroying the region's agricultural diversity and threatening economic security.[6]

The decline of pollinators is not unique to France. Despite their economic worth—they are responsible for pollinating $10 billion worth of crops each year—bees are disap-

pearing all over the world. Domestic honeybees have lost one third of their hives worldwide, and wild species are also declining because of pesticide and insecticide use, development, and invasive species.[7]

Bees are not the only agriculturally important species to be missing in action. Thousands of breeds of plants and animals are being lost each year to wars, pests and diseases, climate change, urbanization, the global marketing of exotic breeding material, and large-scale industrial agriculture. Big mechanized farms cannot manage a variety of crops, and giant food manufacturers require products of standard size and uniformity. As farms become more and more technologically sophisticated, they also become more and more ecologically fragile.

Since the beginning of the last century, 75 percent of the genetic diversity of agricultural crops has been lost. In China, 10,000 varieties of wheat were under cultivation in 1949; by the 1970s only 1,000 were still in production. Just 20 percent—one fifth—of the maize varieties reported in Mexico in 1930 are known today. And farmers in the Philippines once cultivated thousands of rice varieties, but by the 1980s just two varieties took up 98 percent of the growing area. Green Revolution varieties, introduced only four decades earlier, now cover more than half of all riceland in developing countries. According to Patrick Mulvany of the Intermediate Technology Development Group, the world has 7,000–10,000 edible plant species; 100 or so of these are important for the food security of most countries in the world, yet just 4—maize, rice, wheat, and potatoes—provide 60 percent of the world's dietary energy.[8]

Livestock genetic resources are another cause of worry today. (See Table 4–1.) While action to conserve the world's plant resources has been going on for more than a century—the first seed banks were built in

Table 4–1. Selected Food Animal Breeds in Danger of Disappearing

Breed	Importance	Status
Lulu cattle	Lulu cattle in Nepal are well adapted to living in extreme environments and are highly disease-resistant. They require few inputs and are extremely productive, giving up to two liters of milk a day.	These cattle are endangered as a result of rampant crossbreeding because indigenous breeds are seen as inferior to exotic breeds.
South China pig	The South China pig is a hardy breed, adapted to poor feed and highly resistant to heat and direct solar radiation. It is also immune to kidney worm and liver fluke, unlike foreign breeds of pig.	Because of the intensification of factory farming in Malaysia, there are only about 400 of these pigs left.
Mukhatat chicken	Native to Iraq, Mukhatat chickens can be raised in harsh environments with little nutritional requirements.	Fewer than 600 individuals remain.
Criolla Mora sheep	The Criolla Mora are a Colombian sheep that can be traced back to 1548. Used for meat and wool, these sheep are resistant to endoparasite infestation.	Scientists are uncertain how many remain—anywhere from 100 to 1,000 live in the Colombian highlands.
Warsaw grouper	This fish lives in the southwest Atlantic. Popular for its white, flaky meat, the Warsaw grouper can reach weights of over 300 pounds.	Because these territorial fish never leave their immediate habitat, it is easy to catch them. According to scientists, they face an "extremely high risk" of extinction in the wild in the next 10 years.

SOURCE: See endnote 9.

Russia in 1894—livestock have been of concern only in the last few decades. According to FAO, the increasing demand for meat, eggs, milk, and other animal products has forced producers to abandon local breeds in favor of an increasingly limited number of high-producing livestock.[9]

During the last century 1,000 breeds—about 15 percent of the world's cattle and poultry breeds—have disappeared, and about 300 of these losses occurred in the last 15 years. The problem has been greatest in industrial countries, where factory farming has been most intense. In Europe, more than half of all breeds of domestic animals from the turn of the last century have become extinct, and 43 percent of the remaining breeds are endangered. But as developing countries rise up the protein ladder, the genetic stock of livestock breeds is eroding there too, as indigenous breeds of livestock are being replaced with higher-producing industrial breeds. This creeping homogeneity handicaps the ability of farmers everywhere to respond to pests, disease, and changes in climate.[10]

The importance of crop diversity, or the lack of it, became frighteningly clear in the United States a few decades ago. In 1970, more than 80 percent of the U.S. corn crop carried a gene that made plants susceptible to southern leaf blight, a fungus that produces purplish lesions on leaves or black smears on corn ears. The blight reduced yields by as much as 50 percent, contributing to farmers losing almost $1 billion worth of maize in 1970 alone. Most surprising, the cure for the blight did not come from a laboratory but from fields in southern Mexico, where small farmers maintain the genetic diversity of maize (corn) by growing hundreds of different open-pollinated landraces—the genetic parents of modern maize. Scientists were able to locate a variety that was resistant to the blight and crossbreed it with the U.S. variety.[11]

For centuries the Maya and other indigenous farmers of what is now southern Mexico and Central America used genetically rich and diverse landraces to improve their crops. In contrast, most American production farmers grow a small group of nearly genetically identical corn hybrids that require a chemical cocktail of fertilizers and insecticides to survive until harvest. Unfortunately, these Green Revolution technologies have become the norm, replacing native crop varieties and posing a threat to both local and global food security.[12]

In the same way that forests and grasslands rely on a wide variety of plants and animals to be productive, agricultural ecosystems also have depended for millennia on a vast, rich, diverse storehouse of wild and domesticated seeds and animal breeds to propel agricultural productivity. Farmers, herders, and fishers the world over depend on agrobiodiversity—the variety and variability of animals, plants, and microorganisms used directly and indirectly for food and agriculture—to feed themselves and their communities. Through selective breeding and seed saving, farmers have been able to adapt crops and animals to different climates and growing conditions.[13]

According to Jose Esquinas-Alcazar, Secretary of the Commission on Genetic Resources for Food and Agriculture, "genetic resources are the basis of food security." He compares the thousands of different breeds of crops and livestock to LEGO blocks: "Just as children use a variety of different size and color blocks to build a building or castle, we also need all the little pieces of genetic diversity in agriculture to build food security."[14]

Even in wealthy nations, farmers depend on a steady stream of exotic germplasm to develop new varieties that are resistant to pest and disease. The latest crop breeding technologies, including genetic engineering, still rely on existing genes and varieties. And scattered farm fields preserve diversity best, since seed banks, germplasm libraries, and other dead repositories for diversity are susceptible to decay, mechanical failures, and even sabotage.[15]

But crop genetic diversity is not only important for industrial agriculture. In India, members of the Navdanya Movement are responding to the loss of biodiversity—and the threat of corporate ownership of seeds through patents—by protecting local varieties of wheat, rice, and other crops by cataloguing them and declaring them common property. Navdanya has also set up locally owned seed banks, farm supply stores, and storage facilities and has helped establish "Zones for Freedom"—villages that pledge to reject chemical fertilizers and pesticides, genetically engineered seeds, and patents on life. The crop diversity reduces dependence on expensive agrochemicals and other inputs and provides resilience against major pest outbreaks or climatic shifts. And when farmers produce for local as opposed to export markets, their customer base diversifies considerably, encouraging them to plant a wider range of crops. In this way, crop diversity reinforces self-sufficiency.[16]

In this era of "terror alerts," farms that forsake genetic diversity have in effect shed their battle armor. Despite their mammoth technological capabilities, huge sheds crammed with chickens or pigs are more vulnerable than smaller, more diverse farms to the unintended or malicious introduction of disease. (See Box 4–1.) According to Chuck Bassett of the American Livestock Breeds Conservancy, "the loss of the livestock genetic

---

BOX 4–1. CAN FOOD BE A WEAPON OF MASS DESTRUCTION?

Since September 11, 2001, food security has taken on a new meaning. The sheer scale of agriculture, particularly in industrial countries, and its economic importance make it an easy target for terrorist acts. According to Peter Chalk, an expert on "agro-terrorism" at the RAND Corporation, industrial farms are especially attractive targets for a number of reasons. "One thing about terrorists," says Chalk, "is that they choose the path of least resistance. Attacking agriculture is far more simple than using bombs because of inherent vulnerabilities in the system."

One of the biggest vulnerabilities is the U.S. livestock industry. U.S. livestock, according to Chalk, have become progressively more disease-prone in recent years because of the intensive factory-style conditions on farms. And because each farm contains tens of thousands of animals, operators are unable to monitor all the stock on a regular basis, making them unaware of a disease outbreak until it spreads to the entire herd.

Another vulnerability in industrial agriculture is the rapid movement of agricultural products from farms to processing plants to consumers. In the dairy industry, for example, there is a trend toward contract calf-raising, which may involve operations that can have more than 30,000 animals from as many as 80 different farms. These animals travel in and out of farms on a daily basis. "If a disease is introduced, and you miss it, it has already traveled thousands of miles," says Chalk.

The U.S. Department of Agriculture (USDA) notes that if foot-and-mouth disease were introduced in the United States, it could spread to 25 states in just five days. And because processed foods can be disseminated to hundreds of stores within a matter of hours, a single case of chemical or biological adulteration could spread widely.

In the developing world, the lack of food safety regulations or veterinarians trained to spot animal diseases also could make agriculture vulnerable to attack. And as global trade expands, terrorists may have greater opportunities to use food as a weapon of mass destruction.

SOURCE: See endnote 17.

---

resources makes it harder for livestock to survive a disaster, whether it be natural, man-made, or terror-caused. One properly placed vector can wipe out 90 percent of an indoor flock—no problem. In a flock with a more broad genetic spread, that is harder to do."[17]

Food Scares

From brucellosis and foot-and-mouth disease to mycotoxins and late potato blight, for centuries farmers have been plagued by diseases affecting their livestock and crops. Within the last century, however, as agriculture has become bigger and more intensified and has spread into more areas, the nature of these diseases has also changed. Huge factory farms crammed with animals and swollen with manure, monoculture crops replacing diverse cropping systems, recycling of animals and waste into livestock feed, concentration in the slaughtering and processing industry, the misuse of antibiotics—all these hallmarks of industrial agriculture give pathogens greater opportunities to infect every layer of the food chain and, ultimately, affect human health. (See Table 4–2.)[18]

Take avian flu, for example. According to FAO, the spread of avian flu from Pakistan to China may have been facilitated by the rapid scaling up of poultry and pig operations and the massive geographic concentration of livestock in Thailand, Viet Nam, and China. In East and Southeast Asia alone,

Table 4–2. Selected Animal Diseases That Can Spread to Humans

Disease	Description
Avian influenza	Avian flu jumped the species barrier for the first time in 1997, killing six people in Hong Kong. In 2003–04, the virulent H2N51 virus killed at least 30 people.
Nipah virus	In 1997, Nipah was discovered in Malaysia, where it spread from pigs to humans, causing a large outbreak of encephalitis; 93 percent of the people infected had occupational exposure to pigs and 105 people died.
Bovine spongiform encephalopathy (BSE, or mad cow disease)	BSE is caused by feeding cattle the renderings of other ruminants. Since it was first discovered in the United Kingdom in 1986, more than 30 other countries have reported cases of mad cow disease, and variant Creutzfeldt-Jakob disease (vCJD)—the human form of the disease—has killed more than 150 people worldwide.
Foodborne pathogens	Foodborne illnesses are one of the most common health problems worldwide. Campylobacter, pathogenic *E. coli*, and salmonella are the pathogens most frequently associated with contaminated meat and animal products.

SOURCE: See endnote 18.

some 6 billion birds are raised for food, with major groups of them being raised in the region's rapidly growing megacities. This increasing intensity of production of chickens and other livestock in cities and in rural areas, along with their close proximity to people's homes, is beginning to have a number of unexpected consequences that can threaten human health. Since 1997, avian flu has spread from birds to humans at least three times. And in October 2004, the first probable case of human-to-human transmission was reported in Thailand.[19]

The latest outbreak hit in late 2003 and 2004 and spread across Asia, infecting thousands of birds. At least 100 million birds were "depopulated"—killed, in other words—when the disease jumped the species barrier, and most of the people who became infected died. A recent study in China has also shown that with every new outbreak, the virus becomes more and more lethal. International health officials are now concerned that this deadly strain of avian flu has become impossible to wipe out in Asian birds and may some day precipitate a global human flu pandemic. Because this would be fast-moving and eas-

ily transmittable from person to person, experts fear that it could be even more lethal than AIDS. (See also Chapter 3.)[20]

The effects of the flu on bird and human populations alike can be devastating. FAO, the World Health Organization, and the World Organisation for Animal Health report that killing all the birds on farms near an outbreak is one of the only effective ways of controlling the disease. Although experts suspect that the spread of factory farming across Asia, the unsanitary conditions and close concentration of animals in factory farms, and the genetic uniformity of the animals helped facilitate the emergence and spread of avian flu, it is small producers who are the most devastated economically by the disease. Thailand, for example, is the world's fourth largest poultry exporter, and many farmers there will be forced out of business. According to Emmanuelle Guerne-Bleich of FAO, these farmers, who typically have about 50 chickens, use them as an "insurance policy" in times of need—selling them for food, medicine, or other necessities—and are " amongst the worst affected and least able to recover" from the outbreak of avian flu.[21]

Nipah virus is one of the newest zoonoses—diseases that can jump from animals to humans. It is a perfect, albeit complicated, example of what can happen when big agriculture combines with the destruction of fragile ecosystems. Nipah was first discovered in 1997 in a small Malaysian village, site of one of the largest pig farms in the country. Nearby residents began coming down with flu-like symptoms, and more than 100 of them died. Epidemiologists eventually figured out that the disease originated in bats, which spread it to pigs and then to humans. But how?[22]

Scientists speculate that in 1997, forest fires in Borneo and Sumatra precipitated by El Niño forced thousands of fruit bats to search for food in Malaysia. Many of them visited fruit trees towering over newly established large pig farms. There the bats ate fruit, dripping their saliva and half-eaten fruit into stalls where it was eaten by the pigs. Although bats are not sickened by Nipah, in pigs it causes a severe coughing sickness, allowing it to spread efficiently to humans. According to Peter Daszak, Executive Director of Wildlife Trust's Consortium for Conservation Medicine, "the damage to fruit bat habitats and the growth of massive-scale pig farming likely led to the emergence of Nipah virus. Without these large, intensively managed pig farms in Malaysia, it would have been extremely difficult for the virus to emerge." In April 2004, Nipah struck again, this time in Bangladesh, killing up to 74 percent of its human victims. Scientists predict that as industrial agriculture continues to move into tropical environments, the risk of Nipah viruses and other diseases that can jump the species barrier is growing.[23]

Unlike Nipah and avian flu, mad cow disease (bovine spongiform encephalopathy) originated not in the wild but, some experts speculate, in the feed processing plants of the United Kingdom. One way farmers make animals gain weight quickly and cheaply is to feed them the nonedible bits and pieces of other animals. This recycling of sheep and other ruminants back into the food chain at rendering plants using low temperatures to save money likely led to the formation of certain proteins called prions. These destroy the normal proteins in cattle brains, causing animals to stumble, show aggression, and eventually die. The disease can spread to humans who eat infected meat. Since 1986, when BSE was first detected, more than 150 people have died from variant Creutzfeldt-Jakob disease, the human form of mad cow.[24]

Although the practice of feeding meat and bone meal made from ruminants to cattle has been banned in the United Kingdom, it is impossible to predict how many people may have eaten beef infected with BSE or how many people might eventually contract vCJD. Furthermore, scientists do not know the incubation period and whether the risk of developing vCJD depends on the quantity of meat consumed or on the frequency of consumption. Before 1996, meat and bone meal was shipped from the United Kingdom all over the world. At least 12 nations in Africa imported the meal, as did the United States and most European, Middle Eastern, and Asian nations.[25]

A recent study by the French Institute of Health and Medical Research reports that a mad cow disease epidemic in France went completely undetected for years and led to nearly 50,000 severely infected animals entering the food chain. In the United States, after repeated USDA assurances that the risk of BSE was almost nonexistent, the first U.S. case of mad cow disease was discovered in late 2003.[26]

Recently, a new form of mad cow disease was discovered in cattle in Italy. Unlike BSE, this new strain—called BASE, for bovine amyloidotic spongiform encephalopathy—has appeared in cows showing no symptoms. Researchers do not know if BASE can be spread to humans, but they suspect that it may be responsible for some cases of Creutzfeld-Jakob disease that seemed to occur spontaneously. Until they know for sure, scientists are calling for more stringent testing of cows for both BSE and BASE.[27]

Farmers from the American breadbasket to the North China Plain are finding that patterns of rainfall and temperature they have relied on for generations are shifting.

Factory farming also contributes to less publicized health problems, including the rise of foodborne illnesses, one of the most common health problems worldwide. According to the World Health Organization, episodes of foodborne illness could be 300–350 times more frequent than reported. Crowded, unsanitary conditions and poor waste treatment in factory farms exacerbate the rapid movement of animal diseases and foodborne infections. For example, the deadly *E. coli* 0157:H7 pathogen is spread from animals to humans when people eat food contaminated by manure. Live transport of animals can also increase the incidence of animal diseases and foodborne illness. According to FAO, 44 million cattle, sheep, and pigs are traded across the world each year. A 2002 study in the *Journal of Food Protection* found that transporting beef cattle from feedyards to slaughterhouses and packing plants increases the prevalence of salmonella on hides and in feces, which can later end up in food.[28]

Factory farms rely on heavy doses of antibiotics. But drugging animals can have some disastrous consequences. Livestock are often given antibiotics sub-therapeutically—in the absence of disease, that is—as part of their daily ration of feed. Residues from the antibiotics end up in our food and can travel throughout the environment in the waste of livestock, polluting both surface and groundwater. This constant use, or misuse, of drugs—some of which are classes of antibiotics important in human medicine—is leading to antibiotic resistance and making it hard to fight diseases among humans and animals alike.[29]

In addition to diseases, new technologies can also infect crops and livestock, changing their genetic makeup and weakening their ability to survive. Consider genetically modified organisms (GMOs). While proponents claim that the technology will feed the world, advocates of sustainable agriculture fear that GMOs will wipe out native and wild populations of corn, rice, wheat, fish, and other sources of food. According to a recent report by biologist Richard Howard at Purdue University, genetically engineered fish have the potential to replace some wild fish populations. Howard and his colleagues inserted salmon growth genes into medaka, a small Japanese freshwater species that reproduces quickly. They found that the male engineered fish grew bigger than their wild counterparts, chasing the smaller fish away from males during mating time. As a result, the bigger fish were more effective at spreading their DNA. But in an ironic twist, the offspring of the modified fish were less likely to survive into adulthood. Researchers call this the "Trojan-gene effect." If genetically engineered fish escape into the wild and replace native ones, they could eventually drive entire species to extinction.[30]

In the United States, more than two thirds of conventional crops are contaminated with genetically modified material, according to a recent report by the British House of Commons on the spread of GMO crops in North America. Citing data from the Union of Concerned Scientists (UCS), the report says that "contamination [from GMOs]...is endemic to the system." It adds that "heedlessly allowing the contamination of traditional plant varieties with genetically engineered sequences amounts to a huge wager on our ability to understand a complicated technology that manipulates life at the most elemental level."[31]

The problem is that once GMOs contaminate native seeds, there is no way to reverse the process. It changes the very nature of the seeds forever. Pretty soon, the contamination may include traits never intended for eating; "pharmaceutical crops," for example, are engineered to produce vaccines and drugs. Furthermore, contaminating the seed supply removes any safety net the world might have if proponents of GMOs turn out to be wrong. According to the UCS report, "seeds will be our only recourse if the prevailing belief in the safety of genetic engineering proves wrong.... Our ability to change course if genetic engineering goes awry will be severely hampered."[32]

Climate Shifts

High in the Peruvian Andes, five hours by car from Cuzco and six hours by horseback, a new disease has invaded the potato fields in the town of Chacllabamba. Warmer and wetter weather associated with climate change has allowed late blight—the same fungus that caused the Irish potato famine—to creep 4,000 meters up the mountainside for the first time since humans started planting tubers in this region thousands of years ago. In 2003, farmers here saw their crops almost totally destroyed. Breeders are rushing to develop potatoes that retain the taste, texture, and quality preferred by local people and

that also resist the "new" disease.[33]

Like the growers in Chacllabamba, farmers from the American breadbasket to the North China Plain to the fields of southern Africa are already finding that patterns of rainfall and temperature they have relied on for generations are shifting. As farming depends so heavily on a stable climate, this industry will struggle more than others to cope with more erratic weather, severe storms, and changes in growing season lengths. (Ironically, archeologists now belief that the shift to a warmer, wetter, more stable climate at the end of the last Ice Age was vital for humanity's successful foray into food production.)[34]

The possibility that such changes could wreck havoc on the world's food supplies has not escaped the attention of the defense community. In February 2004, the Pentagon released a report that argued climate change could bring the planet to the edge of anarchy as countries develop a nuclear threat to defend and secure dwindling food, water, and energy supplies. The authors, Doug Randall and Peter Schwartz of the California-based future-oriented consulting firm Global Business Network, examined the possibility that global warming and ice cap melting would disrupt oceanic heat transfer and plunge North America and Europe into a mini-Ice Age—an often-discussed scenario supported by evidence in the climate record. "With inadequate preparation, the result could be a significant drop in the human carrying capacity of the Earth's environment," the report noted. In other words, the sudden change in climatic conditions 8,200 years ago that brought widespread crop failure, famine, disease and mass migration of populations could soon be repeated.[35]

The same month the Pentagon report was released, Canadian Environment Minister David Anderson—in an unusual if not unique statement for a government leader—called climate change a bigger threat than terrorism, suggesting that the wheat-growing prairies of Canada and the Great Plains of the United States would eventually no longer produce enough food to support their populations if nothing were done to fight climate change. The bottom line, according to Doug Randall: "We have not been hit with a dramatic climate event since the dawn of modern civilization." On one hand, the modern world's technologies will allow countries like America to "weather" through it. On the other hand, a more populated and built-up planet has more to lose.[36]

As plant scientists refine their understanding of climate change—and the subtle ways in which plants respond—they are beginning to think that the most serious threats to agriculture will not be the most dramatic: a lethal heat wave or severe drought or endless deluge. Instead, for plants that humans have bred for optimal climatic conditions, subtle shifts in temperatures and rainfall during key periods in the crop's life will be most disruptive.

Plant scientists at the International Rice Research Institute in the Philippines are already noting regular heat damage in Cambodia, India, and their own test farms in Manila, where the average temperature is now 2.5 degrees Celsius higher than 50 years ago. "In rice, wheat and maize, grain yields are likely to decline by 10 percent for every one degree [Celsius] increase over 30 degrees," says researcher John Sheehy. "We are already at or close to this threshold." Sheehy estimates that grain yields in the tropics might fall as much as 30 percent over the next 50 years—a period when the region's already malnourished population will increase by 44 percent.[37]

Hartwell Allen, a plant scientist with the USDA and the University of Florida-Gainesville, has found that while a doubling of carbon dioxide and a slightly increased temperature do stimulate seeds to germinate and plants to grow larger and lusher, the higher temperatures are deadly when the plant starts producing pollen. For instance, at temperatures above 36 degrees Celsius during pollination, peanut yields dropped by about 6 percent per degree of increase. Allen is particularly concerned about the implications for places like India and West Africa, where peanuts are a dietary staple and temperatures during the growing season already wander into the upper thirties. "In these regions the crops are mostly rain-fed," he notes. "If global warming also leads to drought in these areas, yields could be even lower."[38]

The world's major plants can cope with temperature shifts to some extent, but since the dawn of agriculture farmers have selected plants that thrive in stable conditions. When climatologists consult global climate models, however, they see anything but stability. As greenhouse gases trap more of the sun's heat in the earth's atmosphere, there is also more energy in the climate system, which means more extreme swings—dry to wet, hot to cold. (This is why there can still be severe winters on a warming planet, and why March 2004 was the third warmest month on record after one of the coldest winters ever.) Climatologists have already observed a number of these projected impacts in most regions: Higher maximum temperatures and more hot days. Higher minimum temperatures and fewer cold days. More variable and extreme rainfall events. Increased summer drying and associated risk of drought in continental interiors. All these conditions will likely accelerate into the next century.[39]

Perhaps nowhere is predictability more central to farming than in unirrigated rice paddies and wheat fields across Asia, where the annual monsoon makes or breaks millions of lives. "If we get a substantial global warming, there is no doubt in my mind that there will be serious changes to the monsoon," said David Rhind, a senior climate researcher with NASA's Goddard Institute for Space Studies at Columbia University. For instance, El Niño events often correspond with weaker monsoons, and El Niños will likely increase with global warming. What is less clear, Rhind said, is the direction of these changes. "My guess is that responses will be much more amplified in all directions."[40]

Cynthia Rosenzweig, a senior research scientist with the Goddard Institute, argues that although climate models will always be improving, there are certain changes we can already predict with a level of confidence. First, most studies indicate "intensification of the hydrological cycle"—big words that essentially mean more droughts and more floods, more variable and more extreme rainfall. Second, Rosenzweig notes that "basically every study has shown that there will be increased incidence of crop pests." Longer growing seasons mean more generations of pests during the summer. Shorter and warmer winters mean that fewer adults, larvae, and eggs will die off.[41]

Third, most climatologists agree that climate change will hit farmers in the developing world hardest. This is partly a result of geography. Farmers in the tropics already find themselves near the temperature limits for most major crops. So any warming is likely to push their crops over the top. "All increases in temperature, however small, will lead to decreases in production," said Robert Watson, chief scientist at the World Bank and former chairman of the Intergovernmental Panel on Climate Change. (By the 2080s, Watson adds, projections indicate that even temperate latitudes will begin to approach this threshold.) "Studies have consistently shown that agricultural regions in the developing world are more vulnerable, even before we

consider the ability to cope," Cynthia Rosenzweig notes. They have less money, more limited irrigation technology, and virtually no weather tracking systems. "Look at the coping strategies, and then it's a real double whammy." In sub-Saharan Africa—ground zero for global hunger, where the number of starving people has doubled in the last 20 years—current problems will undoubtedly be exacerbated by climate change.[42]

"Scientists may indeed need decades to be sure that climate change is taking place," says Patrick Luganda, chairman of the Network of Climate Journalists in the Greater Horn of Africa. "But, on the ground, farmers have no choice but to deal with the daily reality as best they can." Several years ago local farming communities in Uganda could determine the timing of annual rains fairly accurately. "These days there is no guarantee that the long rains will start, or stop, at the usual time," Luganda says. The Ateso people in north central Uganda report the disappearance of *asisinit*, a swamp grass favored for thatch houses because of its beauty and durability. The grass is increasingly rare because farmers have started to plant rice and millet in swampy areas in response to more frequent droughts. (Rice farmers in Indonesia coping with droughts have done the same thing.) Ugandan farmers have also begun to sow a wider diversity of crops and to stagger their plantings to hedge against abrupt climate shifts. Luganda adds that repeated crop failures have pushed many farmers into cities: the final coping mechanism.[43]

New Approaches for New Threats

While threats to food security seem to be multiplying—from HIV/AIDS and climate change to the loss of agricultural diversity and emerging animal diseases—there are no shortage of solutions to ensuring a safe food supply. And while many agricultural officials, scientists, and agribusiness executives will continue to favor technological fixes, it is unlikely that this same emphasis, which generated many of our current problems, will do the trick. Instead, policymakers and farmers are cultivating conceptual and political changes on the ground.

For example, after more than two decades of often bitter negotiations, the International Treaty on Plant Genetic Resources for Food and Agriculture came into force on 29 June 2004. Its aim is to protect agricultural biodiversity and ensure the fair and equitable sharing of its benefits—and ultimately to protect the basis of agriculture and food security. While this achievement is significant, ambiguities in the treaty have some nongovernmental organizations (NGOs) worried that economically powerful nations will be allowed to extract and privatize genetic resources and minimize their contribution to protecting those resources for farmers all over the world. What is missing in the treaty, say some NGOs, is a clear statement of farmers' rights that would protect their ability to save and exchange seeds without restrictions imposed by intellectual property rights.[44]

According to Cary Fowler, Director of Research of the Centre for International Environment and Development Studies at the Agricultural University of Norway, the treaty also fails to spell out a clear role for governments in protecting plant genetic resources and to nail down commitments. Despite these shortcomings, says Fowler, "the treaty provides positive international norms for the conservation and management of plant genetic resources for food and agriculture." It also may reduce the political tensions that have thwarted cooperation, conservation, and research and development in this field for so long.[45]

Farmers are now pushing for a similar treaty to protect domestic animal breeds. In October 2003, leaders of traditional pastoral communities, NGOs, and government representatives met in Karen, Kenya, and drafted the Karen Commitment, calling for animal genetic resources to be protected from patenting and for pastoralists to be recognized for their efforts to conserve and protect domestic animal breeds.[46]

But it will take more than treaties to build food security and protect agricultural diversity. While the scientific community and governments have been engaged in bureaucratic wrangling over the state of the world's agricultural resources, farmers have been quietly cultivating their own genetically diverse crops and animals. According to Pat Mooney of the Action Group on Erosion, Technology, and Concentration, "while the official figures will show big losses of plant and domestic animal diversity, a huge amount is not known by the scientific community. What's lost to scientists isn't always lost to farmers." In fact, through seed saving and selective breeding, farmers are conserving genetic resources on the ground and sharing the fruits of their labor with other farmers at seed fairs and markets.[47]

Farmers, not surprisingly, know best how to "grow" diversity and protect crops and animals from disease and climate on the farm. In northeastern Brazil, for example, community seed banks (CSBs) are being built to help farmers get access to seeds and train them to conserve agricultural biodiversity. Assessment and Services in Alternative Agriculture Projects and other local organizations have trained farmers who by 2000 had organized 220 CSBs, storing more than 80 tons of seeds of the main crop varieties, including 67 varieties of three different bean species. And in Kenya, seed fairs have become an effective and empowering way for women farmers to trade seeds, share knowledge, and improve genetic diversity and food security at the local level.[48]

Ex-situ conservation—keeping animals in zoos, livestock embryos frozen in gene banks, or seeds in seed banks—has been an effective, although costly, approach. But it is not very useful to people who depend on agriculture for their livelihoods. A far more effective and productive way for farmers to preserve livestock breeds and plants is on the farm, especially if farmers raise breeds that have a high monetary value. For example, the multicolored hides of N'guni cattle in South Africa are "in vogue" for furniture coverings. South African hut pigs are also becoming popular because of the large amount of fat they produce to make crackling, or fried pork skin, for the local market. These pigs sell for as much as 1,000 rand ($150), much more than commercial pigs fetch. In the United States, Heritage Foods USA is restoring culinary traditions—and saving North American breeds of livestock from extinction—by developing a market for rare-breed turkeys, geese, and pigs as well as Native American foods.[49]

Beyond the market value, there are other reasons to preserve rare breeds of livestock. Conserving farm animal genetic diversity is a low-cost way of protecting food security in developing countries. Farm animals are not just a source of food. Livestock manure is a valuable resource, keeping soil fertile and productive. Animal power is used on the farm to cultivate and irrigate crops and to transport them at harvest time. Animal meat, hides, wool, and feathers provide important sources of income for rural communities. According to Dr. Jacob Wanyama of the Intermediate Technology Development Group, "it is important to conserve not only animal genetic resources currently or likely to be used in the future for food and agriculture, but also ensure that the people who have conserved them for their livelihoods continue to do

so." For many of the world's poor living in arid and semiarid regions, livestock are the only efficient means of food production.[50]

While governments have been engaged in bureaucratic wrangling over agricultural resources, farmers have been quietly cultivating their own genetically diverse crops and animals.

Keeping livestock healthy and free from disease also strengthens food security. In 2004, the World Organization for Animal Health and FAO agreed to more cooperation to monitor and control highly contagious animal diseases. They have called for more research regarding the transmission of avian flu from birds to pigs and other zoonotic diseases and for major investments in strengthening veterinary services in detecting and reporting the disease. On the farm, many farmers are protecting both human and animal health by reintroducing indigenous breeds of livestock, which tend to be more resistant to disease than nonnative breeds. National governments can reinforce this work by collaborating with these farmers and breeder associations, like the American Livestock Breeds Conservancy, to help them account for a larger part of herds and flocks around the world.[51]

The many variables associated with climate change make coping difficult, but not futile. In essence, farmers will best resist a wide range of shocks by making themselves more diverse and less dependent on outside inputs. When the temperature dramatically shifts, a farmer growing a single variety of wheat is more likely to lose the whole crop than a farmer growing several wheat varieties or even several varieties of plants besides wheat. The additional crops help form a sort of ecological bulwark against blows from climate. More diverse farms will cope better with drought, increased pests, and a range of other climate-related jolts. And they will tend to be less reliant on fossil fuels for fertilizer and pesticides. Climate change might also be the best argument for preserving local crop varieties around the world, so that plant breeders have as wide a range of choices as possible when trying to develop plants that can cope with more frequent drought or new pests.

The Land Institute's Sunshine Farm Project has been farming without fossil fuels, fertilizers, or pesticides as a way to reduce its contribution to climate change.

Planting a wider range of crops, for instance, is perhaps the greatest hedge that farmers can take against more erratic weather. In parts of Africa, planting trees alongside crops—a system called agroforestry, which might include shade coffee and cacao, or leguminous trees with corn—might be part of the answer. "There is good reason to believe that these systems will be more resilient than a maize monoculture," says Lou Verchot, the lead scientist on climate change at the World Agroforestry Centre in Nairobi. Trees send their roots considerably deeper than crops do, allowing them to survive a three- to four-week drought that might damage the grain crop. In addition to this hedge, tree roots will pump water into the upper soil layers where crops can tap it. Trees also improve the soil: their roots create spaces for water flow; their leaves decompose into compost. In other words, a farmer who has trees will not lose everything.[52]

Farmers in central Kenya are using a coffee, macadamia, and cereal mix that results in as many as three marketable crops in a good

years. "Of course, in any one year, the monoculture will yield more money," Verchot admits, "but farmers need to work on many years." These more diverse crop mixes are all the more relevant since rising temperatures will eliminate much of the traditional coffee- and tea-growing areas in the Caribbean, Latin America, and Africa. In Uganda, where coffee and tea account for nearly all agricultural exports, an average temperature rise of 2 degrees Celsius would dramatically reduce the harvest, as all but the highest altitude areas will become too hot to grow coffee.[53]

Farms with trees planted strategically between crops will not only do better in torrential downpours and parching droughts, they will also "lock up" more carbon. Verchot notes that the improved fallows used in Africa can lock up 10–20 times the carbon of nearby cereal monoculture and 30 percent of the carbon in an intact forest. And building up a soil's stock of organic matter—the dark, spongy stuff that gives soils their rich smell and is the form in which soils store carbon dioxide—not only increases the amount of water the soil can hold (good for weathering droughts), it helps bind more nutrients (good for crop growth).[54]

At the Land Institute in Salina, Kansas, where local climatologists suspect that climate change could turn the state's wheat fields into a Dust Bowl–like desert, farmers are not taking the news lying down. The Institute's Sunshine Farm Project raises crops without fossil fuels, fertilizers, or pesticides as a way to reduce its contribution to climate change and to find an inherently local solution to a global problem. As its name implies, the farm runs essentially on sunlight. Homegrown sunflower seeds and soybeans become biodiesel that fuels tractors and trucks. The farm raises nearly three fourths of the feed—oats, grain sorghum, and alfalfa—it needs for its draft horses, beef cattle, and poultry.

Manure and legumes in the crop rotation substitute for energy-gobbling nitrogen fertilizers. A 4.5-kilowatt photovoltaic array powers the workshop tools, electric fencing, water pumping, and chick brooding. In total, the farm eliminated the energy used to make and transport 90 percent of its supplies. Including the energy required to make the farm's machinery lowers the figure to 50 percent, but this is still a huge gain over the standard American farm.[55]

Marty Bender, research director at Sunshine Farm, notes that "carbon farming is a temporary solution." He points to a recent paper in *Science* showing that even if all U.S. soils were returned to their pre-plow carbon content—a theoretical maximum for how much carbon they could lock up—this would be equal to only two decades of U.S. carbon emissions. "That is how little time we will be buying," Bender says, "despite the fact that it may take a hundred years of aggressive, national carbon farming and forestry to restore this lost carbon." Cynthia Rosenzweig of the Goddard Institute also notes that the potential to lock up carbon is limited and that a warmer planet will reduce the amount of carbon that soils can hold: as land heats up, invigorated soil microbes produce more carbon dioxide.[56]

"We really should be focusing on energy efficiency and energy conservation to reduce the carbon emissions by our national economy," Marty Bender concludes. In the United States, as in other nations, although agriculture ranks second among economic sectors contributing to the greenhouse effect, its contribution is less than one tenth that of energy production. For farms to play a significant role in minimizing climate change, changes in cropping practices must happen on a large scale, across large swaths of India and Brazil and China and the American Midwest.[57]

It is in farmers' best interest to make obvious reductions in their own energy use, simply to save money. But the lasting solution to greenhouse gas emissions and climate change will depend mostly on the choices that everyone else makes. For instance, a basic meal—some meat, grain, fruits, and vegetables—using imported ingredients can easily generate four times the greenhouse gas emissions as the same meal with ingredients from local sources. And even if farmers decide that they want to start raising more diverse breeds of livestock and plants, the shift will still depend on shoppers wanting to seek out these foods at the supermarket.[58]

In other words, farmers are not the only ones with a stake in a more secure food system, and they cannot shore up our fields and ranches by themselves. They will need the help of a public committed to farms that can withstand climate change and new diseases and that yield food that is safe to eat. Luckily, this isn't a hard sell.

Toxic Chemicals

In 1984, a malfunction at a pesticide plant in Bhopal, India, triggered the worst chemical disaster in history. More than 27 tons of methyl isocyanate gas leaked out, forming a deadly cloud that killed thousands of people and injured hundreds of thousands more. But this was just one of the tens of thousands of accidental chemical discharges that occur each year. In the United States alone, there were more than 32,000 in 2003. Chemicals are essential to modern economies and serve countless uses. When poorly managed, however, they pose a considerable threat to global security, not just because of accidental releases but because of their potential use by terrorists as well as their effects on the environment and human health.[1]

Chemicals—stored and processed at millions of industrial facilities and transported in countless trucks, trains, and ships—are significant targets for terrorists. According to an analysis by the U.S. Army Surgeon General, in the worst-case scenario an attack on a U.S. chemical plant could kill more than 2 million people. In 2004, in Ashdod, Israel, suicide bombers killed themselves near a citrus packaging plant that uses the pesticide methyl bromide. If they had ruptured the bromide tanks, the release of this poisonous chemical could have killed thousands.[2]

More insidious than these immediate dangers is the fact that many chemicals can threaten long-term security by contaminating both people and the environment. Few of the 70,000 or so chemicals on the market in Europe have been adequately assessed for safety. But several of those that have been tested increase the prevalence of cancer, disrupt hormonal systems, and retard child development. Even people who should have had little exposure to chemicals are affected: 200 toxic chemicals have been detected in the bodies of the Inuit in Greenland—sometimes at concentrations so high that certain tissues and the breast milk of some Inuit could be classified as hazardous waste. Along with increasing rates of disease and death, these toxins can significantly raise health care and social costs, which is especially a problem in countries staggering under other heavy public health burdens.[3]

Perhaps the biggest security risk is that the problems chemicals cause may not arise until it is too late to solve them. In 1962, American scientist Rachel Carson alerted the public to the devastating effects that DDT was having on the environment—including jeopardizing the viability of bird populations and causing health problems in humans. This was after almost 20 years of widespread usage. But it took another 10 years to ban the compound in the United States.[4]

Endocrine-disrupting chemicals may be the next hidden danger. These disrupt hormonal cycles not just in humans, but in many species. This is most starkly revealed by the decline in shellfish and mollusks after extensive use of tributyltin to prevent organisms from attaching themselves to ship hulls. Because of the long time lag between exposure and effects, endocrine disruption could have serious human and societal impacts, including causing reproductive difficulties, if steps are not taken to remove from circulation the chemicals responsible. Already, male sperm counts have dropped significantly in both the United States and Europe, while in many industrial countries testicular cancer has been on the rise—two phenomena possibly connected to increased body burdens of endocrine-disrupting chemicals.[5]

If properly managed, however, chemicals need not pose a security threat. Requiring industrial facilities to provide the public with

Chemical dump, Spain

their annual chemical releases can help reduce the use and emission of toxic chemicals. In the United States, the Toxics Release Inventory provides data on usage and releases. By 2002, releases of 300 chemicals tracked since 1988 had been cut in half. This transparent system has allowed civil society to pressure facilities to use chemicals more efficiently and safely.[6]

In October 2003, the European Commission adopted the REACH proposal—Registration, Evaluation, and Authorization of Chemicals. If passed, this will require manufacturers to register all widely used chemicals (about 30,000). Producers will need to evaluate the safety of those of concern and, for the most toxic ones, seek authorization to continue usage—after proving that no safer alternatives exist. The environmental and health benefits of this legislation promise to be significant. When exposure to three types of chemicals linked to non-Hodgkin's lymphoma was reduced in Sweden, for instance, incidence of this cancer also decreased.[7]

When it is clear that certain chemicals have to be banned, countries must proceed aggressively. A recent success in this comes from the Stockholm Convention on Persistent Organic Pollutants, which entered into force in May 2004. This treaty, ratified by over 70 countries, bans or severely restricts the production and use of 12 of the most hazardous chemicals, including DDT and dioxin, and will mobilize hundreds of millions of dollars to ensure they are phased out.[8]

Even when chemicals are not banned, safer alternatives need to be adopted more rapidly.

Carbon dioxide (CO_2), for example, has many of the properties of organic solvents—chemicals that are frequently very hazardous. CO_2 has replaced methylene chloride in coffee decaffeination and in some cases is replacing perchloroethylene in dry-cleaning operations (both these chemicals are carcinogens). However, shifts to safer chemicals have been limited by higher production costs and by the required investments in new technologies. Governments will need to speed transitions by making less-toxic alternatives more affordable—either directly through subsidies or by pricing hazardous chemicals more accurately (such as through increased chemical taxes to offset elevated environmental and health costs).[9]

Making industrial processes more efficient would help reduce demand for certain chemicals, lessening the risk of accidental or intentional releases while also allowing greater reuse and recycling of any chemicals used. While chemicals will undoubtedly remain abundant, their use and toxicity can be lowered, making them less of a short-term or long-term threat. This will only occur, however, with commitments from political, industrial, and community leaders to safety, efficiency, transparency, and caution.[10]

—*Erik Assadourian*

CHAPTER 5

Managing Water Conflict and Cooperation

*Aaron T. Wolf, Annika Kramer,
Alexander Carius, and Geoffrey D. Dabelko*

Stanley Crawford, a former *mayordomo* (ditch manager) of an *acequia* (irrigation ditch) in New Mexico, writes of two neighbors who "have never been on good terms…the lower neighbor commonly accusing the upper of never letting any water pass downstream to his place and then of dumping trash into it whenever he rarely does." Such rivalries over water have been the source of disputes since the Neolithic revolution, when humans settled down to cultivate food between 8000 and 6000 BC. Our language reflects these ancient roots: "rivalry" comes from the Latin *rivalis*, or "one using the same river as another." Riparians—countries or provinces bordering the same river—are often rivals for the water they share. Today the downstream neighbor's complaint about the upstream riparian is echoed by Syria about Turkey, Pakistan about India, and Egypt about Ethiopia.[1]

Regardless of the geographic scale or the riparians' relative level of economic development, the conflicts they face are remarkably similar. Sandra Postel, director of the Global Water Policy Project, describes the problem in *Pillars of Sand*: Water, unlike other scarce, consumable resources, is used to fuel *all* facets of society, from biology and economy to aesthetics and spiritual practice. Water is an integral part of ecosystems, interwoven with the soil, air, flora, and fauna. Since water flows, use of a river or aquifer in one place will affect (and be affected by) its use in another, possibly distant, place. Within watersheds, everything is connected: surface water and groundwater, quality and quantity. Water fluctuates wildly in space and time, further complicating its management, which is usually fragmented and subject to vague,

Aaron T. Wolf is Associate Professor of Geography in the Department of Geosciences at Oregon State University and Director of the Transboundary Freshwater Dispute Database. Annika Kramer is Research Fellow and Alexander Carius is Director of Adelphi Research in Berlin. Geoffrey D. Dabelko is the Director of the Environmental Change and Security Project at the Woodrow Wilson International Center for Scholars in Washington, D.C.

arcane, or contradictory legal principles.[2]

Water cannot be managed for a single purpose: all water management serves multiple objectives and navigates among competing interests. Within a nation, these interests—domestic users, farmers, hydropower generators, recreational users, ecosystems—are often at odds, and the probability of a mutually acceptable solution falls exponentially in proportion to the number of stakeholders. Add international boundaries, and the chances drop yet again. Without a mutual solution, these parties can find themselves in dispute, and even violent conflict, with each other or with state authorities. Still, water-related disputes must be considered in the broader political, ethnic, and religious context. Water is never the single—and hardly ever the major—cause of conflict. But it can exacerbate existing tensions and therefore must be considered within the larger context of conflict and peace.

From the Middle East to New Mexico, the problems remain the same. So, however, do many of the solutions. Human ingenuity has developed ways to address water shortages and cooperate in managing water resources. In fact, cooperative events between riparian states outnumbered conflicts by more than two to one between 1945 and 1999. In addition, water has also been a productive pathway for building confidence, developing cooperation, and preventing conflict, even in particularly contentious basins. In some cases, water provides one of the few paths for dialogue in otherwise heated bilateral conflicts. In politically unsettled regions, water is an essential part of regional development negotiations, which serve as de facto conflict-prevention strategies.[3]

Key Issues

While the underlying reasons for water-related controversy can be numerous, such as power struggles and competing development interests, all water disputes can be attributed to one or more of three issues: quantity, quality, and timing. (See Table 5–1.)[4]

Competing claims for a limited quantity of water are the most obvious reason for water-related conflict. The potential for tensions over allocation increases when the resource is scarce. But even when pressure on the resource is limited, its allocation to different uses and users can be highly contested. As people become more aware of environmental issues and the economic value of ecosystems, they also claim water to support the environment and the livelihoods it sustains.

Another contentious issue is water quality. Low quality—whether caused by pollution from wastewater and pesticides or excessive levels of salt, nutrients, or suspended solids—makes water inappropriate for drinking, industry, and sometimes even agriculture. Unclean water can pose serious threats to human and ecosystem health. Water quality degradation can therefore become a source of dispute between those who cause it and those affected by it. Further, water quality issues can lead to public protests if they affect livelihoods and the environment. Water quality is closely linked to quantity: decreasing water quantity concentrates pollution, while excessive water quantity, such as flooding, can lead to contamination from overflowing sewage.

Third, the timing of water flow is important in many ways. Thus the operational patterns of dams are often contested. Upstream users, for example, might release water from reservoirs in the winter for hydropower production, while downstream users might need it for irrigation in the summer. In addition, water flow patterns are crucial to maintaining freshwater ecosystems that depend on seasonal flooding.

Conflicting interests concerning water

Table 5–1. Selected Examples of Water-related Disputes

Location	Main Issue	Observation
Cauvery River	Quantity	The dispute on India's Cauvery River sprung from the allocation of water between the downstream state of Tamil Nadu, which had been using the river's water for irrigation, and upstream Karnataka, which wanted to increase irrigated agriculture. The parties did not accept a tribunal's adjudication of the water dispute, leading to violence and death along the river.
Okavango River	Quantity	In the Okavango River basin, Botswana's claims for water to sustain the delta and its lucrative ecotourism industry contribute to a dispute with upstream Namibia, which wants to pipe water passing through the Caprivi Strip to supply its capital city with drinking water.
Mekong River basin	Quantity	Following construction of Thailand's Pak Mun Dam, more than 25,000 people were affected by drastic reductions in upstream fisheries and other livelihood problems. Affected communities have struggled for reparations since the dam was completed in 1994.
Incomati River	Quality and quantity	Dams in the South African part of the Incomati River basin reduced freshwater flows and increased salt levels in Mozambique's Incomati estuary. This altered the estuary's ecosystem and led to the disappearance of salt-intolerant plants and animals that are important for people's livelihoods.
Rhine River	Quality	Rotterdam's harbor had to be dredged frequently to remove contaminated sludge deposited by the Rhine River. The cost was enormous and consequently led to controversy over compensation and responsibility among Rhine users. While in this case negotiations led to a peaceful solution, in areas that lack the Rhine's dispute resolution framework, siltation problems could lead to upstream/downstream disputes, such as those in Central America's Lempa River basin.
Syr Darya	Timing	Relations between Kazakhstan, Kyrgyzstan, and Uzbekistan—all riparians of the Syr Darya, a major tributary of the disappearing Aral Sea—exemplify the problems caused by water flow timing. Under the Soviet Union's central management, spring and summer irrigation in downstream Uzbekistan and Kazakhstan balanced upstream Kyrgyzstan's use of hydropower to generate heat in the winter. But the parties are barely adhering to recent agreements that exchange upstream flows of alternate heating sources (natural gas, coal, and fuel oil) for downstream irrigation, sporadically breaching the agreements.

SOURCE: See endnote 4.

quality, quantity, and timing can occur on many geographic scales, but the dynamics of conflict play out differently at international, national, and local levels. (See Table 5–2.) Whether the dispute is over quality, quantity, and timing, or at the international, national, or local level, however, the key to understanding—and preventing—water-related conflicts can be found in the institutions established to manage water resources.

International Basins

International basins that include political boundaries of two or more countries cover 45.3 percent of Earth's land surface, host

Table 5–2. Conflict Dynamics on Different Spatial Levels

Geographic Scale	Characteristics
International	Disputes can arise between riparian countries on transboundary waters
	Very little violence, but existing tensions between parties are pervasive and difficult to overcome, resulting in degraded political relations, inefficient water management, and ecosystem neglect
	Long, rich record of conflict resolution and development of resilient institutions
National	Disputes can arise between subnational political units, including provinces, ethnic or religious groups, or economic sectors
	Higher potential for violence than at international level
	Rationale for international involvement is more difficult, given national sovereignty concerns
Local (indirect)	Loss of water-based livelihoods (due to loss of irrigation water or freshwater ecosystems) can lead to politically destabilizing migrations to cities or neighboring countries
	Local instability can destabilize regions
	Poverty alleviation is implicitly tied to ameliorating security concerns

about 40 percent of the world's population, and account for approximately 60 percent of global river flow. And the number is growing: in 1978 the United Nations listed 214 international basins (in the last official count). Today there are 263, largely due to the "internationalization" of basins through political changes like the breakup of the Soviet Union and the Balkan states, as well as access to improved mapping technology.[5]

Strikingly, territory in 145 nations falls within international basins, and 33 countries are located almost entirely within these basins. The high level of interdependence is illustrated by the number of countries sharing each international basin (see Table 5–3); the dilemmas posed by basins like the Danube (shared by 17 countries) or the Nile (10 countries) can be easily imagined.[6]

The high number of shared rivers, combined with increasing water scarcity for growing populations, leads many politicians and headlines to trumpet coming "water wars." In 1995, for example, World Bank vice president Ismail Serageldin claimed that "the wars of the next century will be about water." Invariably, these warnings point to the arid and hostile Middle East, where armies have mobilized and fired shots over this scarce and precious resource. Elaborate—if misnamed—"hydraulic imperative" theories cite water as the prime motivation for military strategies and territorial conquests, particularly in the ongoing conflicts between Arabs and Israelis.[7]

The only problem with this scenario is a lack of evidence. In 1951–53 and again in 1964–66, Israel and Syria exchanged fire over the latter's project to divert the Jordan River, but the final exchange—featuring assaults by both tanks and aircraft—stopped construction and effectively ended water-related hostilities between the two states. Nevertheless, the 1967 war broke out almost a year later. Water had little—if any—impact on the military's strategic thinking in subsequent Israeli-Arab violence (including the 1967, 1973, and 1982 wars). Yet water was an underlying source of political stress and one of the most difficult topics in subsequent negotiations.

Table 5–3. Number of Countries Sharing a Basin

Number of Countries	International Basins
3	Asi (Orontes), Awash, Cavally, Cestos, Chiloango, Dnieper, Dniester, Drin, Ebro, Essequibo, Gambia, Garonne, Gash, Geba, Har Us Nur, Hari (Harirud), Helmand, Hondo, Ili (Kunes He), Incomati, Irrawaddy, Juba-Shibeli, Kemi, Lake Prespa, Lake Titicaca-Poopo System, Lempa, Maputo, Maritsa, Maroni, Moa, Neretva, Ntem, Ob, Oueme, Pasvik, Red (Song Hong), Rhone, Ruvuma, Salween, Schelde, Seine, St. John, Sulak, Torne (Tornealven), Tumen, Umbeluzi, Vardar, Volga, and Zapaleri
4	Amur, Daugava, Elbe, Indus, Komoe, Lake Turkana, Limpopo, Lotagipi Swamp, Narva, Oder (Odra), Ogooue, Okavango, Orange, Po, Pu-Lun-T'o, Senegal, and Struma
5	La Plata, Neman, and Vistula (Wista)
6	Aral Sea, Ganges-Brahmaputra-Meghna, Jordan, Kura-Araks, Mekong, Tarim, Tigris and Euphrates (Shatt al Arab), and Volta
8	Amazon and Lake Chad
9	Rhine and Zambezi
10	Nile
11	Congo and Niger
17	Danube

SOURCE: See endnote 6.

In other words, even though the wars were not fought over water, allocation disagreements were an impediment to peace.[8]

While water supplies and infrastructure have often served as military tools or targets, no states have gone to war specifically over water resources since the city-states of Lagash and Umma fought each other in the Tigris-Euphrates basin in 2500 BC. Instead, according to the U.N. Food and Agriculture Organization, more than 3,600 water treaties were signed from AD 805 to 1984. While most were related to navigation, over time a growing number addressed water management, including flood control, hydropower projects, or allocations in international basins. Since 1820, more than 400 water treaties and other water-related agreements have been signed, with more than half of these concluded in the past 50 years.[9]

Researchers at Oregon State University have compiled a dataset of every reported interaction—conflictive or cooperative—between two or more nations that was driven by water. Their analysis highlighted four key findings.[10]

First, despite the potential for dispute in international basins, the incidence of acute conflict over international water resources is overwhelmed by the rate of cooperation. The last 50 years have seen only 37 acute disputes (those involving violence), and 30 of those occurred between Israel and one of its neighbors. Non-Mideast cases account for only 5 acute events, while during the same period 157 treaties were negotiated and signed. The total number of water-related events between nations is also weighted toward cooperation: 507 conflict-related events versus 1,228 cooperative, implying that violence over water is neither strategically rational, hydrographically effective, nor economically viable.[11]

Second, despite the fiery rhetoric of politicians—aimed more often at their own constituencies than at the enemy—most actions taken over water are mild. Of all the events, some 43 percent fall between mild verbal support and mild verbal hostility. If the next levels—official verbal support and official verbal hostility—are added in, verbal events account for 62 percent of the total. Thus almost two thirds of all events are only verbal and more than two thirds of these had no official sanction.[12]

Third, there are more examples of cooperation than of conflict. The distribution of cooperative events covers a broad spectrum, including water quantity, quality, economic development, hydropower, and joint management. In contrast, almost 90 percent of the conflict-laden events relate to quantity and infrastructure. Furthermore, almost all extensive military acts (the most extreme cases of conflict) fall within these two categories.[13]

Fourth, despite the lack of violence, water acts as both an irritant and a unifier. As an irritant, water can make good relations bad and bad relations worse. Despite the complexity, however, international waters can act as a unifier in basins with relatively strong institutions.

This historical record proves that international water disputes do get resolved, even among enemies, and even as conflicts erupt over other issues. Some of the world's most vociferous enemies have negotiated water agreements or are in the process of doing so, and the institutions they have created often prove to be resilient, even when relations are strained.

The Mekong Committee, for example, established by the governments of Cambodia, Laos, Thailand, and Viet Nam as an inter-

governmental agency in 1957, exchanged data and information on water resources development throughout the Viet Nam War. Israel and Jordan have held secret "picnic table" talks on managing the Jordan River since the unsuccessful Johnston negotiations of 1953–55, even though they were at war from Israel's independence in 1948 until the 1994 treaty. (See Box 5–1.) The Indus River Commission survived two major wars between India and Pakistan. And all 10 Nile Basin riparian countries are currently involved in senior government–level negotiations to develop the basin cooperatively, despite fiery "water wars" rhetoric between upstream and downstream states.[14]

The historical record proves that international water disputes do get resolved, even among enemies, and even as conflicts erupt over other issues.

In southern Africa, a number of river basin agreements were signed when the region was embroiled in a series of local wars in the 1970s and 1980s (including the "people's war" in South Africa and civil wars in Mozambique and Angola). Although negotiations were complex, the agreements were rare moments of peaceful cooperation among many of the countries. Now that most of the wars and the apartheid era have ended, water is one of the foundations for cooperation in the region. In fact, the 1995 Protocol on Shared Watercourse Systems was the first protocol signed within the Southern African Development Community. Riparians will go through tough, protracted negotiations in order to gain benefits from joint water resources development. Some researchers have therefore identified cooperation over water resources as a particularly fruitful entry

Box 5–1. Water Sharing Between Israel, Jordan, and the Palestinians

The most severe water scarcity in the world is in the Middle East. The deficit is particularly alarming in the Jordan River basin and the adjacent West Bank aquifers, where Israeli, Palestinian, and Jordanian water claims intersect. In Gaza and the West Bank, the annual availability of water is well below 100 cubic meters of renewable water per person, while Israel has less than 300 and Jordan around 100 cubic meters. A country is generally characterized as water-scarce if the availability falls below 1,000 cubic meters.

Population growth, a result both of high birth rates among Palestinians and Jordanians and of immigration to Israel, puts increasingly severe pressure on the already scarce water resources and raises the risk of water-related conflicts. Israeli settlers in the West Bank and Gaza receive a larger share of the available water than the Palestinians, further complicating the situation.

Despite fears of water-related violence, Israel has maintained basic cooperation with Jordan and the Palestinians over their shared waters. This was true even after the second *intifada* began in September 2000. Low-level water cooperation between Israel and Jordan—under U.N. auspices—extends back to the early 1950s, even though both countries were formally at war. This interaction helped build trust and a shared set of rules and norms, which were later formalized within the peace agreement between Israel and Jordan in 1994. As stipulated in that agreement, a Joint Water Committee for coordination and problem solving was established that helped resolve disagreements over allocations.

A 1995 interim agreement regulates Israeli-Palestinian water issues such as protection of water and sewage systems. The Joint Water Committee and its subcommittees have continued to meet despite the violence of the last years. For the Palestinians, the existing agreement is unsatisfactory from both a rights and an availability perspective. Talks aimed at a final agreement are part of the overall negotiating process and, given the political stalemate and ongoing violence, are not likely to be completed any time soon. Nevertheless, there is agreement between Israel and the Palestinians that cooperation over their shared water is indispensable.

Two main policy recommendations can be drawn from this case. First, water cooperation is intimately linked to politics—a highly complex process influenced by both domestic and international considerations. If donors fail to thoroughly analyze the political context, they are unlikely to understand how water is sometimes subordinate to more important political priorities and used as a political tool.

Second, donor agencies and international organizations can play an important role if they are prepared to provide long-term support for establishing cooperation over shared water. Donors typically want to see tangible results within a short time frame. Yet it is essential to understand that risks are involved, occasional setbacks will occur, and rewards are unlikely to materialize quickly. Donors will need to engage in "process financing" that supports not an ordinary development project with a cycle of 2–4 years but rather a process that can span 10–25 years. In the Israeli–Jordanian case, the U.N. Truce Supervision Organization, which worked as an "umbrella" for discussions on water coordination in spite of the absence of a peace agreement, played a critical role.

Although more conflicts of interest are likely to arise in the future over the waters in the Jordan River basin, water management—properly supported—offers a window of opportunity for broader cooperation in this troubled part of the world.

—*Anders Jägerskog*
Expert Group on Development Issues
Ministry for Foreign Affairs, Sweden

SOURCE: See endnote 14. The views expressed are those of the author and not the Swedish Ministry for Foreign Affairs.

point for building peace. (See Chapter 8.)[15]

So, if shared water does not lead to violence between nations, what is the problem? In fact, complicating factors, such as the time lag between the start of water disputes and final agreements, can cause water issues to exacerbate tensions. Riparians often develop projects unilaterally within their own territories in an attempt to avoid the political intricacies posed by sharing resources. At some point, one of the riparians (usually the most powerful one) will begin a project that affects at least one of its neighbors.

Without relations or institutions conducive to conflict resolution, unilateral action can heighten tensions and regional instability, requiring years or decades to resolve: the Indus treaty took 10 years of negotiations; the Ganges, 30; and the Jordan, 40. Water was the last—and most contentious—issue negotiated in a 1994 peace treaty between Israel and Jordan, and was relegated to "final status" negotiations between Israel and the Palestinians, along with difficult issues like refugees and the status of Jerusalem. During this long process, water quality and quantity can degrade until the health of dependent populations and ecosystems is damaged or destroyed. The problem worsens as the dispute intensifies; the ecosystems of the lower Nile, the lower Jordan, and the tributaries of the Aral Sea have effectively been written off by some as unfortunate products of human intractability.[16]

When unilateral development initiatives produce international tensions, it becomes more difficult to support cooperative behavior. As mistrust between riparians grows, threats and disputes rage across boundaries, as seen in India and Pakistan or Canada and the United States. Mistrust and tensions (even if they do not lead to open conflict) can hamper regional development by impeding joint projects and mutually beneficial infra-

structure. One of the most important sources of water for both Israelis and Palestinians, the Mountain Aquifer, is threatened by pollution from untreated sewage. The existing conflict has impeded donor initiatives to build wastewater treatment plants in Palestine, setting the stage for a vicious circle as groundwater pollution increases regional water scarcity and, in turn, exacerbates the Israeli-Palestinian conflict.[17]

Disputes within Nations

The literature on transboundary waters often treats political entities as homogeneous monoliths: "Canada feels…" or "Jordan wants…." Recently, analysts have identified the pitfalls of this approach, showing how subsets of national actors have different values and priorities for water management. In fact, the history of water-related violence includes incidents between tribes, water use sectors, rural and urban populations, and states or provinces. Some research even suggests that as the geographic scale drops, the likelihood and intensity of violence increases. Throughout the world, local water issues revolve around core values that often date back generations. Irrigators, indigenous populations, and environmentalists, for example, all may view water as tied to their way of life, which is increasingly threatened by new demands for cities and hydropower.[18]

> **Unilateral action can heighten tensions and regional instability, requiring years or decades to resolve.**

Internal water conflicts have led to fighting between downstream and upstream users along the Cauvery River in India and between Native Americans and European settlers. In 1934, the landlocked state of

Arizona commissioned a navy (it consisted of one ferryboat) and sent its state militia to stop a dam and diversion project on the Colorado River. Water-related disputes can also engender civil disobedience, acts of sabotage, and violent protest. In the Chinese province of Shandong, thousands of farmers clashed with police in July 2000 because the government planned to divert agricultural irrigation water to cities and industries. Several people died in the riots. And from 1907 to 1913 in California's Owens Valley, farmers repeatedly bombed a pipeline diverting water to Los Angeles.[19]

National instability can also be provoked by poor or inequitable water services management. Disputes arise over system connections for suburban or rural areas, service liability, and especially prices. In most countries, the state is responsible for providing drinking water; even if concessions are transferred to private companies, the state usually remains responsible for service. Disputes over water supply management therefore usually arise between communities and state authorities. (See Box 5–2.) Protests are particularly likely when the public suspects that water services are managed in a corrupt manner or that public resources are diverted for private gain.[20]

Local Impacts

As water quality degrades or quantity diminishes, it can affect people's health and destroy livelihoods that depend on water. Agriculture uses two thirds of the world's water and is the greatest source of livelihoods, especially in developing countries, where a large portion of the population depends on subsistence farming. Sandra Postel's list of countries that rely heavily on declining water supplies for irrigation includes eight that currently concern the security community:

Bangladesh, China, Egypt, India, Iran, Iraq, Pakistan, and Uzbekistan. When access to irrigation water is cut off, groups of unemployed, disgruntled men may be forced out of the countryside and into the city—an established contributor to political instability. Migration can cause tensions between communities, especially when it increases pressure on already scarce resources, and cross-boundary migration can contribute to interstate tensions. (See Chapter 2.)[21]

Thus, water problems can contribute to local instability, which in turn can destabilize a nation or an entire region. In this indirect way, water contributes to international and national disputes, even though the parties are not fighting explicitly about water. During the 30 years that Israel occupied the Gaza Strip, for example, water quality deteriorated steadily, saltwater intruded into local wells, and water-related diseases took a toll on the residents. In 1987, the second *intifada* began in the Gaza Strip, and the uprising quickly spread throughout the West Bank. While it would be simplistic to claim that deteriorating water quality caused the violence, it undoubtedly exacerbated an already tenuous situation by damaging health and livelihoods.[22]

An examination of relations between India and Bangladesh demonstrates that local instabilities can spring from international water disputes and exacerbate international tensions. In the 1960s, India built a dam at Farakka, diverting a portion of the Ganges from Bangladesh to flush silt from Calcutta's seaport, some 100 miles to the south. In Bangladesh, the reduced flow depleted surface water and groundwater, impeded navigation, increased salinity, degraded fisheries, and endangered water supplies and public health, leading some Bangladeshis to migrate—many, ironically, to India.[23]

So, while no "water wars" have occurred, the lack of clean fresh water or the competi-

---

Box 5–2. Conflict over Water Services Management: The Case of Cochabamba

Issues of water supply management can lead to violent conflict, as demonstrated by the confrontations that erupted in 2000 in Cochabamba, Bolivia's third largest city, following the privatization of the city's water utility. Cochabamba had long suffered from water scarcity and insufficient, irregular provision of water services. Hoping for improved services and higher connection rates, in September 1999 the Bolivian government signed a 40-year concession contract with the international private water consortium Aguas del Tunari (AdT).

By January 2000, drinking water tariffs increased sharply; some households had to pay a significant share of their monthly income for water services. Consumers felt they were simply paying more for the same poor services and responded with strikes, roadblocks, and other forms of civil protest that shut the city down for four days in February 2000.

While increased water bills triggered the protests, some people also opposed a law threatening public control of rural water systems. Long-standing water scarcity had encouraged the development of well-established alternative sources of supply. In rural municipalities surrounding Cochabamba, farmer cooperatives drilled their own wells and used an informal market for water based on an ancient system of property rights. Under the concession contract, AdT was granted the exclusive use of water resources in Cochabamba, as well as any future sources needed to supply city consumers. It was also granted the exclusive right to provide water services and to require potential consumers to connect to its system. The rural population feared they would lose their traditional water rights and AdT would charge them for water from their own wells.

Farmers from surrounding municipalities joined the protest in Cochabamba, which spread to other parts of Bolivia. Months of civil unrest came to a head in April 2000, when the government declared a state of siege for the whole country and sent soldiers into Cochabamba. Several days of violence left more than a hundred people injured and one person dead. The protests eased only after the government agreed to revoke AdT's concession and return the utility's management to the municipality.

Performance continues to be unsatisfactory, however. Many neighborhoods have only occasional service, and the valley's groundwater table continues to sink. Although many view the concession's cancellation as a victory for the people, it did not solve their water problems. Meanwhile, AdT filed a complaint against the Bolivian government in the World Bank's trade court, the International Centre for Settlement of Investment Disputes, in Washington, D.C. According to the *San Francisco Chronicle*, the consortium is demanding $25 million in compensation for the canceled contract. The case is still pending.

SOURCE: See endnote 20.

---

tion over access to water resources has occasionally led to intense political instability that resulted in acute violence, albeit on a small scale. Insufficient access to water is a major cause of lost livelihoods and thus fuels livelihood-related conflicts. Environmental protection, peace, and stability are unlikely to be realized in a world in which so many suffer from poverty.[24]

Institutional Capacity: The Heart of Water Conflict and Cooperation

Many analysts who write about water politics, especially those who explicitly address the issue of water conflicts, assume that scarcity of such a critical resource drives people to conflict. It seems intuitive: the less water there is, the more dearly it is held and the more likely it is that people will fight over it. Recent research on indicators for transboundary water conflict, however, did not find any statistically significant physical parameters—arid climates were no more conflict-prone than humid ones, and international cooperation actually increased during droughts. In fact, no single variable proved causal: democracies were as susceptible to conflict as autocracies, rich countries as poor ones, densely populated countries as sparsely populated ones, and large countries as small ones.[25]

When Oregon State University researchers looked closely at water management practices in arid countries, they found institutional capacity was the key to success. Naturally arid countries cooperate on water: to live in a water-scarce environment, people develop institutional strategies—formal treaties, informal working groups, or generally warm relations—for adapting to it. The researchers also found that the likelihood of conflict increases significantly if two factors come into play. First, conflict is more likely if the basin's physical or political setting undergoes a large or rapid change, such as the construction of a dam, an irrigation scheme, or territorial realignment. Second, conflict is more likely if existing institutions are unable to absorb and effectively manage that change.[26]

Water resource management institutions have to be strong to balance competing interests and to manage water scarcity (which is often the result of previous mismanagement), and they can even become a matter of dispute themselves. In international river basins, water management institutions typically fail to manage conflicts when there is no treaty spelling out each nation's rights and responsibilities nor any implicit agreements or cooperative arrangements.[27]

Similarly, at the national and local level it is not the lack of water that leads to conflict but the way it is governed and managed. Many countries need stronger policies to regulate water use and enable equitable and sustainable management. Especially in developing countries, water management institutions often lack the human, technical, and financial resources to develop comprehensive management plans and ensure their implementation.

Moreover, in many countries decision-making authority is spread among different institutions responsible for agriculture, fisheries, water supply, regional development, tourism, transport, or conservation and environment, so that different management approaches serve contradictory objectives. Formal and customary management practices can also be contradictory, as demonstrated in Cochabamba, where formal provisions of the 1999 Bolivian Water Services Law conflicted with customary groundwater use by farmers' associations.[28]

In countries without a formal system of water use permits or adequate enforcement and monitoring, more powerful water users can override the customary rights of local communities. If institutions allocate water inequitably between social groups, the risk of public protest and conflict increases. In South Africa, the apartheid regime allocated water to favor the white minority. This "ecological marginalization" heightened the black population's grievances and contributed to social instability, which ultimately led to the end of the regime.[29]

Institutions can also distribute costs and benefits unequally: revenues from major water infrastructure projects, such as large dams or irrigation schemes, usually benefit only a small elite, leaving local communities to cope with the resulting environmental and social impacts, often with little compensation. (See Box 5–3.)[30]

The various parties to water conflicts often have differing perceptions of legal rights, the technical nature of the problem, the cost of solving it, and the allocation of costs among stakeholders. Reliable sources of information acceptable to all stakeholders are therefore essential for any joint efforts. This not only enables water-sharing parties to make decisions based on a shared understanding, it also helps build trust.[31]

A reliable database, including meteorological, hydrological, and socioeconomic data, is a fundamental tool for deliberate and farsighted water management. Hydrological and meteorological data collected upstream are crucial for decisionmaking downstream. And in emergencies such as floods, this information is required to protect human and environmental health. Tensions between different water users can emerge when information is not exchanged. Disparities in stakeholders' capacity to generate, interpret, and legitimize data can lead to mistrust of those with better information and support systems. In the Incomati and Maputo River basins, the South African monopoly over data generation created such discomfort in downstream Mozambique that the basins' Piggs Peak Agreement broke down, and Mozambique used this negotiation impasse to start developing its own data.[32]

Moving Toward Cooperative Water Management

Although there are many links between water and conflict, and competing interests are inherent to water management, most disputes are resolved peacefully and cooperatively, even if the negotiation process is lengthy. Cooperative water management mechanisms—probably the most advanced

approach—can anticipate conflict and solve smoldering disputes, provided that all stakeholders are included in the decisionmaking process and given the means (information, trained staff, and financial support) to act as equal partners. Cooperative management mechanisms can reduce conflict potential by:

- providing a forum for joint negotiations, thus ensuring that all existing and potentially conflicting interests are taken into account during decisionmaking;
- considering different perspectives and interests to reveal new management options and offer win-win solutions;
- building trust and confidence through collaboration and joint fact-finding; and
- making decisions that are much more likely to be accepted by all stakeholders, even if consensus cannot be reached.[33]

In international river basins, water management institutions typically fail to manage conflicts when there is no treaty spelling out each nation's rights and responsibilities nor any implicit agreements.

On the local level, traditional community-based mechanisms are already well suited to specific local conditions and are thus more easily adopted by the community. Examples include the *chaffa* committee, a traditional water management institution of the Boran people in the Horn of Africa, or the Arvari Parliament, an informal decisionmaking and conflict-resolution body based on traditional customs of the small Arvari River in Rajasthan, India. On the international level, river basin commissions with representatives from all riparian states have been successfully involved in joint riparian water resources management. Especially in transboundary basins, achieving

---

Box 5–3. Harnessing Wild Rivers: Who Pays the Price?

Since World War II, some 45,000 large dams have been built, generating an estimated 20 percent of the world's electricity and providing irrigation to fields that produce some 10 percent of the world's food. Yet for the 40–80 million people whose lives and livelihoods were rooted in the banks and valleys of wild rivers, dam development has profoundly altered the health, economy, and culture of communities and entire nations.

Because dams are generally situated near the ancient homes of indigenous nations, it is ultimately rural and ethnic minorities far from the central corridors of power who are typically forced to pay the price. Ill-considered development plans, forced evictions, and resettlement with inadequate compensation generate conditions and conflicts that threaten the security of individual and group rights to culture, self-determination, livelihood, and life itself.

These dynamics are illustrated in the case of the Chixoy Dam in Guatemala, which provides 80 percent of that nation's electricity. It was planned and developed by INDE (the National Institute for Electrification) and largely financed with loans from the Inter-American Development Bank and the World Bank. Designs were approved and construction was begun without notifying the local population, conducting a comprehensive survey of affected villages, or addressing compensation and resettlement for the 3,400 mostly Mayan residents. The military dictatorship of Lucas Garcia declared the Chixoy Dam site and surrounding region a militarized zone in 1978.

Some villagers accepted resettlement offers but found poorer quality housing, smaller acreage, and infertile land. Others refused to move and instead attempted to negotiate more equitable terms. Tensions escalated as the government declared remaining villagers subversive, seized community records of resettlement promises and land documents, and killed community leaders. Following a second

military coup in March 1982, General Rios Montt initiated a "scorched earth" policy against Guatemala's Mayan population. As construction on the dam was completed and floodwaters began to rise, villages were emptied at gunpoint and homes and fields burned. Massacres ensued, including in villages that provided refuge to survivors. In the village of Rio Negro, for instance, 487 people—half the population—had been murdered by September 1982.

Following the 1994 Oslo Peace Accords ending Guatemala's civil war, a series of investigations broke the silence over the massacres. In 1999 a United Nations–sponsored commission concluded that more than 200,000 Mayan civilians had been killed, that acts of genocide were committed against specific Mayan communities, and that the government of Guatemala was responsible for 93 percent of the human rights violations and acts of violence against civilians.

Today, the issue is far from settled. The failure to provide farm and household land of equivalent size and quality for those resettled has produced severe poverty, widespread hunger, and high malnutrition rates. Communities that were excluded from the resettlement program also struggle with an array of problems. Dam releases occur with no warning, and the ensuing flashfloods destroy crops, drown livestock, and sometimes kill people. Most inhabitants of former fishing villages, their livelihoods destroyed, have turned to migrant labor. Upstream communities saw part of their agricultural land flooded, and access to land, roads, and regional markets was cut off. No mechanism exists for affected people to complain or negotiate assistance.

Chixoy Dam–affected communities have met to discuss common problems and strategies, testified before truth commissions, and, with help from national and international advocates, are working to document the dam's impact. In September 2004, some 500 Mayan farmers seized the dam, threatening to cut

---

Box 5–3. (continued)

power supplies unless they were compensated for land and lives lost.

In a growing number of instances, the efforts by dam-affected peoples to document experiences and protest injury, damage, and loss have succeeded in producing some measure of remedy. In Thailand, where the Pak Mun Dam destroyed fisheries and the livelihood of tens of thousands, a decade of protests prompted the government to decommission the dam temporarily. Affected villagers conducted research on the impact of the dam on their lives and the Mun River ecosystem, documenting the return of 156 fish species to the river after floodgates were opened and the subsequent revitalization of the fishing economy and village life. These assessments played a key role in the decision to operate the dam on a seasonal basis.

At a second dam on the Mun River, the Rasi Salai, displaced people established a protest village in 1999, refusing to leave while the reservoir waters submerged their encampment. Their nonviolent protest and their willingness to face imminent drowning struck a chord in the nation. In July 2000, the Rasi Salai floodgates

were opened to allow environmental recovery and impact assessments, and they remain open to this day.

In documenting the many failures to address rights and resources properly, dam-affected communities have taken the lead in challenging the assumptions that drive development decisionmaking and in demanding institutional accountability. Their demands for "reparations" are much more than cries for compensation. They are demands for meaningful remedy, which means that free, prior, and informed consent of residents is obtained before financing is approved and dam construction initiated, that scientific assessments and plans are developed with the equitable participation of members of the affected community, that governments and financiers respect the rights of indigenous peoples to self-determination—including the right to say no, and that new projects are not funded until any remaining problems from past projects are addressed.

—Barbara Rose Johnston,
Center for Political Ecology, Santa Cruz, California

SOURCE: See endnote 30.

---

cooperation has been a drawn-out and costly process. Recognizing this, the World Bank agreed to facilitate the Nile Basin Initiative negotiation process for 20 years.[34]

Capacity building—to generate and analyze data, develop sustainable water management plans, use conflict resolution techniques, or encourage stakeholder participation—should target water management institutions, local nongovernmental organizations, water users' associations, or religious groups. On the international level, strengthening less powerful riparians' negotiating skills can help prevent conflict. On the local level, strengthening the capacity of excluded, marginalized, or weaker groups to articulate

and negotiate their interests helps involve them in cooperative water management. The Every River Has Its People Project in the Okavango River basin, for instance, aims to increase participation by communities and other local stakeholders in decisionmaking and basin management through educational and training activities.[35]

Preventing severe conflicts requires informing or explicitly consulting all stakeholders, such as downstream states or societies, before making management decisions. The process of identifying all relevant stakeholders and their positions is crucial to estimating, and consequently managing, the risks of conflict. Without extensive and regular public participation, the general public might reject infrastructure project proposals. For example, the decision to build the Hainburg Dam on the Danube River was announced in 1983 after only limited public participation. Environmental groups and other civil society organizations, supported by the general public, occupied the project site and managed to stop the dam's construction. Subsequently, the site became a national park.[36]

The crux of water disputes is still about little more than opening a diversion gate or garbage floating downstream.

Cooperative water management is a challenging issue that requires time and commitment. Extensive stakeholder participation might not always be feasible; in some cases, it may not even be advisable. On any scale of water management, if the level of dispute is too high and the disparities are too great, conflicting parties are not likely to reach consensus and might even refuse to participate in cooperative management activities. In such cases, confidence and consensus-building measures, such as joint training or joint fact-finding, will support cooperative decisionmaking.

Conflict transformation measures involving a neutral third party, such as mediation, facilitation, or arbitration, are helpful in cases with open disputes over water resources management. Related parties, such as elders, women, or water experts, have successfully initiated cooperation when the conflicting groups could not meet. The women-led Wajir Peace Initiative, for example, helped reduce violent conflict between pastoralists in Kenya, where access to water was one issue in the conflict. In certain highly contentious cases, such as the Nile Basin, an "elite model" that seeks consensus between high-level representatives before encouraging broader participation has enjoyed some success in developing a shared vision for basin management. Effectively integrating public participation is now the key challenge for long-term implementation of elite-negotiated efforts.[37]

Water management is, by definition, conflict management. For all the twenty-first century wizardry—dynamic modeling, remote sensing, geographic information systems, desalination, biotechnology, or demand management—and the new-found concern with globalization and privatization, the crux of water disputes is still about little more than opening a diversion gate or garbage floating downstream. Yet anyone attempting to manage water-related conflicts must keep in mind that rather than being simply another environmental input, water is regularly treated as a security issue, a gift of nature, or a focal point for local society. Disputes, therefore, are more than "simply" fights over a quantity of a resource; they are arguments over conflicting attitudes, meanings, and contexts.

Obviously, there are no guarantees that the future will look like the past; the worlds of water and conflict are undergoing slow but steady changes. An unprecedented number of people lack access to a safe, stable supply of water. As exploitation of the world's water supplies increases, quality is becoming a more serious problem than quantity, and water use is shifting to less traditional sources like deep fossil aquifers, wastewater reclamation, and interbasin transfers. Conflict, too, is becoming less traditional, driven increasingly by internal or local pressures or, more subtly, by poverty and instability. These changes suggest that tomorrow's water disputes may look very different from today's.

On the other hand, water is a productive pathway for confidence building, cooperation, and arguably conflict prevention, even in particularly contentious basins. In some cases, water offers one of the few paths for dialogue to navigate an otherwise heated bilateral conflict. In politically unsettled regions, water is often essential to regional development negotiations that serve as de facto conflict-prevention strategies. Environmental cooperation—especially cooperation in water resources management—has been identified as a potential catalyst for peacemaking. (See Chapter 8.)[38]

So far, attempts to translate the findings from the environment and conflict debate into a positive, practical policy framework for environmental cooperation and sustainable peace show some signs of promise but have not been widely discussed or practiced. More research could elucidate how water—being international, indispensible, and emotional—can serve as a cornerstone for confidence building and a potential entry point for peace. Once the conditions determining whether water contributes to conflict or to cooperation are better understood, mutually beneficial integration and cooperation around water resources could be used more effectively to head off conflict and to support sustainable peace among states and groups within societies.

Resource Wealth and Conflict

Abundant natural resources—such as oil, minerals, metals, diamonds, timber, and agricultural commodities, including drug crops—have fueled a large number of violent conflicts. Resource exploitation played a role in about a quarter of the roughly 50 wars and armed conflicts of recent years. More than 5 million people were killed in resource-related conflicts during the 1990s. Close to 6 million fled to neighboring countries, and anywhere from 11–15 million people were displaced inside their own countries.[1]

The money derived from the often illicit resource exploitation in war zones has secured an ample supply of arms for various armed factions and enriched a handful of people—warlords, corrupt government officials, and unscrupulous corporate leaders. But for the vast majority of the local people, these conflicts have brought a torrent of human rights violations, humanitarian disasters, and environmental destruction, helping to push these countries to the bottom of most measures of human development.[2]

In places like Afghanistan, Angola, Cambodia, Colombia, and Sudan, the pillaging of resources allowed violent conflicts to continue that were initially driven by grievances or secessionist and ideological struggles. Revenues from resource exploitation replaced the support extended to governments and rebel groups by superpower patrons that largely evaporated with the cold war's end. Elsewhere, such as in Sierra Leone or the Democratic Republic of the Congo, predatory groups initiated violence not necessarily to gain control of government, but as a way to seize control of a coveted resource.[3]

Commercial resource extraction can also be a source of conflict where governance is undemocratic and corrupt. The economic benefits accrue only to a small domestic elite and to multinational companies, while the local population shoulders an array of social, health, and environmental burdens. All over the world, indigenous communities confront oil, mining, and logging firms. Violent conflict has occurred in places like Nigeria (more than 1,000 people were killed there in 2004), Colombia, Papua New Guinea's Bougainville island, and Indonesia's Aceh province.[4]

Finally, tensions and disputes arise as major consumers of natural resources jockey for access and control. The history of oil, in particular, is one of military interventions and other forms of foreign meddling, of which the Iraq invasion is but the latest chapter. As demand for oil becomes more intense, a new set of big-power rivalries is now emerging.[5]

The United States, Russia, and China are backing competing pipeline plans for Caspian resources, and China and Japan are pushing mutually exclusive export routes in their struggle for access to Siberian oil. In Africa, France and the United States are maneuvering for influence by deepening military ties with undemocratic regimes in Congo-Brazzaville, Gabon, and Angola. China is seeking a greater role for its oil companies, particularly in Sudan, and working to increase its political clout in Africa and the Middle East. U.S. soldiers patrol the oil-rich, violence-soaked Niger Delta with their Nigerian counterparts and help protect a Colombian export pipeline against rebel attacks.[6]

Resource-rich countries often fail to invest adequately in critical social areas or public infrastructure. But resource royalties help their leaders maintain power even in the absence of popular legitimacy—by funding a system of patronage and by beefing up an internal security apparatus to suppress challenges to their power.[7]

A number of conflicts—in Sierra Leone,

L. Lartigue/USAID

Diamond miners, Sierra Leone

Liberia, and Angola—have finally come to an end, but others burn on. In the Democratic Republic of the Congo, foreign forces that invaded in 1998 have withdrawn, yet fighting among various domestic armed factions continues, and elaborate illegal networks and proxy forces have been set up that continue to exploit natural resources.[8]

The enormous expansion of global trade and financial networks has made access to key markets relatively easy for warring groups. They have little difficulty in establishing international smuggling networks and sidestepping international embargoes, given a degree of complicity among certain companies and often lax customs controls in importing nations.[9]

Over the past five years or so, awareness of the close links between resource extraction, underdevelopment, and armed conflict has grown rapidly. Campaigns by civil society groups and investigative reports by U.N. expert panels have shed light on these connections, making it at least somewhat more difficult for "conflict resources," such as diamonds, to be sold on world markets. To discourage illicit deals, revenue flows associated with resource extraction need to become more transparent, but governments, companies, and financial institutions often still shirk their responsibilities.[10]

Commodity-tracking regimes are equally important. In the diamond industry, national certification schemes and a standardized global certification scheme have been established. But the resulting set of rules still suffers from a lack of independent monitoring and too much reliance on voluntary measures. Efforts are also under way by the European Union to establish a certification system for its tropical timber imports—up to half of which are connected to armed conflict or organized crime.[11]

Natural resources will continue to fuel deadly conflicts as long as consumer societies import materials with little regard for their origin or the conditions under which they were produced. Some civil society groups have sought to increase consumer awareness and to compel companies to do business more ethically through investigative reports and by "naming and shaming" specific corporations. Consumer electronics companies, for instance, were pressured to scrutinize their supplies of coltan, a key ingredient of circuit boards, and to ask processing firms to stop purchasing illegally mined coltan.[12]

Promoting democratization, justice, and greater respect for human rights are key tasks, along with efforts to reduce the impunity with which some governments and rebel groups engage in extreme violence. Another goal is to facilitate the diversification of the economy away from a strong dependence on primary commodities to a broader mix of activities. A more diversified economy, greater investments in human development, and help for local communities to become strong guardians of the natural resource base would lessen the likelihood that commodities become pawns in a struggle among ruthless contenders for wealth and power.

—*Michael Renner*

The Private Sector

In an address to the United Nations Security Council in April 2004, U.N. Secretary-General Kofi Annan highlighted the important role that private companies can play—good or bad—in the world's most conflict-prone countries: "Their decisions—on investment and employment, on relations with local communities, on protection for local environments, on their own security arrangements—can help a country turn its back on conflict, or exacerbate the tensions that fuelled the conflict in the first place."[1]

In recent years, grassroots campaigners and U.N. panels have documented the alleged complicity of multinational companies in a wide range of conflict situations—from human rights abuses in oil-rich Sudan and Nigeria, to the trafficking of diamonds and timber from the Congo and Sierra Leone, to the misuse of financial services for arms purchases and terrorist acts. In light of these reports, corporations are increasingly aware that in addition to fueling violence, investments in a conflict situation can seriously taint a company's reputation, and may even become a legal liability.[2]

In one prominent case, the Canadian petroleum company Talisman Energy was forced to sell its oil interests in Sudan following accusations that it had contributed to the 20-year-long civil war. Beginning with the completion of an export pipeline in 1999, crude oil produced by the Talisman-led consortium contributed as much as $500 million a year to government revenues. These payments were alleged to have contributed to a doubling of the government's defense budget in the same period and thus to the "scorched earth" campaign to clear people out of the country's oil fields. In at least one reported instance, helicopter gunships and other military aircraft used the consortium's landing strip as a staging point for attacks on civilians.[3]

In March 2003, having been targeted in a class action suit in New York, Talisman sold its share in the oil consortium to the Indian energy firm ONGC Videsh. Yet even as this initiated a boom in Talisman's share value, the company's retreat from Sudan posed a complex dilemma. On the one hand, it demonstrated to the oil industry that questionable investments or activities could affect a company's reputation and lower its stock value (by up to 15 percent in Talisman's case). On the other hand, the withdrawal of top multinational investments from unstable countries could ultimately reduce international scrutiny of these places, lessening pressure on remaining firms to adhere to minimum social and environmental standards.[4]

There are also instances where the private sector has been instrumental in helping bring hostilities to a close. In Sri Lanka, an attack on the international airport in July 2001 marked a turning point in the decades-long conflict between the Sinhalese majority and separatist Tamils. Prominent business leaders from both sides formed Sri Lanka First to build grassroots support against the war. The group helped coordinate a million-person demonstration in September, and during the subsequent election it campaigned on behalf of legislators who favored a negotiated settlement. These actions helped move the Tamil separatists and the government toward a cease-fire in early 2002.[5]

Companies should play a role in reducing conflict rather than contributing to it. To do so, however, they will need to develop guidelines for managing social risks, strengthening transparency and accountability, and forging collaborative relationships—thus enabling managers to navigate difficult

Esso Photo

Building the Chad-Cameroon pipeline

situations more responsibly.

First and foremost, the consequences of business and development projects must be better understood. By analyzing the likely impacts of conflict on company operations, as well as the impacts of corporate activities on local communities and the broader social fabric, companies would have the opportunity to refocus their core business operations, social investment activities, and public policy strategies on the goal of minimizing harm. To spur their adoption, governments could require export credit agencies (ECAs) and other lenders to conflict-prone areas to make such assessments a condition for preferential access to finance. Similarly, the World Bank's private-sector lending arms and the ECAs could establish guidelines for the assessments, similar to those they use for the environment.[6]

Increasing the transparency of corporate actions will also be essential. The nongovernmental Publish What You Pay initiative seeks to ensure transparency of extractive project royalties and other payments to governments. And the U.K. government–led Extractive Industries Transparency Initiative calls on host governments to be more transparent about the use of these revenue streams. Boosting the capacity of civil society in host countries to hold governments accountable for how these funds are spent is the other necessary building block.[7]

Clear and internationally agreed norms of legal accountability for corporate complicity in gross human rights violations, war crimes, and violations of U.N. sanctions are needed.

Corporate accountability could be upheld through the International Criminal Court or through domestic civil courts using mechanisms like the Alien Tort Claims Act in the United States. While voluntary codes of conduct that address human rights and corruption—such as the U.N. Global Compact—are valuable starting points, a degree of enforceability based on internationally agreed minimum standards is critical.[8]

Private-sector actors can also form valuable partnerships with governments, development agencies, and civil society organizations in areas of ongoing or potential conflict. These can enhance corporate sensitivity and legitimacy while reducing risk, thus increasing overall investment. Multistakeholder assurance groups set up under the supervision of the World Bank, for example, have strengthened the accountability of governments and project operators for delivery of social programs and mitigation of project impacts in the case of the Chad-Cameroon and Baku-Tblisi-Ceyhan pipelines.[9]

The price of getting private-sector investments wrong has reached unprecedented heights. Corruption, patronage, and war profiteering are destabilizing countries and causing unjustified human suffering. But if ethics, regulation, and incentives support the shift, responsible business can become a leading force for peace.

—*Jason Switzer, International Institute for Sustainable Development*

CHAPTER 6

Changing the Oil Economy

*Thomas Prugh, Christopher Flavin,
and Janet L. Sawin*

Oil has become so central to modern civilization that language strains to convey its importance; the common metaphors for its role—linchpin, lifeblood, prize—seem tired and inadequate. Oil saturates virtually every aspect of modern life, and the well-being of every individual, community, and nation on the planet is linked to our oil-based energy culture. Even as oil has become indispensable, however, its continued use has begun to impose unacceptable costs and risks.

The costs and risks of using oil can be grouped into three broad categories. First, oil threatens global economic security because it is a finite resource for which no clear successor has been developed and because the gap between supply and demand appears to be growing, making the world vulnerable to serious economic shocks. Second, oil's value as a commodity undermines civil security by compromising efforts to achieve peace, civil order, human rights, and democracy in many regions. Third, oil threatens climate stability because its use, which is accelerating, accounts for a major share of global greenhouse gas

emissions and because its overwhelming dominance of the transportation fuel market makes it difficult to replace. In short, where oil once helped ensure human security, it now makes us more vulnerable.

A Strategic Commodity

To understand how oil has gone from an asset to a liability, we must begin with its place in modern life. Consider a typical citizen living in a city or suburb in the industrial world—call him Mr. Lee—who embarks on an ordinary Saturday morning of errands. He rises to the sound of a clock radio, showers, puts in his contact lenses, and dresses in a track suit and sneakers. In the kitchen, Mr. Lee swallows some antihistamines to treat his head cold and wolfs down a bowl of cereal, then brushes his teeth, slips on a nylon windbreaker, and heads off through the morning drizzle to the stores. Either car or tram is an option; today, he takes the car.

He stops first at his favorite music store, where he parks, opens his umbrella, and dashes for the entrance. Inside, he browses awhile and settles on a couple of CDs, paying for them with a credit card. Then it's down the street to the sports shop—pausing on the way for a quick pastry from the local baker—where he buys a new tennis racket and a can of tennis balls for Mrs. Lee's birthday. On the way home, Mr. Lee stops at a camera store for a new digital camera, also for his wife, and a blank cassette for their video camera. Then he calls Mrs. Lee on his cell phone to see if they need anything from the pharmacy; she asks him to pick up some hand lotion and her favorite lipstick.

With minor adjustments, this vignette might apply to the lives of hundreds of millions of people in Singapore, Berlin, New York, or elsewhere in the industrial world. But imagine how the picture would change if one element—oil—were erased from the frame.

To begin with, both cars and sprawling suburbs are creatures of cheap oil and would be far less common. But even at a finer grain, our story would change radically. The following items it mentions or implies are at least partly made from petroleum: radios, shower curtains, shampoo, contact lenses, toothbrushes and toothpaste, drugs and pill capsules, fabrics, shoes, automobiles (the carpets and upholstery, insulation, fan belts, battery cases, safety glass and seatbelts, speakers, tires, dashboards, paint, antifreeze), umbrellas, CDs, tennis rackets and balls (and the cans they come in), credit cards, ballpoint pens, cameras, film, cell phones, and countless cosmetics. Mr. Lee's snack pastry is a stand-in for the huge role oil plays in agricultural production—from the manufacture and fueling of farm machinery to the use of oil as a fertilizer input and in processing, packaging, and transport. Then there are the furnishings and floor coverings in the Lee family's house, the roof over their heads, and the roads Mr. Lee drives on—literally thousands of things.

In many cases, there are no ready substitutes for oil in the manufacture of these items.[1]

Clearly, oil is important as a feedstock; in the United States, for instance, feedstock uses account for roughly one fifth of oil consumption. But oil is even more important as a source of energy. Energy has become an enormous presence in the world economy and in the lives of billions. Few people grasp just how critical it is—or that the sheer abundance of energy is what defines life in industrial nations and distinguishes it from traditional ways. Those ways were marked by bondage to a trickle of solar energy, essentially human and animal muscle power fueled by plants. People made minor uses of coal and oil before the Industrial Revolution, but that era literally transformed the world's energy economy. The general pattern can be seen in the energy history of the United States. (See Figure 6–1.)[2]

The sheer abundance of energy is what defines life in industrial nations and distinguishes it from traditional ways.

Although wood was the chief non-muscle-based source of energy throughout pre-modern times (and remains so for billions in the developing world), once fossil fuels became widely available in the late nineteenth century—first coal, then oil and natural gas—they quickly came to dominate energy budgets in those nations with ready access to them. Both per capita and total energy consumption skyrocketed, particularly with the introduction of technologies such as automobiles and electric power plants that were adapted to the new fuels' advantages.

Today, the global consumption of useful energy per person is about 13 times higher than in pre-industrial times—even though total population has increased tenfold since

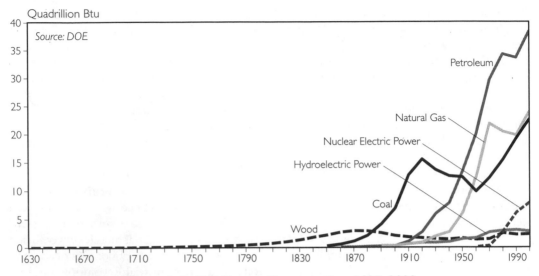

Figure 6–1. U.S. Energy Consumption, 1635–2000

1700. (Per capita consumption is far higher than the global average in industrial nations, of course, and much lower in developing countries.) Oil—easily extracted, flexible, and energy-dense—is the most highly prized energy source of all. It is the world's single largest source of energy, accounting for about 37 percent of global energy production. And it holds a dominant place in the global economy. (See Box 6–1.) Oil's value and availability as a source of transportation fuels mean that oil accounts for nearly all transportation-sector energy consumption. The combustion of oil also accounts for 42 percent of all emissions of carbon dioxide (CO_2), the chief human-caused greenhouse gas.[3]

In this "culture of energy consumption," unique in human history, the health, welfare, prosperity, and prospects of billions of people—their security and that of their nations— are directly influenced by oil's price and availability. Oil has arguably become the most important strategic commodity ever. In a globalized economy, it binds all the world's economies and peoples in a shared matrix. But that matrix is under growing strain. The world's enslavement to oil creates threats that add up to a compelling argument for the end of the current energy regime.[4]

Oil and Global Economic Security

Oil dependence means economic vulnerability. Oil price spikes can trigger both inflation and recession, with real impacts on personal incomes and employment. In the heavily oil-dependent United States, oil price increases have preceded 9 of the 10 recessions since World War II.[5]

The key actors on the oil stage—importing and exporting nations—enjoy much the same relationship as junkies and pushers: neither can easily do without the other. This addiction theme is familiar, but it is not just a conceit. In chemical dependency studies, the classic definition of addiction has three aspects: tolerance, which is the tendency to use more of a substance to achieve the desired effect; withdrawal, the experience of unpleas-

<table>
<tr><td>

BOX 6–1. SOME MEASURES OF OIL'S CENTRAL PLACE IN THE ECONOMY

Two of the top 10 U.S. corporations by sales (ExxonMobil and ChevronTexaco), and 3 of the top 20 (the first two plus ConocoPhillips), are oil companies. The top 10 U.S. oil companies had revenues of nearly $430 billion in 2002. In 1999, 6 of the 10 largest corporations around the world (and 9 of the top 20) were oil companies or their soulmates—automobile companies.

Oil companies are large and profitable because of the strong global demand for oil. Most oil is used in transportation, and the largest category of vehicle is the automobile (including, at least in the United States, light trucks and sport utility vehicles). The world's automobile fleet grew from 53 million in 1950 to 539 million in 2003. Automobile production likewise soared from 8 million in 1950 to more than 41 million in 2003. This trend is likely to continue as developing countries motorize; in China, for example, over 2 million cars were sold in 2003 (80 percent more than in 2002), and the fleet is projected to number 28 million by 2010.

Although air travel accounts for a much smaller share of total oil consumption, it too has increased dramatically, especially following the introduction of commercial jet aircraft in the late 1950s. Air travel volume has increased by a factor of over 100 since 1950, from 28 billion to 2,942 billion passenger-kilometers in 2002.

SOURCE: See endnote 3.

</td></tr>
</table>

cent of the energy budget in France, 39 percent in the United States, 49 percent in Japan, 51 percent in Thailand, and 77 percent in Ecuador. Even these raw numbers understate the dependency, since in many countries oil provides virtually all transportation fuel. Global consumption has generally risen over time—except when price spikes trigger "withdrawal" bouts of economic malaise—despite increasingly troublesome pollution, greenhouse gas emissions, and other problems.[6]

Although industrial countries still use most of the world's oil, developing nations are on average more dependent on oil as a share of total energy use (excluding biomass) and use much more oil in proportion to the size of their economies. Many developing countries import virtually all their oil and are therefore more vulnerable to price shocks than many industrial nations. The International Energy Agency (IEA) has estimated that if the $20-per-barrel price increase in 2004 were sustained, it would reduce economic growth in 2006 by 1.0 percent in the United States and 1.6 percent in Europe, but 3.2 percent in India and 5.1 percent in the most highly indebted nations, mostly in Africa. Such price increases translate directly into human costs in poor countries, since rising food transport costs can affect the diets of impoverished city dwellers, and higher kerosene prices may mean doing without cooking fuel.[7]

Oil-exporting nations, in their own way, are just as dependent. Many rely heavily on a continuous stream of oil revenues because they have failed to use past income from exports to diversify their economies. In some cases, much of the oil income has been diverted to enrich elites and to pay for sophisticated military buildups. In Saudi Arabia, the Sa'ud dynasty subsidizes thousands of "princes" with royal stipends of as much as

ant effects when use is curtailed; and continued use of a substance despite adverse consequences.

All three are visible in the modern world's relationship to oil. Oil accounts for 36 per-

$270,000 a month. James Woolsey, former director of the U.S. Central Intelligence Agency, notes that in the 1990s the "Muslim Middle East," with a population of 260 million, had smaller non-oil exports than Finland, which has 5 million people.[8]

The world received a sharp reminder of its dependence on oil in 2004, when two decades of relative calm ended with a price surge from about $33 per barrel early in the year to over $50 in October—the highest price, adjusted for inflation, since the mid-1980s. (See Figure 6–2.) The spike angered drivers, rocked the world's stock markets, and put a nascent global economic recovery at risk. Several factors contributed to the leap in prices, including sabotage of oil facilities in Iraq and Saudi Arabia, political unrest in the oil fields of Nigeria, and hurricane damage to oil infrastructure in the Gulf of Mexico. But a more elemental force was also at work: supply and demand.[9]

Oil consumption accelerated with the end of the 2001–02 recession, rising 1.5 million barrels per day in 2003 and another 2.3 million barrels per day in 2004, to reach a new record of over 82 million barrels a day by October 2004. Consumption rose in scores of developing and industrial countries. The United States alone contributed one quarter of the increase, while China's consumption rose even more rapidly—from 5.2 million barrels a day in 2002 to an estimated 6.6 million barrels a day in 2004, with most of the demand being met by soaring imports. (See Figure 6–3.)[10]

Such post-recession surges are not unusual. But what was unusual was that oil producers were unable to meet the stronger demand. As prices rose well above their target levels early in 2004, the Organization of the Petroleum Exporting Countries (OPEC) raised production quotas and repeatedly assured the world that they were doing all they could to raise output. Neither action did anything to stem the rise in prices.[11]

From the mid-1980s until 2003, there had generally been enough spare production capacity to allow OPEC producers to keep prices within their target range of $22–28 per barrel by raising output when necessary. In fact, OPEC's main problem during the period was keeping oil prices from getting too low, which happened in 1998, when they briefly fell to $12 per barrel. During most of the 1980s and 1990s, rising oil production outside the Persian Gulf in countries such as Norway, Nigeria, and Brazil was sufficient to meet growing demand, and

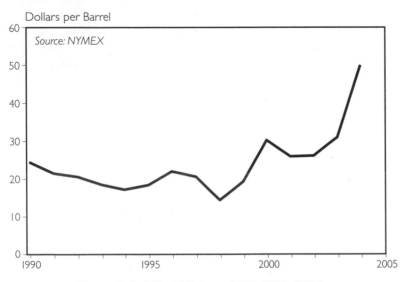

Dollars per Barrel

Source: NYMEX

Figure 6–2. World Price of Oil, 1990–2004

OPEC was able to maintain a substantial spare production capacity, mostly in the Persian Gulf and especially in Saudi Arabia. But production growth outside the Persian Gulf has slowed substantially in recent years. Part of the shortfall was made up by higher production in Russia; the rest came from Saudi Arabia and other Persian Gulf countries. Today, virtually all spare production capacity is gone, leaving the oil market sensitive to everything from a weather report in the Caribbean to a strike in Nigeria.[12]

Does the imbalance in supply and demand represent a short-term challenge, or something more fundamental? The conventional view, echoed by everyone from the U.S. Geological Survey to the IEA, is that there is plenty of oil left to be produced and that slightly higher prices will open the floodgates. They blame higher oil prices on everything from the surge in demand to oil companies' failure to invest, all of which they believe the market will soon correct. Most of these analysts think that officially reported world oil reserves (over a trillion barrels) will allow production to increase for decades, boosted by new technology that will allow oil to be extracted from ever-more-inaccessible reservoirs and from unconventional oil shale and tar sands. These assumptions lead many government forecasters to predict that world oil production will keep on rising. The IEA, for example, projects global production will reach 121 million barrels a day in 2030.[13]

Such projections are pure fantasy, according to a growing number of dissident analysts who believe the recent price rise is an early sign that producers are just not finding enough oil to keep up with demand growth. Led by former Amoco geologist Colin Campbell and others with affiliations ranging from the U.S. Geological Survey to the Iranian National Oil Company, these geologists have looked closely at oil discoveries over the past half-century and concluded that even as exploration increases and technology advances, oil is being found in smaller and smaller quantities and in ever-more-remote regions.[14]

Indeed, Campbell and his colleagues point to data showing that world oil discoveries peaked in the early 1960s and have fallen in each subsequent decade, so that annual discoveries are now running at less than one fifth their peak level. Production has outrun discovery for the past three decades, they argue—and the gap continues to grow. While they acknowledge the role that enhanced technology has played in increasing production, the oil pessimists believe that it simply

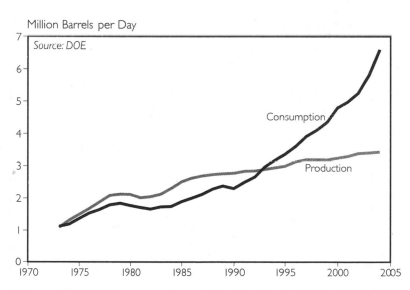

Million Barrels per Day

Source: DOE

Consumption

Production

Figure 6–3. Oil Consumption and Production in China, 1973–2004

allows a bit more oil to be extracted from a given field and, by increasing efficiency, actually accelerates the rate at which a given field is depleted. Global production must inevitably begin to decline—by 2007, according to the latest modeling by Campbell and his colleagues, and slightly earlier or later according to other forecasts.[15]

Exhibit A for these analysts is the United States, where the world's first large oilfield was discovered at Spindletop, Texas, in 1901. U.S. oil production peaked in 1970 and has been falling ever since, only temporarily slowed by record-high oil prices in the late 1970s and early 1980s. U.S. continental oil production fell by half—from 9.4 million barrels a day in 1970 to 4.7 million barrels a day in 2004. Production in Alaska peaked at 2 million barrels per day in 1988 and is now at less than 1 million barrels daily. (See Figure 6–4.) Campbell and his allies point out that U.S. oil discoveries peaked in the 1930s, four decades ahead of the downturn in production, and that the world is now passing the fortieth anniversary of the global peak in oil discoveries.[16]

Many other countries are following the same path. Production has plateaued or declined in 33 of the 48 largest producers, including 6 of OPEC's 11 members. Production is already declining in the United Kingdom and Indonesia, for example, and is no longer increasing significantly in Norway, Mexico, or Venezuela. Moreover, the cornucopian visions of the optimists were shaken in March 2004 when Royal Dutch Shell, the world's second largest private oil company, conceded it had been artificially inflating its reserve estimates for years. This fed growing skepticism about the industry's robust reserve figures. Another index of looming constraints is the recent rise in long-term oil futures prices; for instance, the BP Royalty Trust contract for 2020 nearly doubled from $49 per barrel in May 2003 to over $93 in August 2004.[17]

Earlier oil-peak conjectures have proved premature. But some mainstream analysts think the pessimists may be right this time. PFC Energy, a respected Washington-based oil forecasting firm, released a study in September 2004 estimating that world oil production would peak by roughly 2015 at no more than 20 percent above the current level. Based on a country-by-country analysis of reserve and production trends, the new study concludes that there is just not enough oil being found to sustain growing production. PFC Senior Director

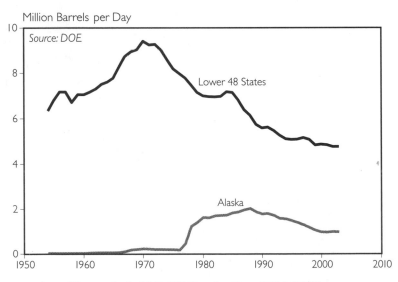

Figure 6–4. U.S. Oil Production, 1954–2003

Michael Rogers believes that oil companies are not increasing their exploration budgets, even at today's high prices, because the fields have gotten so small they are not worth the effort.[18]

No one can predict the precise date at which world oil production will peak, simply because there are far too many uncertainties: demand growth rates, price movements, technological developments, and the political stability of oil-producing countries. These factors have always complicated projections, but now they have been joined by an additional uncertainty concerning the real state of oil reserves in the Persian Gulf, particularly Saudi Arabia.

Matthew Simmons, a Houston-based investment banker to the oil industry, has pored over technical papers published by experts from the Saudi national oil company and concluded that their vaunted world-class oil fields are in trouble. The Saudis' huge official reserve figures (claimed to be 40 percent of the world total) were abruptly raised during the 1980s—a move that was widely seen within the industry as a means of increasing the Saudi share of OPEC production quotas. Simmons found that even today just six fields, all of them more than 30 years old, account for nearly all Saudi production. The Saudi papers suggest that high levels of production are being maintained by pumping large amounts of salt water into the fields. "They're effectively sweeping the reservoirs until the easily recoverable oil is gone," Simmons says. "There isn't any Act II." He believes that far from doubling, as the IEA assumes, Saudi Arabia's oil production could begin falling within a decade.[19]

If so, the future could be even more chaotic than the more pessimistic petroleum geologists expect. The Saudis vehemently dispute these forecasts. But even optimists agree that meeting continued oil demand growth will entail a substantial increase in dependence on supplies from the Middle East—the world's most unstable region. Tens of billions of dollars in foreign investment will be needed to increase Persian Gulf oil production, but that investment will only be forthcoming if oil companies perceive the region to be sufficiently stable. The lack of security is the reason that in 2004 Iraqi oil production ran well short of even the constrained levels of Saddam Hussein's final years in power, contrary to U.S. Pentagon forecasts that production would rise in the year after Saddam was deposed.[20]

Oil production has plateaued or declined in 33 of the 48 largest producers, including 6 of OPEC's 11 members.

Amid the uncertainties, it seems likely that the world energy economy has entered a period of prolonged turbulence. The rate of oil demand growth of the past decade cannot be met much longer, yet this reality has emerged just as China and India—home to 2.5 billion people—and other countries are entering an oil-intensive stage of economic development and moving to stake their claims on the world's oil reserves. With competition for oil intensifying, the peak of oil production, whenever it comes, will almost certainly trigger a period of soaring prices. The aggregate effects on the global economy—on transport, agriculture, and industry—are likely to be grave. Just how grave will depend on many things, but especially on governments' political will to restrain consumption and otherwise find a path away from oil. These measures will need to be developed and in place within a generation, probably sooner. If we wait until then to begin, it will be far too late.

Oil and Civil Security

Oil's long history is one of competition, corruption, political repression, maneuvering for access, and open conflict. At the global scale of great-power contention, this history began in 1912, when the British Royal Navy began switching its ships from Welsh coal to oil in pursuit of greater speed and range. Lacking domestic sources of oil, Britain deployed its fleet abroad to secure dependable supplies, thus beginning a deep entanglement in Middle Eastern politics that ended only with the 1956 Suez crisis, when the United States finally displaced the United Kingdom as the dominant power in the Middle East.[21]

The United States consumes one quarter of global oil production, and the Persian Gulf still supplies one fifth of U.S. imports.

World War I hinted at oil's strategic value—a French official called it "the blood of victory"—as it became vital for warships, defense factories, and new weapons such as tanks and combat aircraft. During World War II, the industrial economies and the mechanized militaries of all the major combatants demanded secure access to oil for fuel and lubricants, but only the United States and the Soviet Union enjoyed plentiful domestic supplies. Lack of such supplies prompted the Japanese invasion of Southeast Asia and the German invasion of the Soviet Union, and the ultimate failure of those initiatives contributed to the defeat of the Axis powers.[22]

The British were instrumental in creating the country of Iraq, partly with an eye toward controlling the flow of oil from the region, and U.S. oil firms joined the British in searching for oil in the Middle East in the 1920s.

But World War II marked a new phase of deep U.S. involvement in the region as the geological significance of Persian Gulf oil reserves became clearer to U.S. policymakers in the early 1940s. President Franklin Roosevelt and Saudi King Ibn Sa'ud met on board a U.S. warship in early 1945 to negotiate the beginning of an official relationship that continues to this day. The Saudis gained a powerful patron able to protect them from their many regional enemies, and the United States laid "the cornerstone of [its] postwar industrial machine."[23]

At war's end, U.S. oilfields still supplied about two thirds of the world's oil. But an explosion of postwar economic growth drove demand skyward. The United States became a net importer of oil in 1948 and since then has become increasingly import-dependent—as have nearly all industrial nations. The vast oil reserves of Saudi Arabia and other Persian Gulf nations became more vital than ever, in part because their exploitation allowed U.S. resources to be conserved. The result, as the United States wielded its growing influence in the area, was a web of U.S.-linked business relationships that allowed the efficient extraction and export of oil through the Arabian American Oil Company (better known as Aramco) and other firms. Production in Saudi Arabia and the surrounding Gulf states skyrocketed, and the royal family and its allies grew rich.[24]

U.S. and other western oil holdings were expropriated in the 1960s and 1970s, but Saudi dependence on U.S. and European engineers and managers continued. Oil money allowed the Saudis to buy a great deal of Western military hardware as well, from army boots to fighter jets and radar systems. And the U.S. government, anxious to protect the region's oil from the Soviet Union, Iran, and any other challengers—and to improve its balance of payments—was eager to sell.

Oil, however, was and remains far too important to be left to the market. It has always provoked a "realpolitik" approach to international relations: hard-nosed, even ruthless measures to ensure access to oil. As early as 1946, American economist Herbert Feis argued that "American interests must have actual physical control of, or at least assured access to, adequate and properly located source of supply." That bald language means a willingness to use military force—a willingness first expressed in the 1950s, when Presidents Harry Truman and Dwight Eisenhower explicitly assured Ibn Sa'ud of the U.S. commitment to act against threats to Saudi sovereignty.[25]

For at least 30 years the United States has had military contingency plans to seize key Middle Eastern oilfields if necessary to secure the flow of oil—plans stimulated by the Arab oil embargo of 1973–74, which ironically was the first time oil itself was used as a weapon against western interests. After the embargo was lifted, Secretary of State Henry Kissinger described for *Business Week* the appropriate circumstances for using military force in defense of the flow of oil. Even more openly, in January 1980 President Jimmy Carter announced in his last State of the Union address that any attempt to control the Persian Gulf would be taken as "an assault on the vital interests of the United States" and would be "repelled by any means necessary, including military force." The "Carter Doctrine" was effectively invoked in 1991 when the United States military drove Iraq from the Kuwaiti oil fields it had occupied a few months earlier.[26]

The Carter Doctrine is still part of U.S. policy. The stakes are higher than ever: the United States consumes one quarter of global oil production, and while it has diversified its sources of imported oil in recent years, the Persian Gulf still supplies one fifth of U.S. imports. Moreover, important U.S. allies, including Japan and many West European nations, depend heavily on oil from the region, and Persian Gulf output has helped stabilize world oil prices at relatively low levels over the years, to the economic benefit of importing nations.[27]

Any loss of production—especially the loss of Saudi Arabia's output—would have devastating consequences for the entire world economy. In this context, the recent wars in the Gulf region may be seen as further exercises in applying the Carter Doctrine. In April 2001, an energy policy report to the U.S. Vice President prepared by a think tank with Republican Party ties noted that at a time of tightening oil supplies and declining excess production capacity, Iraq had become a key "swing" producer and a destabilizing influence on oil supplies. To the extent that it allowed the securing of control over Iraq's oil reserves (10 percent of the global total) and production capacity, the 2003 invasion was meant not only to prevent Saddam Hussein from exerting pressure on world oil prices but to capture that power for the United States.[28]

By one mid-range estimate, the direct cost to U.S. taxpayers of maintaining a military presence designed to secure Middle Eastern oil flows from 1993 through 2003 was $49 billion a year. These costs—not paid at the pump—do not include additional appropriations specifically for the two U.S.-led wars in Iraq. And needless to say, they do not include the human costs—soldiers' loss of life and limb, the grief of loved ones—incurred in military actions.[29]

Western actions to secure reliable oil supplies have extended to other oil-rich parts of the world as well. These measures include major military expenditures as well as alliances of convenience with countries and political leaders whose values, aims, and methods may be anti-democratic, repressive, or even murderous. As the current major-power guarantor of political stability and oil availability in the Middle East, the United States has long aided or allied itself with many repressive regimes, including Saudi Arabia, Iran, and, at one time, Iraq. Ties of this sort are perpetually under revision as regional political circumstances shift and as the pursuit of more diversified sources of oil encourages the development of new relationships.[30]

Oil vividly illustrates the tendency for resource wealth to support corruption and conflict rather than growth and development.

In recent years, the United States has cultivated links (civil or military aid, including bases for U.S. troops in some instances) to a number of central Asian countries—including Afghanistan, Azerbaijan, Kazakhstan, Kyrgyzstan, Pakistan, Turkmenistan, and Uzbekistan—that either have important undeveloped oil reserves or straddle key potential pipeline routes. Much of this activity represents U.S. initiatives in a three-way struggle with Russia and China to secure access to the oil and gas resources of the region, a competition that will keep arms and other aid flowing abundantly to the region's regimes and undoubtedly keep tensions high. According to Human Rights Watch, the regimes in all these countries are marked by significant human rights abuses, such as arrests or harassment of political opposition, repression of journalists, corruption, police brutality, election violence and fraud, and lack of religious freedom.[31]

In addition to great-power maneuvering, military interventionism, and alliances of convenience, oil is associated with a variety of other actions that undermine civil security. For example, oil vividly illustrates the "natural resource curse"—the tendency for resource wealth to support corruption and conflict rather than growth and development. The effects have been evident in a number of countries, including the United States. In addition to Saudi Arabia and other Persian Gulf nations, the natural resource curse can be seen at work in Angola, Cameroon, Colombia, Ecuador, Equatorial Guinea, Indonesia, Nigeria, the Republic of the Congo, Sudan, and Venezuela, among others. The "curse" is often aided and abetted by corporations acting with the knowledge of (and frequently in collusion with) national governments in pursuit of resources. The measures taken to secure access to and extract those resources have been known to abridge the rights of indigenous peoples and despoil or even poison their traditional homelands.[32]

Oil and other mineral wealth also appear to impede the establishment and preservation of democracy because the wealth allows governments to reduce agitation for democracy by keeping taxes low and spending high. And when that does not work, it lets them deploy strong security forces to suppress dissent. This is a particularly important point in view of the evidence that terrorism is more likely a response to lack of political rights and opportunities than it is to poverty. More generally, countries that depend on oil revenues tend to be more authoritarian, more corrupt, more conflict-prone, and less developed than countries with diversified economies. They also spend more on their militaries and have larger shares of their people mired in poverty. Oil, a toxic fluid in more ways than one, is such a mixed blessing that it has been called the devil's tears.[33]

The most recent oil-related threat to civil security is terrorism. The best-known part of this story emerged after the terrorist attacks of September 11, 2001. Most of the aircraft hijackers were Saudi citizens—one of the ironic legacies of U.S. and Saudi support for the radical Muslim fighters who fought and defeated the Soviets in Afghanistan in the 1980s. This support, marshaled in part by the U.S. Central Intelligence Agency and the Saudi royal family, created a pool of tens of thousands of Muslim radicals, including Osama bin Laden, from which al Qaeda drew its members. The Saudi contribution included funds for a network of charities that built thousands of mosques and schools run according to the tenets of Wahhabism, a fundamentalist strain of Islam, as well as paramilitary training camps and terrorist recruitment operations. With the Soviets out of Afghanistan, the growing American military presence in Saudi Arabia and elsewhere in the Middle East following the 1991 Iraq war helped provoke the radicals to commit a string of assaults on U.S. interests, including the embassies in Kenya and Tanzania, the USS *Cole*, and the World Trade Center.[34]

Saudi support for the charities amounted to an estimated $70 billion over the years. Given the dearth of Saudi income from any source other than sales of crude oil, those monies in effect make western consumers complicit in acts of terror directed primarily against westerners. In this and the other ways already mentioned—the military adventurism, alliances with thuggish regimes, resource conflicts—the dependence of industrial countries, particularly the United States, on oil imposes a heavy burden of geopolitical risk and moral culpability.[35]

Oil and Climate Security

Fifty pages into a 2004 White House report on climate change, a diligent reader will find this statement: "Comparison of index trends in observations and model simulations shows that North American temperature changes from 1950 to 1999 were unlikely to be due only to natural climate variations." With this unassuming sentence, U.S. government officials finally stepped in line with the global consensus on climate change: that Earth is warming and that human actions, mainly deforestation and the burning of fossil fuels (oil, coal, and natural gas), are the major causes. Oil alone accounts for over two fifths of total emissions of carbon dioxide, the chief human-caused greenhouse gas.[36]

The consensus on climate change has been building for some time. As early as 1988, scientists noted that human tinkering with the climate amounted to "an unintended, uncontrolled, globally pervasive experiment whose ultimate consequences could be second only to a global nuclear war." Each successive report of the Intergovernmental Panel on Climate Change, the most authoritative body synthesizing the vast research on this subject, has argued the case for human influence on climate in stronger and stronger terms. Earth's temperature at the surface is about 0.6 degrees Celsius higher than it was a century ago, and greenhouse gas concentrations are rising as emissions continue to increase. A growing number of leaders around the world now warn that climate change is, in the words of U.K. Chief Scientific Advisor David King, "the most severe problem that we are facing today—more serious even than the threat of terrorism."[37]

Climate change, whether incremental (the most likely scenario) or abrupt, is likely to trigger regional droughts, famines, and weather-related disasters that could claim thousands or millions of lives, exacerbate existing tensions, and contribute to diplomatic or trade disputes. In the worst case, further incremental warming will raise sea levels more and reduce the capacities of Earth's natural systems, which could threaten the very survival of low-lying island nations, destabilize the global economy and geopolitical balance, and incite violent conflict.[38]

Human civilization itself was made possible only because the climate has been relatively stable over the past several thousand years. But that climate stability—unusual over geologic time scales—is at risk. The concentration of CO_2 in Earth's atmosphere is now higher than at any time in the past 400,000 years, and the rate of increase is accelerating. In June 2004, a new, more accurate supercomputer modeling system revealed that global temperatures could rise more rapidly than previously projected.[39]

As CO_2 concentrations build and the planet warms further, the effects are likely to include more severe and frequent storms, floods, and droughts; prolonged and more frequent heat waves; the spread of diseases such as malaria and dengue fever; and ocean-water acidification, coral bleaching, and sea level rise. This will further stress Earth's carrying capacity, which is already pushed to its limit by some estimates. Existing threats to security will be amplified as the impacts of climate change affect regional water supplies, agricultural productivity, human and ecosystem health, infrastructure, financial flows and economies, and migration patterns. Uncertainty about the future availability of essential resources will only heighten such threats.[40]

Global poverty will likely increase as the climate changes, threatening people's homes and livelihoods through increased storms, droughts, disease, and other stressors. This may in turn impede development, increase national and regional instability, and intensify income disparities between rich and poor countries. Such impacts could in turn lead to military confrontations over distribution of the world's wealth or could feed terrorism

or transnational crime.[41]

Countries with large populations, such as China and India, may find themselves in especially serious trouble from multiyear droughts and floods, and the resulting surges in food imports could dramatically increase food prices around the world. Tightened food supplies could trigger internal unrest and increase the use of food as a weapon by exporting countries. Altered rainfall patterns could heighten tensions over the use of shared water bodies and increase the potential for violent conflict over water resources. Such shifts in motives for confrontation and in the positioning of essential resources could alter the balance of power among nations, causing global political instability.[42]

Whether or not conflicts result from climate change will depend greatly on societies' vulnerability to stress and their ability to adapt to or mitigate impacts. Environment-related violence tends to occur in weak and undemocratic states and tends to be internal rather than inter-state. Thus it is most likely that conflicts would take the form of domestic insurgencies or civil wars. Yet climate change could lead to waves of migration as agricultural productivity declines, fresh water becomes scarcer, or sea levels rise, threatening international stability as well. Historically, migration to urban areas has stressed limited services and infrastructure, inciting crime or insurgency movements, while migration across borders has frequently led to violent clashes over land and resources.[43]

The stresses already imposed by climate change make it vital to address current vulnerabilities. Poverty must be reduced through sustainable development so that people can better cope with the changes wrought by global warming. Renewable energy must play a major role on this front, because it can help to alleviate poverty and reduce the risk of conflicts over nonrenewable energy and water resources and because distributed energy systems lessen susceptibility to natural disasters. In addition, environmental management and conservation can reduce vulnerability to climate change impacts by creating more-resilient ecosystems.

Some further warming of the climate is unavoidable. But the more extreme the ultimate warming, the graver the consequences—so it is critical to take every possible measure to reverse the growth of emissions. That means moving as swiftly as possible to a global post-carbon energy economy—an economy that does not release further carbon into the atmosphere.

Climate change already claims more lives annually than terrorism does: a study by the World Health Organization and the London School of Hygiene and Tropical Medicine estimated that perhaps 160,000 people die each year due to the ancillary effects of climate change, such as malaria and malnutrition. As oil contributes significantly to that toll, severing the oil connection will enhance global security.[44]

The Fork in the Road

For global economic security, for national security and personal safety, and for the stability of the world's climatic system, an end to our dependence on oil is necessary. But is it possible? The short answer is yes—in time. The key questions are, How much time do we have? And what sort of transition will it be?

The change should be as rapid as we can make it. Estimates vary regarding how much time is reasonable: a 1995 Shell scenario suggested that renewable energy sources might supply half of the world's energy by 2050, while a Stockholm Environment Institute study argued that a concerted effort could deliver a system relying almost completely

on renewables by 2100, even allowing for steady growth in energy use.[45]

It is important to remember that the timeframe is policy-driven and that, among the security-related problems of oil, nothing improves with time; everything gets worse. Global economic security is increasingly imperiled by the growing strain on oil supplies and the world's intensifying reliance on Middle Eastern oil reserves. The civil security issues will simply continue to fester, claiming lives and undermining development, as long as oil remains an inordinately valuable and unequally distributed commodity. And climate change imposes ever-graver stresses the longer we fail to curb carbon emissions from the use of oil and other fossil fuels.

Climate change could lead to waves of migration as agricultural productivity declines, fresh water becomes scarcer, or sea levels rise.

As for the kind of transition we make, that will depend heavily on the choices made by individuals and, especially, governments. The role of governments cannot be overstressed; the major existing energy companies must be involved in the transition, yet they already have billions of dollars tied up in assets that they cannot simply abandon. Only governments can create the incentive structures to encourage the required investment. Thoughtful government choices thereby increase the odds of making the transition to a new energy era that can relieve some of the major stresses of the current regime without undue economic and social disruption. Bad choices—including "business as usual," or simply letting events run their course—seem likely to lead to an era in which the economic, civil, and climate traumas of the current energy regime are only exacerbated.

Environmentalists have been critical of overdependence on oil and other fossil fuels for years, but there is now considerable agreement in many quarters on the need to overhaul the world's energy economy. It is widely argued that strict energy independence for the United States and other major oil-importing nations is impossible as long as oil remains any significant part of the global energy economy. This may be true, depending on how the new energy economy evolves, particularly with respect to transportation strategies. But the likely mix of options chosen as the successor to the oil economy—renewables, demand management, efficiency, and others—will encourage flexibility, reliability, and reduced vulnerability worldwide by involving less centralized energy systems, use of a greater variety of energy technologies and fuels, and diversification of sources.[46]

Total worldwide investment in renewables topped $20 billion in 2003 and the market could reach $85 billion per year in a few years.

This transition points to the growing prominence and market penetration of renewable energy technologies, discussed briefly in this section. One important element, however, is not a technology at all but rather a policy shift toward intelligent conservation. Far from "freezing in the dark," conservation practiced as demand management involves a wide range of social and economic arrangements that reduce or eliminate the need for energy in the first place without losing the material benefits of energy consumption. It can be as simple as living within walking distance of shopping and work, making it unnecessary to drive or ride the subway to buy

groceries or earn a living.

At the institutional level, community zoning laws can shape development patterns so as to facilitate such living arrangements. Among electric and natural gas utility companies, demand management (also called demand-side management, or DSM) takes the form of programs to make consumers more aware of their energy consumption and waste and show how both of these might be reduced, financial assistance to increase the use of energy-efficient technologies, load management (incentives to use energy at times when it is in less demand), and other options. The aim is to get more use out of the energy consumed rather than simply to increase consumption.[47]

DSM programs can save enormous amounts of energy (and money). In the United States, electric utilities avoided the construction of nearly 30,000 megawatts of capacity in 1996, the peak year of such programs. In economic terms, this meant that utilities did not have to finance and build—and ratepayers did not have to pay for—several large power plants, reducing fuel consumption and pollution. Yet these promising programs were torpedoed and began to lose effectiveness in the late 1990s when U.S. electricity generation was deregulated and utilities once more were rewarded for generating and selling commodity electricity rather than supplying services—a powerful example of how government policy can shape the energy regime.[48]

A related approach is energy efficiency, which involves the deployment of improved technologies that use less energy to accomplish the same tasks. Compact fluorescent lamps, which last longer and produce equivalent light with far less current than ordinary incandescent bulbs, are a familiar example. Millions of homeowners were introduced to both DSM and compact fluorescents by utility programs that made it cheap and easy to acquire the bulbs. Another example is advances in automotive engine technologies—partly mechanical (double overhead cams, three or four valves per cylinder rather than two, variable valve timing) and partly electronic (sophisticated computers that control fuel metering, spark timing, and so on)—and their wider use in more and more cars. The march of these advances was driven by decades of high gas prices in Europe and by the oil price crises of the 1970s and 1980s, which awoke U.S. motorists to the value of good fuel economy. Government policies, in the form of fleet fuel mileage standards, helped sustain progress—at least for awhile.

Such improvements can make an enormous difference in a nation's energy budget. Efficiency gains in transportation and other sectors helped the U.S. economy raise its overall energy productivity (the amount of economic output per unit of energy used) by 64 percent between 1975 and 2000. Oil productivity improved a stunning 93 percent, even though government policies were not consistently sympathetic to that goal. When they did promote efficiency, especially in 1977–85, net oil imports dropped by half and imports from the Persian Gulf plummeted by 87 percent.[49]

Have all the obvious efficiency improvements already been adopted? Hardly. It's been estimated that raising the year-2000 average fuel economy of the U.S. light-vehicle fleet by 3.25 miles per gallon would have saved as much oil as was imported from the Persian Gulf that year. When fuel efficiency was a prized goal in the early 1980s, it took less than three years to achieve such an improvement. A 2002 study from the National Academy of Sciences estimated that the efficiency of the U.S. fleet could nearly be doubled cost-effectively with no loss of safety

or performance—and that study did not consider the gains from wider use of emerging gas/electric hybrid technologies. Other new technologies, involving advances not just in powertrain design but in automobile body and frame construction as well, promise further potential gains.[50]

As for other sectors, efficiency improvements may be the most cost-effective way to reduce business and industrial carbon emissions. One study estimated that most buildings and factories could slash their electricity consumption by at least one quarter, with the savings paying for the investment in under four years. Since the commercial and industrial sectors use large amounts of electricity—over 60 percent of all electricity generated in the United States, for example—and most of that electricity is still made by burning fossil fuels (mainly coal), the carbon savings potential is significant—and cheap.[51]

Using less energy and using it more efficiently are necessary—but not sufficient—to build a sustainable energy regime. A world that wishes to be carbon-free must also plan on making its energy from renewable sources, possibly with some transitional contribution from fossil-fuel systems incorporating carbon capture and storage technology. (See Box 6–2.) Fortunately, this is no longer just a fevered dream of environmentalists; the modern age of renewables is already here. Wind- and solar-generated electricity are the fastest-growing sources of energy in the world; total worldwide investment in renewables topped $20 billion in 2003 and the market could reach $85 billion per year within a few years. The theoretical potential of energy from renewable sources is greater than total global energy consumption by a factor of roughly 18, even using current wind, solar, biomass, and geothermal technologies.[52]

Transportation may be the most chal-

BOX 6–2. CARBON CAPTURE: FOSSIL FUEL REPRIEVE OR RED HERRING?

The chairman of Shell Transport made a startling observation recently: "No one can be comfortable at the prospect of continuing to pump out the amounts of carbon dioxide that we are at present," said Lord Oxburgh, "with consequences we really can't predict, but are probably not good." That statement, progressive for an oil company executive, would seem to demand a turn to renewables.

Perhaps—unless the carbon released by burning fossil fuels can be corralled and herded back into the ground where it came from. This option, called carbon capture and storage (CCS), would leave more of a role for fossil fuels in the energy economy of the future—and would take the pressure off coal, oil, and natural gas companies and countries with large fossil fuel reserves, while possibly reducing the urgency of converting to renewable energy.

Carbon capture is aggressively promoted in some quarters; coal companies in particular have a huge stake in its success because, with oil on the decline, most future carbon emissions would come from burning coal. But the prospects for CCS are not clear. It is already in limited use; for example, the Norwegian oil company Statoil has been injecting captured CO_2 into offshore aquifers since 1996 and now injects a million tons a year to avoid carbon taxes. Carbon dioxide has also long been reinjected into oilfields to enhance oil recovery. One Japanese study concludes that suitable land-based reservoirs alone could store 280 years' worth of global carbon emissions at 1990 rates, and that—in Japan at least—CCS is both practical and cost-competitive with domestic energy conservation and renewable energy measures. Other estimates suggest that there is enough underground storage capacity to hold several decades' worth of CO_2 at current emission rates.

Doubts remain, however. The biggest is cost: about $150 per ton of carbon using current technologies, which would increase the cost of electric power by 2.5–4¢ per kilowatt-hour. Current technologies are also suitable only for large point sources of CO_2, such as coal-fired power plants, which account for less than one third of global carbon emissions. They are thus useless in cleaning up the planet's vast and expanding fleet of motor vehicles, which emit 42 percent of the global total. And most critically, nobody knows how long injected CO_2 will stay put, even if undisturbed—much less when subjected to earthquakes or other stresses. Studies to date have monitored injection sites for only a few years, and even very low rates of leakage could re-release enough CO_2 to pose major problems within decades.

Unless they could be firmly resolved, such worries suggest that relying on CCS for anything but a temporary, transitional contribution to reducing atmospheric carbon concentrations would be risky.

Further research may help answer some of these questions. The Intergovernmental Panel on Climate Change plans to release a major report on CCS in 2005.

SOURCE: See endnote 52.

---

lenging sector to convert, because the machines and the infrastructure are almost exclusively designed for energy-dense and easily handled liquid fuels made from oil. Biomass fuels such as ethanol and biodiesel, however, are technically proven and can be economically competitive with gasoline and diesel. Studies suggest that ethanol from cellulosic biomass (plant residue and waste) could displace roughly a quarter to a third of U.S. gasoline demand at a feedstock price of $50 per ton. The relatively conservative International Energy Agency estimates that, in terms of technical potential, ethanol could account for half or more of global transport fuels by 2050.[53]

The ultimate goal is to switch the global energy economy from carbon to hydrogen.

Hydrogen is not a greenhouse gas, and burning it releases no carbon at all. When it is run through a fuel cell, the products are electricity, heat, and water vapor. Hydrogen fuel cells are an old technology that people became enthused about a few years ago. Some of the excitement has ebbed, however, as sober analyses of the transitional challenges have emerged. The long-term potential of hydrogen is great, but hydrogen is an energy carrier rather than a fuel per se. Capitalizing on it may depend on technical progress in efficient renewable electricity generation to make hydrogen from water. (Using nuclear power as the electricity source, despite proponents' hopes, is too expensive and poses grave security problems.) Such a closed-loop system would be carbon-neutral and could be sustained as long as the sun shines—for several billion years, in other words.[54]

Hydrogen pessimists do not dispute the desirability of a hydrogen energy economy, but they believe the transition will take longer than the optimists think and will require bridging technologies. One such technology is gasoline- and diesel-electric hybrid automobiles and light trucks. Fuel cells are not yet technically ready or cheap enough for vehicle applications, but hybrids can be nearly as efficient as fuel cells (and twice as efficient as internal combustion engines alone) when evaluated on a "well-to-wheels" basis. Moreover, hybrids are facilitating the debut of fuel-cell vehicles by providing a laboratory for the development of critical electronic controls and power management systems.[55]

If renewable energy technologies are so attractive, why don't we already get more energy from them? The market presence of renewables so far is small in absolute terms, despite their extraordinary recent growth rates. The reasons have little to do with technology, however, and more to do with the regulatory and policy environment. Most societies, including the United States, have long been wedded to nonrenewable fuels and large, centralized generating facilities—fossil-fired and nuclear power plants, huge hydroelectric dams—and have supported them with enormous subsidies over the years.

Estimates of those subsidies vary widely because of differing assumptions and definitions; subsidies may include everything from tax breaks and depletion credits to R&D funding. (And in the United States, the Price-Anderson Act limits nuclear industry liability for catastrophic accidents—a priceless asset, because no nuclear plant could be insured or built without it.) Nevertheless, the estimates, both partisan and scholarly, are instructive: in Europe, national and European Union subsidies to the fossil fuel and nuclear power industries were estimated in 1997 at nearly $15 billion a year. A 2004 report put them at 29 billion euros ($36 billion) in 2001. Two separate studies of U.S. subsidies put the total at $5 billion (for fossil fuels alone) and $36 billion a year. Only a small fraction of subsidies anywhere have supported renewables. Imagine the progress that could be made—in raising the relatively low efficiency of solar cells, for example—if $20–30 billion a year were steered into intensive research on renewables or into production incentives and rebates to increase their market penetration.[56]

Besides redressing the subsidies imbalance, governments can and should be proactive in accelerating renewables' growth via regulatory reform. The regulatory environment in many countries has not been stable or conducive enough to renewables investment. Rapid adoption of renewables can mitigate some of the worst effects of tightening oil supplies and climate change, but the risks and uncertainties of the new technologies could choke off the substantial investment required unless investors are both informed about renewables' promise and convinced that the regulatory environment is stable and inviting.

The experience of several countries—especially Germany and Japan, which in only a few years have transformed themselves into the world's leaders in wind power and photovoltaics, respectively—yields clear guidance for governments in four major policy categories in addition to the overhaul of subsidy policies.[57]

First, governments must ensure that renewables actually have access to energy markets. The most effective tactic to date has been pricing laws, which guarantee fixed, minimum prices for electricity and require utilities to provide access to grids. Quota systems, which mandate a minimum share of generation or capacity for renewables, have also worked; examples include the renewable portfolio standards in place in several states in the United States and ethanol quotas in Brazil.

Second, investors and consumers alike are often poorly informed about renewable resource availability and potential, new developments in technology, and existing incentives to build renewable capacity or install equipment. Governments, nongovernmental groups, and industry should cooperate to dispel this ignorance, as well as to ensure that a skilled workforce is available to build, install, and maintain renewable energy systems.

Third, public participation in policymaking, in project development, and in ownership has been shown to increase political support and raise the odds of success, whether the project is a wind turbine farm in Denmark or a solar mini-grid project in Nepal.

Finally, industry standards, permit requirements, and building codes are important to ensure that inferior hardware does not enter the marketplace and destroy consumer and investor confidence, that public concerns about siting are addressed, and that new buildings are designed to be compatible with renewables.

These measures, in effect, are the core of the operating manual for the transition from the oil economy to the renewable economy. As noted earlier, the best current estimates of when scarce oil supplies and rising prices will begin to bring the oil era to a close suggest an upper bound of about 30 years from now. Completely replacing the world's existing energy infrastructure (worth about $10–12 trillion) also takes 30–40 years and will require an estimated $16 trillion. Diverting the lion's share of that routine investment—let alone additional funds—into renewables would give the world a new energy economy in timely fashion.[58]

We know how to do this, and we know it must be done soon. Can governments and ordinary citizens summon the will to act? There are many hopeful signs, such as European renewable energy laws. And in 2004, China pledged that by 2010 it would generate 10 percent of its electric power from "new" renewable sources (that is, not including large hydroelectric dams such as the Three Gorges project). In countries with less-than-progressive national policies, some local governments are taking up the challenge; in the United States, for instance, many states offer rebate programs and other incentives to encourage homeowners to install rooftop solar power equipment.[59]

Building on these and other efforts to transform the oil-dominated global energy regime is now urgent. Oil was once a lifeline for civilization, but now it is a noose. We have arrived at a fork in the road. One path leads to the likelihood of calamitous loss of a prime energy source before the world has prepared for it, and thus to an economically precarious, hotter, and more dangerous planet. The other path leads us away from oil before crisis creates panic and toward a world of abundant, clean, and stable energy, more widely available to more of the world's people than ever before. Put another way, our choice is to be bereft of oil or to be free of it.

Nuclear Energy

Today there are 438 commercial nuclear power plants operating in 30 countries. These reactors supply 16 percent of global electricity, an important consideration in a world where energy demand is growing rapidly. But they also pose a tempting target for terrorists, and damaging them—whether intentionally or by accident—could have catastrophic effects. Further, as nuclear materials are unearthed from mines and distributed across the globe, they contaminate soil, air, and water, damaging the environment and people's health.[1]

Simulated attacks on nuclear power plants have shown that many reactors are poorly secured. In both the United States and Russia, government agencies have launched mock attacks on reactors, only to find that power plant defenses are often inadequate to prevent infiltration and the planting of fake bombs. Twenty-seven of the 57 simulated attacks in the United States during the 1990s revealed significant vulnerabilities that could have caused reactor "core damage" and "radiological release." Even environmental groups have been able to simulate attacks on power plants successfully. In 2003, to expose the plant's vulnerability, Greenpeace activists stormed the United Kingdom's Sizewell power plant and scaled the reactor without resistance.[2]

Sabotage of nuclear reactors is not the only threat. As history has shown, construction flaws and human errors can have disastrous effects when not caught in time. Since the dawn of the nuclear age, there have been hundreds of nuclear accidents. While most have been relatively minor, a few have been catastrophic—the worst of which occurred in Chernobyl, Ukraine, in 1986. A reactor meltdown caused at least 6,000 deaths, as well as elevated rates of thyroid cancer, significant environmental damage,

and the eventual resettlement of more than 370,000 people.[3]

There have also been many near misses. In 2002, for example, at the Davis-Besse power plant in Ohio, boric acid ate a hole through a 17-centimeter reactor vessel head. If it had gotten through the remaining half-centimeter of steel that contained the coolant, a meltdown could have occurred. As a 2004 Union of Concerned Scientists study warned, many of the 103 U.S. nuclear power plants are now entering the last phase of life, which increases the probability of reactor failure and possibly disaster.[4]

Even if there are no attacks or accidents, nuclear materials threaten global security more subtly. In 2002, some 65,000 tons of uranium were used in the world's commercial power plants—36,000 tons of which were extracted from uranium mines. These mines often threaten surrounding communities, creating dust and tailings that can spread radioactive contamination. In Kyrgyzstan, for instance, at least 2 million tons of uranium waste currently sit in 23 tailing ponds in Mailuu-Suu. Left over from Soviet uranium mining and milling operations, this waste is at risk of spilling into the local river and could easily contaminate the Fergana river valley and its 6 million inhabitants.[5]

Some of the most radioactive waste does not sit near old mines but on-site at nuclear power plants in the form of spent fuel. Often, like the reactors, it is inadequately secured and poses security risks. While finding secure long-term storage facilities to safeguard this waste is essential, the challenge will be finding suitable sites that will remain geologically stable for hundreds of thousands of years— which is how long uranium remains hazardous. Currently the United States is planning to build a repository in Yucca

British Energy

Spent fuel rods in a storage pond at a British nuclear plant

Mountain in the state of Nevada. Critics question the appropriateness of this site, however: it is geologically unstable, and water in the repository could corrode the storage casks and contaminate regional groundwater stores.[6]

One immediate way to reduce the nuclear threat would be to decommission as many nuclear weapons as politically feasible and convert them into fuel. There is a dual benefit to converting highly enriched uranium into nuclear fuel: it lessens the potential of weapons-grade uranium falling into terrorists' hands and it reduces the need to mine more uranium, thus slowing the influx of new nuclear materials into circulation. Already the United States and Russia are converting warheads through the Cooperative Threat Reduction Initiative. Over the past 10 years, about 8,000 Russian nuclear warheads have been dismantled and converted into nuclear fuel—providing half of the uranium needed to run U.S. nuclear power plants.[7]

While its advocates often claim that nuclear energy will help reduce the threat of climate change, they rarely incorporate the entire fuel cycle into their considerations. According to the Oeko-Institut, when indirect emissions are included nuclear power produces from one-and-a-half to three times the carbon dioxide per kilowatt-hour that wind power does. Add to this the pollution, health effects, and safety risks of this energy source, and nuclear power becomes less and less a reasonable option.[8]

To completely eliminate the threat that nuclear energy poses, nuclear power plants will need to be phased out entirely. While this may seem impossible in some countries for instance, in France 78 percent of the electricity comes from nuclear power—Belgium, Germany, Sweden, and Spain are planning to eliminate this energy source over the next 20–30 years. Yet this may be more of a countertrend than trend: around the world, 28 new reactors are under construction and another 35 are being planned, including in some countries that have not built new plants in decades. In 2002, Finland's parliament voted to build a new reactor—the first in 20 years. The likely 8-percent rise in nuclear capacity over the next five years will increase the circulation of nuclear materials, in turn adding to security threats, pollution, and damage to health.[9]

Admittedly, facilitating the phaseout of nuclear energy will be a considerable challenge, as it is one of the most protected industries in the world. But by terminating the massive subsidies that the industry receives, removing government-paid catastrophe insurance and insurance exemptions, and factoring in the environmental and social costs of nuclear power, policymakers can make the price of nuclear power reflect its true costs.

—*Erik Assadourian*

CHAPTER 7

Disarming Postwar Societies

Michael Renner

As described throughout *State of the World 2005*, security is influenced by a far broader set of factors than the narrow confines of traditional thinking would suggest. This does not mean, however, that issues related to weapons and combatants are irrelevant. Military spending diverts scarce funds from social and environmental programs that—adequately funded—are the underpinnings of a more stable world. And the legacies of past wars can be powerful obstacles to creating more secure and peaceful societies. This is as true for the environmental repercussions of armed conflict—from plundered forests and devastated irrigation systems to the spread of depleted uranium weapons—as it is for the enormous array of weapons that have been produced and traded across the planet.

The weapons that military budgets purchase are ostensibly acquired to enhance security, but they may well have the opposite effect. Across the spectrum of armaments, there is an urgent need to promote disarmament in order to reduce the likelihood that weapons fuel wars, trigger crime waves, or

become available to extremists.

Although this chapter focuses on conventional small-caliber weapons, other types of arms are of course also of great concern. Nuclear weapons, chemical and biological weapons, landmines, and heavy conventional weapons—tanks, jets, missiles, and warships—are a continuing threat to the economic and physical security of all nations. The nuclear "haves" give no indication of seriously moving toward disarmament. Meanwhile, Israel, India, Pakistan, and possibly North Korea have joined the nuclear weapons "club," for example, and Iran may soon as well. As long as nations pursue "civilian" nuclear programs, there will always be some ambiguity about ultimate intent and always a danger that materials might be diverted into a weapons program or stolen by terrorist groups.

There is also growing concern that chemical weapons or the precursor materials needed to manufacture them might fall into terrorists' hands. Fortunately there is far greater progress toward universal disarmament in this case, driven by a treaty as well as by a sense that existing stocks pose a greater danger to their owners than to any enemy. For biological weapons, however, there is much greater ambiguity: although an international convention outlaws such arsenals, there is no way to know whether states abide by this treaty, and efforts to create a verification mechanism foundered on U.S. opposition.

Anti-personnel landmines have taken a heavy toll for decades. They continue to kill indiscriminately long after a conflict has come to an end and they cannot distinguish between soldiers and civilians. Estimates indicate that a million people have been killed or maimed by landmines since 1975; some 80 percent of them were civilians. Since the passage of an international treaty banning anti-personnel mines in 1999, however, significant headway has been made in recent years, with production and exports dropping sharply and stockpiles being destroyed at a rapid pace. Yet according to the International Campaign to Ban Landmines, six countries still use anti-personnel mines, and Russia and Myanmar use them on a regular basis. In addition, armed opposition forces continue to use mines in 11 countries. So there is a long way to go before landmines cease to be a problem.[1]

The production and trade of large conventional weapons continues more or less unabated. After the end of the cold war, governments did conclude a treaty that regulated several types of such armaments in Europe, and huge numbers of them were dismantled. But many others were sold off to other governments, with little prospect that these nations will consent to limits on heavy conventional arms.

Meanwhile, smaller conventional weapons—the focus of this chapter—have been responsible for most of the killing in recent decades during and after armed conflicts. The easy availability of these light weapons impedes peace-building after armed conflicts have come to an end and facilitates criminal, personal, and political violence even in countries not at war. True peace and stability in war-torn regions will remain elusive unless soldiers and other fighters are disarmed, demobilized, and reintegrated into society.

A Global "Wild West"

In countries all across the globe there are hundreds of millions of low-tech, inexpensive, sturdy, easy-to-use "small arms and light weapons." This category includes a broad spectrum of weapons held by both the military and civilians—such as handguns and hunting rifles, assault rifles, and machine guns.[2]

Small arms are the weapons of choice in most of today's conflicts—battles fought within rather than between countries. It is estimated that about 300,000 people are killed by small arms each year in armed conflicts. But when guns are endemic, they also affect societies that are formally at peace. So another 200,000 people are killed annually in gun-related violence, and 1.5 million are wounded. In the hands of repressive governments or ruthless warlords, small arms can contribute to massive human rights violations—uprooting or "disappearing" people, silencing political opponents, and intimidating civil society.[3]

The dispersal of guns to private armies and militias, insurgent groups, criminal organizations, vigilante squads, and private citizens feeds a cycle of violence that in turn causes even greater demand for weapons. Political violence pits governments against insurgent forces fighting to overthrow the government or to secede; communal violence involves different ethnic, religious, or other identity-based groups; and criminal violence involves drug traffickers, organized crime groups, or petty individual crime. Beyond injuries and the loss of life, the wide availability of small arms creates a climate of fear and lawlessness that can undermine political stability, disrupt economic activity, and threaten to unravel past development accomplishments.

Although high levels of gun ownership do not automatically translate into violence, the easy availability of guns does make a difference. This is particularly true in societies where social and economic inequality are pronounced, poverty is endemic, unemployment leads young people with an uncertain future to join gangs or militias, the social fabric is under severe strain, strong ethnic or other intergroup animosities persist, or the legitimacy of political institutions is in question.

Small arms violence can have fatal consequences for human development, disrupting already overstretched health care and education systems.

Societies that have emerged from long years of war often continue to experience considerable tension and violence, fed by unresolved grievances, a culture of violence, and large stocks of leftover weapons. Recently demobilized combatants, often poorly equipped to make a living in the civilian world, may turn to banditry. Southern Africa and several Central American countries, for instance, experienced a seamless transition from politically motivated to criminal violence in the early 1990s, enduring killings that rivaled the number of people who perished during armed conflict. More recently, countries like Sri Lanka are confronting similar problems; as the civil war appeared to be drawing to a close, widely available weapons of war triggered a crime wave.[4]

Latin America stands out in this global picture because of its high rate of firearm deaths. In Venezuela, there are 21 gun homicides each year per 100,000 inhabitants, in Jamaica 17, in Brazil 14, and in war-torn Colombia 50, compared with 30 per 100,000 in South Africa but 3.5 in the United States and 0.2 in Germany. Children are increasingly involved in drug trafficking syndicates and street violence. In Rio de Janeiro, for instance, some 12,000 children and teenagers are involved in the narcotics trade, and in El Salvador at least 25,000 children belong to gangs.[5]

Small arms violence can have fatal consequences for human development, disrupting already overstretched health care and education systems. Throughout sub-Saharan Africa, for example, immunization and vaccination efforts have been curtailed. In the Democratic Republic of the Congo, one third of all children 5–14 years old were not in school in 1999–2000, with much higher proportions in areas most affected by ongoing violence. In Jamaica, 30 percent of girls surveyed said they were afraid to go to school because of the threat of firearms-related crime.[6]

Small arms can contribute to a precipitous decline and even collapse of economic activity, quite apart from the outright destruction or deterioration of physical infrastructure. Civil wars, banditry, and other forms of armed violence may cause a breakdown of the basic trust and confidence that are essential for trade and other economic transactions. Investors tend to stay away from countries where personal safety and the security of physical assets are at risk. In rural areas, endemic violence may compel farmers to abandon their harvests. In Angola, for example, agriculture's share of gross domestic product (GDP) fell from 23 percent in 1991 to 6 percent by 2000. Among pastoral groups in eastern Africa, the influx of high-caliber weapons has made traditional cattle rustling practices much more deadly.[7]

Coping with crime and violence diverts a growing proportion of investment, foreign aid, and domestic budgets to unproductive purposes. So much may be spent on police, law enforcement, private security guards, and other forms of "security" that few resources remain for social services. This is the case in South Africa, where the police budget for 2000–01 was 26 percent larger than the health budget. In Latin America, public and private security expenses ate up an estimated 13–15 percent of the region's combined GDP in the mid-1990s—surpassing its welfare expenditures.[8]

Sell, Loot, Smuggle

Because they have long been considered essential for state and personal security, small arms have been produced, traded, and stockpiled for many decades, with little thought given to the repercussions for public safety, development, and livelihood. No one knows how many of these guns exist, and most governments and companies are still highly reticent to make detailed information available. Using conservative assumptions, the Geneva-based *Small Arms Survey*—the most authoritative public source of information—puts global stocks of military- and civilian-style weapons at 639 million. For most of the world's regions and countries, only rough and often partial estimates are available for the number of small arms in the hands of civilians, police, state armed forces, and insurgent groups. (See Table 7–1.)[9]

The utility of aggregated numbers alone is limited, of course, because they include a tremendous range of different types of weapons. Military-style weapons, estimated at more than 240 million, have the greatest firepower. The most ubiquitous among these are assault rifles, of which some 90–122 million have been produced worldwide. But civilian-type firearms—far more numerous—may actually present a greater challenge with regard

to crime and urban violence. Yet numbers alone are not good predictors of the likely consequences. Societal context and intent are critical. At least some countries not at war have far greater arsenals of firearms than countries wracked by organized violence. A weapon in the hand of a ruthless individual or of groups such as warlords, rebels, or crime syndicates is far more likely to be used. Even limited quantities of small arms can cause death, injury, and havoc on a large scale.[10]

Global production of small arms is estimated at 7.5–8 million units per year, of which 7 million are civilian-type firearms and the remainder military-style weapons. Annual production of military-caliber small arms ammunition alone is believed to be in the range of 10–14 billion rounds—or roughly one-and-a-half to two bullets for every living person on earth. Production of military-style weapons may be increasing in the wake of the Afghanistan and Iraq invasions and major rearmament programs in the United States, Russia, China, and parts of Europe. If so, this would be a reversal of trends in recent years.[11]

The United States, Russia, and China are the dominant producers, but there are at least another 27 medium-sized producer nations—15 in Europe, 6 in Asia, 3 in the Middle East, plus Canada, Brazil, and South Africa. All in all, at least 1,249 companies in 92 countries are involved in production. Not included in these statistics are insurgent and opposition groups in several nations that are able to produce simple small-caliber weapons. In addition, illicit small-scale production appears to be fairly widespread, taking place in at least 25 countries, including Chile, Ghana, South Africa, Turkey, Pakistan, and the Philippines.[12]

Despite growing efforts at data collection and transparency, the world is still far from having a clear and coherent picture of even authorized trade in small arms, let alone gray

Table 7–1. Rough Estimates of Stockpiles of Small Arms, Selected Countries and Regions

Country or Region	Estimated Stocks
Latin America	45–80 million (37–72 million civilian, 1.8 million police, 7 million military)
United States	238–276 million (civilian, police, and military)
European Union (EU)	67 million (15–16 million military and police)
Non-EU Europe (including Russia)	At least 13–14 million firearms (civilian only)
Sub-Saharan Africa	29 million (23 million civilian; 600,000 insurgent groups; remainder held by armed forces, police)
South Africa	4.5 million (civilian only)
Iraq	At least 7–8 million (military and civilian-style weapons), possibly many more
Yemen	5–8 million (military and civilian)
India	48 million (40 million civilian; 7 million military; 600,000 police; 100,00 insurgents)
Pakistan	23 million (20 million civilian; 3 million military; 400,000 police)
China	30 million (at least 27 million military, 3 million police); insufficient information on civilian ownership

Note: These numbers represent, in most instances, very rough order-of-magnitude estimates.
SOURCE: See endnote 9.

or black market deals. The United States, Italy, Belgium, Germany, Russia, Brazil, and China are the largest exporters, while the United States, Saudi Arabia, Cyprus, Japan, South Korea, Germany, and Canada are leading importers. The *Small Arms Survey* puts the total legal international trade at $4 billion a year, or about half of the estimated value of total production; the illicit trade is believed to be somewhere under $1 billion.[13]

But ultimately it is difficult if not impossible to draw a clear dividing line between these categories, as a "legal" weapon can easily become an illegal one. A significant proportion of even the legal international trade is done secretly. And numerous trading networks allow clandestine supplies by government agencies, black market sales by private arms merchants, and unauthorized transfers from original to secondary recipients. Adding in such factors as the theft or loss of many small arms, once the weapons are produced there is virtually no telling in

whose hands they will ultimately end up.[14]

Trade in secondhand arms began to flourish after the cold war, when armies in North America, Europe, and the former Soviet Union gave away much of their excess equipment to other countries or sold it at bargain rates. Casting off surplus stocks may make sense from a narrow cost-benefit point of view, saving the money and effort required to dismantle and destroy the weapons and providing a source of revenue. But if these weapons go to undesirable or irresponsible customers, the longer-term cost may be considerably higher than anticipated.[15]

While Norway and Germany have essentially stopped exporting excess weapons or ammunition, the United Kingdom, Russia, and the United States are still largely following old practices. In the late 1990s, for example, Washington sold some 320,000 surplus arms to "friendly" governments. (Surplus weapons there are also routinely transferred to domestic law enforcement agencies or sold to the public.)[16]

In a number of developing countries, arms purchases have been financed through the sale of commodities or through direct barter for natural resources, animal products, or drugs. In Liberia, Sierra Leone, and Angola, for instance, revenues from diamonds, oil, timber, and wildlife products helped both the government and rebel forces to buy arms. Cambodian rebels paid for their activities by selling timber and gems. And in Afghanistan, anti-Soviet *mujahideen* and later the Taliban supported themselves through the opium trade.[17]

Other important sources of weapons flows are the capture of arms by insurgent forces, the wholesale looting of military and police depots, and continuous "leaks" from government arsenals as soldiers steal and sell off weapons. There have been reports, for instance, of guns from Argentine military and police arsenals rented out to domestic gangs and smuggled to crime-plagued Rio de Janeiro. There is also evidence that small arms are being diverted from Saudi government stocks to a variety of terrorist organizations, including al Qaeda. In the Philippines, the vast majority of weapons in the hands of insurgents are from the country's police and military depots.[18]

Some cases involve huge quantities of weapons. Following the fall of the Siad Barre dictatorship in Somalia, several hundred thousand weapons were pillaged from army arsenals in 1991–92, fueling the rise of warlords. In 1997, a popular revolt in Albania led to the ransacking of military and police depots. Some 643,000 small arms were stolen, many smuggled to ethnic Albanians in neighboring Kosovo and Macedonia, where fighting flared up later. Small arms are in fact often transferred illicitly from one hotspot of the world to another. (See Table 7–2.) When a conflict in one country comes to an end, ongoing conflicts elsewhere represent a tempting and lucrative market for leftover weapons.[19]

One of the most tremendous and rapid transfers of small arms occurred after the collapse of Saddam Hussein's regime in Iraq, when government arsenals were looted. When the head of the occupation forces, Paul Bremer, disbanded the Iraqi army in May 2003, it not only accelerated the flow of weapons into civil society, it created a class of people who had suddenly lost their livelihoods but were well armed. An estimated 4.2 million guns were lost from military arsenals, adding to at least 3 million firearms already privately owned by civilians. This flood of weapons helped arm militias maintained by rival regional, religious, and factional leaders, and it triggered rising crime and killings among an impoverished and desperate population. It may yet undermine stability in neighboring countries if a significant portion flows across hard-to-control borders. There are already reports of Iraqi AK-47s and rocket-propelled grenades turning up in Saudi Arabia.[20]

Aside from such spectacular losses, there are countless incidents of small volumes of weapons being lost or diverted. The cumulative impact may be no less insidious. At least 1 million firearms are lost or stolen worldwide each year, mostly from private individuals, although the total is likely to be much higher than that.[21]

A Limited Response to Date

Tackling the proliferation of small arms requires a multitude of approaches. The experience of Colombia over the last four decades illustrates the problems facing governments and societies that are awash in small arms. (See Box 7–1.) Among the new approaches needed are greater transparency, tighter export controls to guard against illicit shipments, more cooperation among national

Table 7–2. Selected Examples of Small Arms Transfers from Hotspot to Hotspot, 1970s to 2002

Source	Recipients
Viet Nam	Following the end of the war in Viet Nam, leftover U.S.-supplied arms and ammunition were acquired by Cuba and then by Nicaragua's Sandinista government and Salvadoran insurgents.
Palestine Liberation Organization	Soviet-made ammunition confiscated by Israeli forces was transferred to Nicaragua's Contra rebels by the U.S. Central Intelligence Agency in the 1980s.
Nicaragua and El Salvador	U.S. armaments pumped into Central American civil wars in the 1980s later became part of a regional black market, with arms going to Mexico, Colombia, Peru, and Salvadoran gangs in the United States.
Afghanistan	Two thirds of $6–9 billion worth of weapons intended for anti-Soviet fighters during the 1980s ended up in Pakistan, India's Punjab region, Kashmir, Tajikistan, Sri Lanka, Myanmar, and Algeria.
Lebanon	Weapons left over from civil war in the 1970s and 1980s were transferred to Bosnia in the early 1990s.
Mozambique and Angola	Arms originally supplied to anti-government forces in Mozambique and Angola by South Africa's apartheid regime were later smuggled back into South Africa, feeding a tremendous crime wave.
Cambodia	Weapons outflows have been traced to the Philippines, Aceh (in Indonesia), Sri Lanka, India, and Kashmir.
Southeast Asia	Rebels in India's Assam region received weapons from other insurgents, including Sri Lanka's Tamil rebels, the Kachins in Myanmar, the Khmer Rouge in Cambodia, and Kashmiri groups.
Sudan and Somalia	An influx of arms from war zones into Kenya has transformed low-key skirmishes among cattle herders into ever more deadly confrontations.
Greece and Turkey	Kurdish PKK rebels transferred Stinger missiles to Sri Lanka's Tamil rebels. The missiles were manufactured in Greece under U.S. license.
Liberia	Ex-combatants are allegedly smuggling AK-47s and other weapons to neighboring countries in exchange for consumer goods.

SOURCE: See endnote 19.

customs agencies, codes of conduct and embargoes to prevent transfers to questionable users, a reduction in the number of weapons in circulation through gun buyback programs and other collection methods, and destruction of surplus stocks.[22]

European nations have expanded their sharing of data in the context of the Organization for Security and Co-operation in Europe (OSCE). And in December 2003 the members of the Wassenaar Arrangement on Export Controls for Conventional Arms and Dual-Use Goods and Technologies—in effect, an arms suppliers club—decided to exchange information regularly. Still, greater transparency among government agencies is not the same as a willingness to make this type of information public.[23]

During the last decade, export controls have become increasingly strict and governments have begun to exercise greater caution in their sales to countries with armed conflicts or human rights violations. An array of regional agreements and mechanisms address-

BOX 7–1. COLOMBIA: OBSTACLES TO PEACE

Judging by the massive human impact, the huge economic cost, and the connections between widespread violence, black market economies, social inequality, and human rights violations, the worst conflict in Latin America is undoubtedly the one in Colombia. It has caused a serious humanitarian crisis and has the potential to spread violence and instability throughout the Andean and Amazonian regions.

This conflict is rooted in pervasive inequality, social exclusion, and endemic violence reaching back several decades. Repression and the growing concentration of wealth and power spurred the rise of leftist guerrilla groups in the 1960s. In response, the military aided the emergence of paramilitary groups that became notorious for civilian massacres.

Negotiations currently under way between Autodefensas Unidas de Colombia (AUC, the United Self Defense Forces of Colombia, which is the largest "umbrella" organization of paramilitary groups) and Álvaro Uribe's government may lead to the demobilization of some armed groups. But the proposed amnesty for AUC members is a step backward in terms of letting people go unpunished for villainous acts. And the negotiations may end up legalizing the ownership of land and other assets obtained illegally.

Peace efforts over the last two decades demonstrate that demobilization and reintegration of combatants alone are insufficient. The cycle of violence will not be broken if structural problems that are both a cause and consequence of the conflict remain unresolved. Key aspects of a strategy for peace and re-establishing political legitimacy include building respect for the rule of law and human rights, ending the impunity with which armed groups carry out violence, and establishing an independent and effective judiciary. Attacks on civilians have been an integral part of the strategies pursued by various armed groups—including murders, disappearances, torture and kidnappings, sexual abuse of women, attacks on protected personnel such as medics, and the forced

displacement of people. Colombia has one of the highest numbers of internally displaced persons in the world.

The ongoing militarization of Colombian society represents a huge obstacle to peace. Plan Colombia, adopted by the government and supported by the United States, seeks to mobilize the entire population either as informants or as members of local paramilitary groups. The democratic state is increasingly becoming subordinated to the logic of war, and the distinction between combatants and civilians is blurred, if not lost altogether. Leaders of trade unions, women's groups, and human rights organizations and the like are frequently the target of persecution.

Colombia has one of the world's highest levels of inequality of landownership, which is worsened by forced displacement. The lack of employment opportunities for the urban poor adds pressure to the problem. According to the United Nations, 55 percent of Colombians live in poverty; 27 percent are in extreme poverty. It is thus essential to promote structural reforms that ensure greater access to land and other productive activities and to adopt an economic policy directed toward creating employment, including jobs for those who have been demobilized. Strengthened health and education policies would contribute to reducing poverty and to a better distribution of wealth. This requires a state with more resources. Yet the current fiscal system collects little and favors the wealthy—only a little more than 400,000 people in a population of some 44 million even pay income tax.

While drug-trafficking was not the cause of the conflict, it has turned into its main driving force. The conflict has become more complex due to the growth of paramilitary groups that finance themselves through extortion and drug trafficking. Their responsibility for the forced expulsions of peasants and their ties to some units of the armed forces have been clearly demonstrated.

BOX 7–1. (continued)

U.S. policy toward illegal drugs, as expressed in Plan Colombia, has increasingly become militarized. Major components include military aid, uniformed and private-sector military advisors, and massive aerial spraying of illicit crops. More than 350,000 hectares have been fumigated since 2000. The negative effects of aerial spraying on the environment and human health have engendered broad opposition. Peasants bear the greatest burden of Plan Colombia, while the international drug-trafficking and money laundering networks remain essentially unencumbered. Years of attempts to eliminate coca production in Colombia appear to have only led to a geographical shift in drug-cultivating areas. There is no visible reduction in the availability of drugs in western markets.

An alternative policy would steer clear of criminalizing peasant growers of coca crops, substitute aerial spraying with manual eradication projects, promote alternative development projects that take into account the social and economic causes that push peasants toward illicit crops, and work to ensure better market access for other crops.

—Manuela Mesa, Centro de Investigación para la Paz (Peace Research Center, Madrid)

SOURCE: See endnote 22.

ing arms manufacturing, transfers, and stockpile management are now in place:

- In 1993, the OSCE adopted Principles Governing Conventional Arms Transfers, followed by a Document on Small Arms and Light Weapons, laying out criteria on arms manufacture and export, brokering, and stockpile and surplus weapons management. To encourage higher standards, the OSCE published a *Handbook of Best Practices* in 2003.
- In November 1997, members of the Organization of American States signed the Inter-American Convention Against the Illicit Manufacturing of and Trafficking in Firearms, Ammunition, Explosives and Other Related Materials, the first binding agreement.
- In 1998, the European Union approved a Code of Conduct on Arms Exports stipulating that arms should not be sent to countries where there is a clear risk that they might be used for external aggression or internal repression.
- In October 1998, West African heads of

state proclaimed a moratorium on the import, export, and production of all small arms within the region. Repeated violation of this poses a challenge, however.
- In 2001, the Southern African Development Community adopted a Protocol on Firearms, Ammunition and Related Materials aimed at creating regional controls on possession and against trafficking.
- In 2002, the South Eastern Europe Clearinghouse for the Control of Small Arms and Light Weapons was created in cooperation with the U.N. Development Programme (UNDP).
- In April 2004, representatives of 11 African nations in the Great Lakes and Horn of Africa regions signed the Nairobi Protocol for the Prevention, Control and Reduction of Small Arms and Light Weapons, obliging them to take concrete steps toward curbing the manufacture, trafficking, and possession of illegal small arms.[24]

The effectiveness of these efforts remains somewhat limited, unfortunately, because almost all of them are politically but not legally binding and thus hard to enforce, because the focus is on illicit arms (mostly ignoring state-sanctioned arms transfers), and because there are no express requirements for arms-exporting states to respect international human rights or humanitarian law. Still, these agreements do signal growing commitments on the part of governments and are part of a process that is creating new norms and standards.[25]

Questionable practices still do occur, however. In 2003, for instance, the United Kingdom decided to go ahead with exports to Indonesia in spite of the EU Code of Conduct. Although rising worries about security and public safety in this era of terrorism would presumably spur greater caution about transfers, the "war on terror" has in some ways led to the opposite outcome. A recipient government's willingness to join the "anti-terror" coalition often trumps its human rights abuses and other considerations.[26]

On the global level, small arms control received fresh impetus from a 2001 U.N. conference that led to a Programme of Action. Although it fell short of the expectations of concerned civil society groups, the program does encourage governments to adopt suitable laws and regulations and to report on measures they are taking in a number of areas: improving stockpile management, controlling transfers and regulating brokering, improving marking and tracing of arms as well as recordkeeping, establishing criminal penalties for trafficking, collecting and disposing of surplus arms, and increasing public awareness.[27]

A global civil society coalition, the International Action Network on Small Arms, is keeping pressure on governments to follow through on their commitments and to close the gap between rhetoric and action. Meanwhile, a number of governments, international organizations, and nongovernmental

organizations (NGOs) come together in what has been dubbed the "Geneva Process" for regular, informal consultations on this issue. Progress in implementing the Programme of Action is particularly slow in North Africa, the Middle East, and parts of Asia. To spur things along, the Regional Human Security Center in Jordan organized a workshop on national and regional measures, and an NGO network was founded in November 2002.[28]

Crime-plagued Brazil adopted a Disarmament Statute in December 2003 that seeks, among other goals, to ban the carrying of firearms in public.

Another effort at the global level is the Firearms Protocol to the existing Convention against Transnational Organized Crime, which the U.N. General Assembly adopted in May 2001. The protocol aims to promote cooperation among governments in preventing and countering illicit manufacturing and trafficking in firearms, components, and ammunition by developing harmonized international standards. Yet by September 2004, only 52 states had signed it (the United States, France, and Russia notably have not); the 26 ratifications to date still fell considerably short of the 40 needed to put the protocol into force.[29]

A number of governments have intensified their efforts to tighten domestic gun control. Efforts have particularly focused on possession of automatic and semi-automatic firearms. One of the most ambitious initiatives is found in Canada, which passed a new law in January 2003. And crime-plagued Brazil—where 300,000 people have been killed in urban violence in the past decade—adopted a Disarmament Statute in December 2003 that seeks, among other goals, to ban the carrying of firearms in public, raise the minimum age of legal gun possession to 25, and conduct a national referendum in 2005 on banning all firearms and ammunition sales. In Thailand, prime minister Thaksin Shinawatra wants to make the country gun-free in five to six years, beginning with a ban on new gun sales. In the United States, by contrast, the political influence of the gun lobby prevents any far-reaching measures. The power of the National Rifle Association was demonstrated once again in September 2004 when the Bush administration and Congress declined to extend a 1994 ban on assault weapons in defiance of strong popular support for the ban.[30]

Finally, international arms embargoes—although typically not limited to or specifically focused on small arms—are another effort to control the illicit flow of weapons. Several governments and rebel groups have been targeted by U.N. arms embargoes since 1990. (See Table 7–3.) In addition, the European Union has adopted embargoes aimed at Libya, China, Myanmar, several Yugoslav successor states, the Democratic Republic of the Congo, Sudan, Nigeria, Indonesia, and Zimbabwe. But these efforts are frequently violated, and greater resources for monitoring and enforcement are needed.[31]

Efforts to limit and control the flow of small arms are but one side of the coin; the other side is reducing the number of arms in circulation. One of the most pressing tasks is to collect arms left over at the end of civil wars. Since 1990, there have been at least 17

Table 7–3. International Arms Embargoes, 1990 to Present

Target Country	Entry into Force	Duration	Observation
Iraq	August 1990	Continuing	
Former Yugoslavia	September 1991	June 1996	
Somalia	January 1992	Continuing	
Libya	March 1992	April 1999	
Liberia	November 1992	Continuing	The original embargo was terminated in March 2001 but was replaced with one imposed for different reasons.
Haiti	June 1993	1994	
Angola (UNITA rebels)	September 1993	Continuing	
Rwanda (rebels)	May 1994	Continuing	Applied only to Rwandan rebel groups after August 1995.
Afghanistan (Taliban-held territory)	December 1996	January 2002	Initiated as a voluntary embargo; became mandatory in December 2000. After January 2002, limited to Osama bin Laden and members of al Qaeda and the Taliban.
Former Yugoslavia, Kosovo	March 1998	September 2001	
Sierra Leone (RUF rebels)	October 1997	Continuing	Applied only to RUF rebels after June 1998.
Eritrea and Ethiopia	February 1999	May 2001	Initiated as a voluntary embargo; became mandatory in May 2000.

SOURCE: See endnote 31.

major U.N. and non-U.N. peacekeeping operations—in Central America, the Balkans, different parts of sub-Saharan Africa, Afghanistan, and Cambodia—whose mandate included disarming former soldiers and rebel fighters.[32]

Typically, at the end of a conflict there is no firm or reliable inventory of the weapons in the possession of combatants, so it is difficult to establish a baseline and assess how much disarmament is actually taking place. During peace negotiations, each side has an interest in inflating the number of weapons under its control in order to win concessions from opponents. When it comes to actual disarmament, however, the protagonists tend to downplay their holdings, turning in only a portion of the arms and retaining those in superior condition. Two important unanswered questions are whether coercive measures should be taken if some actors renege on commitments and whether civilians in addition to ex-fighters should be disarmed.[33]

Timing is critical. Disarmament is best undertaken once the political situation is favorable, funding is adequate, and enough peacekeeping forces have been deployed. But often it takes too long to create favorable conditions, and experience shows that the protagonists' willingness to be disarmed tends to diminish over time, irrespective of commitments on paper. And once arms are collected, someone must decide what to do with them. Peacekeepers often turn them over to a reconstituted national army (one that integrates government and rebel soldiers) instead of destroying them. Yet, weak controls over these arms and the fact that many soldiers earn low salaries are a virtual invitation to steal and sell arms, causing additional problems.[34]

In addition to arms collection efforts in the context of peacekeeping operations, a variety of gun buy-back programs have been

launched, encouraging individuals to turn arms in voluntarily in return for monetary or in-kind compensation. Frequently, governments establish an "amnesty" period during which unlicensed or otherwise illegal firearms can be turned in without fear of prosecution. Following El Salvador's civil war, for instance, a goods-for-guns program run by the Patriotic Movement Against Crime collected close to 5,000 weapons in 1996–97, and New York–based Guns for Goods sponsored the exchange of food and clothing vouchers for guns in three cities. In both cases, lack of funding limited effectiveness. In neighboring Nicaragua, cash and food incentives and an Italian-sponsored microenterprise program yielded 64,000 weapons in 1992–93 (with another 78,000 confiscated). In post-conflict Mozambique, the Christian Council initiated a Transformation of Arms into Ploughshares project in 1995 (underwritten in part by Germany and Japan); it allowed people to exchange weapons for cows, plows, sewing machines, and bicycles.[35]

In addition to arms collection efforts in the context of peacekeeping operations, a variety of gun buy-back programs have been launched.

Important lessons have been learned from these various efforts. Monetary compensation in return for guns is one way to go, but it may provide an incentive to steal guns in order to turn them in for cash, thus stimulating illegal activities. Pricing can be a crucial factor: at compensation levels too far below the black market value, few firearms will be turned in; prices that are too high, on the other hand, will stimulate the black market. But particularly in developing countries, where many ex-fighters can be expected to return to rural areas, programs that provide food or agricultural implements are more appropriate than offering cash for weapons. Generally speaking, buy-back schemes will tend to be more successful if they are embedded in broader community programs.[36]

UNDP is promoting "weapons for development" programs in more than 15 countries in the Balkans, sub-Saharan Africa, and Central America. In Albania, where an estimated 200,000 weapons looted from government depots were still in circulation, a UNDP project in 2002–04 had communities compete for funding for development projects by handing in weapons. In Cambodia, weapons-for-development efforts have attracted EU financial support. Once the virtually exclusive preserve of defense and foreign ministries, weapons collection programs supported by international donors are increasingly integrating disarmament and developmental considerations. Since the late 1990s, the British, Canadian, German, and Japanese governments have developed cross-sectoral responses that combine

Table 7–4. Selected Small Arms Collection Programs, 1989–2003

Region/ Country	Period	Organized or Implemented by	Small Arms Collected[1]	
			Weapons	Rounds of Ammunition
Africa				
Mali	1995–96	UNDP	3,000	—
Mozambique	1995–2003	Mozambique, South African governments	34,903	11.4 million
Liberia	1996–97	West African/U.N. peacekeepers	17,287	1.4 million
Sierra Leone	1999-2002	West African/U.N. peacekeepers	26,000	935,495
Angola	2002	UNITA rebel group	25,000	—
South Africa	since 1995	government	260,000	—
The Americas				
Nicaragua	1989–93	U.N. peacekeepers	159,833	250,000
El Salvador	1992–93, 1996–99	U.N. peacekeepers, anti-gun NGOs	28,927	4.1 million
Brazil	2001–02	government, NGOs	110,000	—
Argentina	2001–02	government, NGOs, U.N.	12,766	7,200
Asia-Pacific				
Australia	1996–98	government	643,726	—
Cambodia	1998–2002	government	119,000	—
Pakistan	2001–02	government	141,180	848,407
Thailand	2003	government	100,000+	—
Europe				
Croatia	1996–97	U.N. peacekeepers	21,929	1.8 million
Britain	1996–97	government	185,000	—
Kosovo	1999	NATO peacekeepers	38,200	5 million
Albania	1997–2002	U.N. agencies	200,377	—
Bosnia	1999–2002	NATO peacekeepers	96,230	6.6 million

[1]Some of these weapons were subsequently destroyed.
SOURCE: See endnote 38.

different types of expertise.[37]

Overall, substantial numbers of weapons have been collected in recent years through a variety of methods. (See Table 7–4.) There is now increasing recognition that the weapons are best destroyed in order to prevent them from being stolen. The United Nations, for instance, sponsored a Global Gun Destruction Day in July 2002, with events designed to have public impact taking place in Argentina, Brazil, Bosnia, Serbia, South Africa, Indonesia, the Philippines, and other countries. In Brazil, the NGO Viva Rio was instrumental in having 100,000 guns destroyed in 2001—the largest number destroyed in a single day anywhere in the world.[38]

The biggest quantities of weapons destroyed, however, have involved surplus government stocks from police or military holdings, eliminating more than 8 million small arms since 1990. (See Table 7–5.) Russia, Ukraine, and Bulgaria may soon destroy another 3.2 million unwanted arms. (Russia was planning to decommission 1 million small arms along with 140 million rounds of ammunition between 2002 and 2005 but is still considering export sales instead of destruction.)[39]

Many of the developments toward greater collection and destruction of weapons over the past decade are encouraging. But much

Table 7–5. Major Surplus Small Arms Destruction Efforts, 1990–2003

Country	Time Period	Number of Weapons Destroyed
Germany	1990–2003	2.2 million
China	1999–2001	1.3 million
Russia	1998–2002	890,000
United States	1993–96	830,000
Australia	1997–98	644,000
South Africa	1998–2001	315,000

SOURCE: See endnote 39.

greater progress is necessary in order to truly tackle the scourge of small arms. Considering that there are an estimated 8 million or so new weapons each year, production still surpasses destruction by at least a factor of 10.

From Combat to Civilian Life

In countries recovering from armed conflict, the process of demobilizing soldiers and other combatants is an enormous challenge. Ensuring that weapons are not dispersed to new conflicts is only one dimension of the problem; the other is preventing ex-combatants from becoming agents of discontent and instability. Reintegrating them into civilian life is a monumental task at a time when warfare has destroyed a large portion of public infrastructure, economic activity remains handicapped, and national treasuries are depleted. Though political violence may finally be absent, social and criminal violence are often increasing.

Demobilization typically involves the temporary encampment of former fighters in areas where they can be disarmed, provided with food and medical care, and given some basic training and orientation to help them master civilian life. In poor countries, the resources available are often barely sufficient, and thus encampment sites are typically inadequate—lacking proper accommodations, sanitation, and enough food and water. Delays stemming from lack of funds or political and bureaucratic obstacles can sometimes come close to derailing the whole demobilization process. Following the encampment stage, the capacity and willingness of communities to absorb returning combatants and their dependents is critical for the success of reintegration efforts. One obstacle is that in civil wars the civilian population bears the brunt of the violence and is likely to resent those thought responsible for their ordeals.[40]

Finding a new livelihood is often quite difficult. Even where short-term jobs and benefits materialize, longer-term employment is far from assured. Many former combatants face tremendous difficulties because they have limited or inappropriate education and skills, as well as little experience with the ways of the civilian world. Training in civilian skills is often either not available or inadequate. Jobs are scarce. As a result, the temptation to engage in banditry, drug trafficking, or other criminal activities to survive may be hard to resist—particularly as it tends to be more lucrative than the precarious life of a subsistence farmer or day laborer. Others may decide to sell off any weapons they kept in order to supplement otherwise meager incomes, feeding a rampant black market in surplus arms.[41]

Reintegration of ex-combatants needs to go hand in hand with the broader reconstruction of society, including reconciliation and the building of political processes and institutions that can prevent renewed insta-

---

BOX 7–2. ANGOLA: THE CHALLENGE OF RECONSTRUCTION

Angola is a test case for the complex reconstruction challenges of countries emerging from many years of warfare. Out of a total population of 13 million, the war killed or maimed roughly 1 million people; there are about 4 million internally displaced persons and 500,000 refugees in neighboring countries. The devastating civil war that lasted from 1975 to 2002 pitted the Movimento Popular para a Libertação da Angola (MPLA, the Popular Movement for the Liberation of Angola) against the União Nacional para a Independencia Total da Angola (UNITA, the National Union for the Total Independence of Angola).

Angola is rich in mineral and agricultural resources. It is the second largest oil producer in sub-Saharan Africa—with production of 900,000 barrels per day, and expected to overtake Nigeria by 2008—and it has the fourth largest diamond reserves in the world. But with oil and diamonds fueling the war, agriculture and industry were either destroyed or neglected. And during the war, loans from international banks were secured against future oil production, mortgaging Angola's future. The country currently ranks 166th out of 175 countries on UNDP's Human Development Index: 70 percent of the people live in poverty; life expectancy, at approximately 40 years, is one third below the average of developing countries; one in four

children dies before the age of five; 60 percent of people lack access to clean water; and more than half the children do not go to school. Some 1.5 million people depend on international food assistance for survival.

The death of UNITA commander Jonas Savimbi in February 2002 opened the door to the signing of the Luena Accords and the demobilization of UNITA. But Cabinda, where the Front for the Liberation of the Cabinda Enclave fights for secession, is still at war. This province produces 60 percent of Angola's oil but suffers from even greater poverty than the country as a whole. Portions are now "off-limits" for human rights groups and foreign observers.

The World Bank and U.N. agencies are now providing support to more than 400,000 former combatants and their dependents. Efforts focus on farming, vocational training, job creation and placement, and micro-credit support. Reintegration is due to be completed by December 2006. But this support is not enough. The situation for women is particularly critical, since their status as ex-combatants has never been recognized. Likewise, most refugees and displaced people returned without any assistance and remain highly vulnerable. Disarmament has had limited impact in reducing the number of small arms (mostly in the possession of civilians), estimated at about 4 million. Vast bility and violence. The recent experience of Angola provides an example of the many difficulties involved. (See Box 7–2.)[42]

In some cases, continued instability compounds the challenge, as the situation in Afghanistan suggests. Economic stagnation and unemployment, political factionalism, and the continued power of regional warlords present a thorny situation. Further magnifying these problems are the slow and inadequate disbursement of international aid and the much faster-than-expected return of refugees that is straining resources and generating significant animosity. Ongoing U.S. military operations have actually helped the warlords consolidate power while alienating much of the population.[43]

The experience with demobilization in different parts of the world over the last 10–15 years is decidedly mixed. (See Table 7–6.) Undoubtedly, however, the practical understanding of what it takes to make postconflict demobilization and reintegration processes work has increased significantly. [44]

BOX 7–2. (continued)

areas are still inaccessible due to landmines (variably estimated at 2–6 million), which impedes return and the resumption of farming. Demining will likely take about 10 years.

A clientele network pervades all sectors of Angolan society. Access to jobs, goods, services, and resources strongly depends on these relationships and is marked by class, ethnic, and regional divides. In many municipalities, *comunas*, and *aldeias*, the state is virtually absent, lacking both the political will and the means to deliver social services. The MPLA's control of the state machinery allows it to appropriate resources, control wealth, and create a predatory autocracy, blurring the distinction between ruling party and state. Yet this is a symptom of Angolan political culture more generally: other parties would likely replicate this system if they were in power.

UNITA has transformed itself into a political party, but reconciliation—so badly needed—is not part of the public debate. The situation remains volatile, and acts of political violence have occurred. Especially in rural areas, there is a climate of fear. Tensions may rise in the run-up to general elections, which may be held in 2006. Fear continues to curtail the participation of broad sectors of the population in public affairs, which is fueled by a political culture that confuses criticism with subversion and by public officials who use the argument of patriotism and sovereignty to silence opposition voices.

The elections could signify a turning point for the country. But they will only be meaningful if people's basic needs are met so that concern about daily survival does not overshadow political debate, if voter registration progresses sufficiently, and if greater freedom of information is created, ending the state's almost complete media monopoly. Only then can the budding yet still weak civil society make itself heard. Otherwise, elections will only serve to legitimize the MPLA's authoritarian system.

Angola has largely disappeared from the international agenda because its war is over. There is a danger that the international community will support the government as long as other countries are assured a continuous supply of oil. But donor support should be conditioned on progress toward transparency, good governance, and democratic behavior. In addition, oil companies and other foreign investors must be held accountable for their activities. These are critical tasks for civil society both inside and outside of Angola.

—*Mabel González Bustelo, Centro de Investigación para la Paz (Peace Research Center, Madrid)*

SOURCE: See endnote 42.

Table 7–6. Selected Demobilization Experiences in Countries Emerging from War, 1992 to Present

Country	Observation
El Salvador	Unemployment as high as 50 percent made reintegration exceedingly difficult for many of the 40,000 soldiers and guerrillas demobilized in 1992. Heavily armed gangs formed by some ex-soldiers and disaffected youth are responsible for murders, kidnappings, robberies, and arms and drug trafficking.
Nicaragua	Much of the land, health care, and economic aid promised to 88,000 demobilized Sandinista and Contra combatants and their families failed to materialize. Severe hardships led former combatants to turn to banditry and gun-running. It took years to achieve a settlement with all the groups.
Mozambique	Insufficient and delayed international funding limited reintegration programs such as vocational training, public work schemes, and provision of seeds and agricultural implements. The lack of jobs led to a rise in crime and violence.
Sierra Leone	By January 2002, a total of 72,490 combatants had completed the disarmament process. Long-term reintegration depends on revitalizing the economy, jobs, and well-designed community-based projects. It is unclear how many ex-combatants have found a new livelihood.
Liberia	A peace accord ended the civil war in August 2003. By summer 2004, 49,000 out of 60,000 former government and rebel fighters had been disarmed, and about 7,000 had received vocational training. But aid shortfalls imperil reintegration, resettlement of refugees and internally displaced people, and reconstruction.
Afghanistan	Poor economic conditions hinder reintegration of ex-combatants. A U.N. pilot project helps 20,000 ex-combatants and their communities through vocational and professional training, income-generating micro-credit and micro-enterprise development schemes, and public-private investment partnerships. The International Labour Organization is providing training needed for construction work.
Sri Lanka	Given major unemployment problems, reintegrating large numbers of ex-fighters could exacerbate social tensions. Deserters from the Sri Lankan armed forces are responsible for increasing crime in the south.

SOURCE: See endnote 44.

A number of factors are critical for a successful outcome. Government capacity and political will are most decisive. But other key factors include the influence wielded by outside actors; coordination among national donor agencies, international aid and development organizations, and civil society groups; political cost-benefit perceptions among combatants, local communities, and other actors; and economic opportunities for those seeking to reintegrate.[45]

Child soldiers have particular needs. Many of them never had a "normal" childhood, and some know nothing but organized violence. Family, friends, and community have typically been ravaged by war; schools are often destroyed or abandoned, meaning that many child soldiers lack the literacy and other skills necessary for civilian life. Altogether, more than a half-million children—most aged 15 to 18, but some far younger—have been recruited into government armed forces and a wide range of non-state armed groups in more than 85 countries worldwide. More than 300,000 of these minors are thought to be actively involved in fighting in some 33 ongoing or recent conflicts. While some children have been recruited forcibly, others are driven by poverty—particularly a lack of education and jobs, alienation, and discrimination—to join. Because small arms are simple to operate and lightweight, their proliferation has facilitated the growth of the ranks of child soldiers.[46]

UNICEF has played a key role in running counseling, literacy, and vocational programs and seeking to reunite children with their families in Sierra Leone, Burundi, Liberia, Angola, Sri Lanka, and the Philippines. Efforts are under way to outlaw the recruitment of minors. In May 2000, the Optional Protocol to the Convention on the Rights of the Child, which is on the involvement of children in armed conflict, was opened for signature. It entered into force in February 2002, and 78 states have ratified it so far. The protocol raises the minimum age for direct participation in hostilities, for compulsory recruitment, and for any recruitment by nongovernmental armed groups from 15 to 18 years.[47]

Demobilization and reintegration programs frequently suffer from a dearth of financial support, lacking the wholehearted commitment of donor states. In general, it has proved easier to secure funding for disarmament than demobilization; the reintegration component—which tends to have less visibility and requires longer-term commitments—has been particularly shortchanged.[48]

As important as it is to ensure that ex-combatants are not a continuing threat to society, it is equally crucial not to provide assistance to them at the expense of other groups who have suffered from conflict (such as refugees and internally displaced people), often at the hands of those being demobilized. It therefore makes sense to design integrated programs that benefit communities broadly. After all, only thriving communities will ultimately ensure that former combatants, or others, do not reach for guns to settle scores and grievances. In this context, the importance of providing post-conflict justice, supporting reconstruction, and promoting reconciliation can hardly be overstated.[49]

Traditional security theoreticians and practitioners tend to assume that dealing with armaments is key. Those promoting an alternative view of security emphasize the importance of nonmilitary considerations. But in the end, this is not an either-or situation; it is not a chicken-and-egg question. In the absence of meaningful development that provides jobs, livelihoods, and reasonable hope for the future—and in the absence of sustainable solutions to grievances and political disputes—people are likely to resort to force in order to bring about change or seek redress. And as long as weapons are abundant and disputes remain unresolved, it will be difficult or impossible for meaningful development to go forward. In the interest of human development, disarmament needs to proceed; in the interest of disarmament and security, sustainable development is indispensable.

Nuclear Proliferation

Several grave dangers arise from the approximately 28,000 nuclear weapons held by eight states around the world. Most dangerous is the wide availability of highly enriched uranium and plutonium, the fissile materials at the cores of nuclear weapons. These materials have become more accessible to terrorists because of the collapse of the Soviet Union and the poor security at nuclear stockpiles in the former Soviet republics and dozens of other countries with nuclear power.[1]

There is also a danger that some nations could acquire nuclear weapons by exploiting inadequacies in the Nuclear Non-Proliferation Treaty (NPT). As the treaty now stands, countries can acquire technologies that bring them to the brink of nuclear weapon capability without explicitly violating the agreement; they can then leave the treaty without penalty.

Finally, there are rising doubts about the sustainability of the nonproliferation regime. This is most disturbing in nations with the technological ability to develop nuclear weapons that have made a political decision not to. Some Brazilian and Japanese leaders, for example, have openly suggested that their countries reconsider nuclear weapon options. Recent revelations that South Korean scientists have produced a small quantity of highly enriched uranium also raise concerns. Some of the failures to contain proliferation result from flaws in the nonproliferation regime itself; many others stem from leaders' unwillingness to enforce commitments and resolutions earnestly passed.[2]

There are, however, positive trends to build on. Since the signing of the NPT in 1968, many more countries have given up nuclear weapon programs than have started them. There are fewer nuclear weapons in the world and fewer nations with these programs than there were 20 years ago. The United

States and Russia continue to cooperate on dismantling and securing nuclear weapons and materials from the cold war. Still, the bilateral Moscow Treaty lacks verification measures and either party can back out easily at the end of its term; it thus fails to build on earlier arms control treaties, START I and II.[3]

Libya's decision to forgo and verifiably dismantle its clandestine nuclear weapons capabilities is an important success. Libya follows in the footsteps of South Africa—the first country to build nuclear weapons and then, in 1993, to give them up—and should serve as a model for other delinquent nations.[4]

On the other hand, India, Pakistan, and presumably Israel joined the nuclear "club." The Iraq crisis, Pakistani peddling of nuclear technologies, North Korean maneuvering, and worries about Iran have heightened international awareness of the dangers posed by proliferation. In the face of these dangers, the European Union has forged a new resolve to combat proliferation, working hard to curb programs in Libya and Iran and adopting a unified strategy that requires full compliance with nonproliferation norms in all future trade and cooperation agreements.[5]

To build on these successes and to prevent new threats, the world needs a new strategy. The strategic aim must now be universal compliance with the norms and terms of a deepened nuclear nonproliferation regime. The United States must take the lead in developing this global plan. To do so, the next administration will have to work hard to overcome Washington's loss of credibility due to false claims of Iraqi weapons of mass destruction. It must also reverse course on current U.S. nuclear policy by committing to the Comprehensive Test Ban Treaty and to a verifiable Fissile Materials Cutoff Treaty, as well as by ending all research and development of

U.S. Airforce

The Peacekeeper intercontinental ballistic missile

new nuclear weapons.

Compliance means more than signatures on treaties or declarations of fine intent—it means actual performance. And universal means that all actors must comply with the norms and terms that apply to them. This includes states that have joined the NPT and those that have not. It also includes corporations and individuals. The burden of compliance extends not only to states seeking nuclear weapon capabilities through dual-use fuel cycle programs or those abetting proliferation through technology transfers, but also to states with nuclear weapons that are not honoring their pledges.

Five obligations form the core of the universal compliance strategy. Their successful fulfillment will answer the most pressing problems. Each of these general objectives requires subsidiary national and international policies, resources, and institutional reforms. Some of the necessary steps require new laws and voluntary codes of conduct; others need only the will to live up to existing commitments.[6]

First, non–nuclear weapon states must reaffirm commitments to never acquire nuclear weapons. This commitment must evolve to proscribe the further national acquisition of facilities that can produce materials directly usable in nuclear weapons (separated plutonium and highly enriched uranium).

Second, states must secure all nuclear materials, maintaining robust standards and mechanisms for securing, monitoring, and accounting for all fissile materials in any form. Such mechanisms are necessary both to pre-

vent nuclear terrorism and to create the potential for secure nuclear disarmament. Options for the safe long-term disposal of fissile materials must be developed.

Third, nations must establish enforceable prohibitions against individuals, corporations, and states that help others secretly acquire the technology, material, and know-how needed for nuclear weapons.

Fourth, steps must be undertaken to devalue the political and military currency of nuclear weapons. All states must honor their obligations to end nuclear explosive testing and must diminish the role of nuclear weapons in security policies and international politics. They must also identify and strive to create the conditions necessary to verifiably eliminate all nuclear arsenals.

Fifth, states must commit to fostering diplomacy-based conflict resolution strategies. Those that possess nuclear weapons must use their leadership to resolve regional conflicts that compel, or excuse, some states' pursuit of security by means of nuclear, biological, or chemical weapons.

Political leaders must forge a bold, new nuclear security strategy—one that secures and eliminates nuclear materials before terrorists can steal them and that reinforces a badly damaged nonproliferation regime before new nuclear states emerge. With active informed citizen involvement, with international collaboration, and with real leadership, this is an achievable goal.

—*Joseph Cirincione, Carnegie Endowment for International Peace*

Chemical Weapons

The September 11th attacks in the United States and the subsequent "war on terrorism" have raised public and government attention about the importance of accelerating anti-terrorism and weapons nonproliferation programs. Fortunately, all the major powers have agreed to abolish one major class of weapons of mass destruction: chemical weapons. Iraq's use of such weapons against its Kurdish citizens in Halabja in March 1988 and the use of sarin gas by the Japanese terrorist group Aum Shinrikyo in the Tokyo subway in March 1995 illustrated the gruesome potential of these deadly weapons.[1]

Currently, only six countries worldwide possess declared stocks of chemical weapons—Albania, India, Libya, Russia, South Korea, and the United States. Russia and the United States have over 98 percent of these stockpiles. One of the greatest liabilities today is the possibility of diversion and terrorist threat. Over the past decade, bilateral and multilateral on-site inspections of stockpile sites have illustrated how vulnerable some, if not all, are to infiltration, theft, and possible diversion to national and subnational groups.[2]

Some two decades ago, the United States and the Soviet Union unilaterally and reciprocally agreed to abolish their large and aging arsenals. This commitment was strengthened in 1993 when these two joined 128 other countries in an internationally binding Chemical Weapons Convention. The convention entered into force in April 1997, obliging the four signatories holding acknowledged stockpiles at the time—United States, Russia, India, and South Korea—to abolish their arsenals by 2007, with the option of extending this to 2012. As of September 2004, 165 countries had ratified or acceded to the treaty, committing to halt all research, development, production, use, and transfer of chemical weapons and to destroy all stockpiles, abandoned weapons, and production facilities. Among the 29 countries that have not yet ratified or acceded to the convention are Egypt, Iraq, Israel, Lebanon, North Korea, Somalia, and Syria.[3]

Destruction of chemical weapons stockpiles is proceeding, albeit slowly. The United States began an active destruction program in the early 1990s and has so far destroyed more than 8,000 tons—26 percent of its declared arsenal of approximately 31,500 tons. Russia lags far behind, having destroyed only about 800 tons, some 2 percent of its declared arsenal of approximately 40,000 tons. The pace should quicken in the next five years, however, as two new facilities to handle destruction come online. Both India and South Korea are making good progress in eliminating their much smaller arsenals, and Albania and Libya will soon begin programs as well.[4]

The keys to success have become apparent. First and foremost, it has become clear that all stakeholders—including local and regional governments and nearby communities—must be involved in the overall process of weapons destruction. Russia learned this hard fact in 1989 after a local community opposed and stopped its initial (secret) plan to destroy its chemical weapons at a centralized facility in Chapeyevsk. While the United States has had a more inclusive process, with local outreach offices and Citizens' Advisory Commissions, opposition still arises from citizens, state regulators, and governors' offices if there is insufficient planning and discussion. What is needed is recognition that social and economic development, technical assistance, and demilitarization must go hand-in-hand for a project to succeed.[5]

It is also important that the potential risks and impacts to public health and the

Russian Munitions Agency

Artillery shells with VX nerve agent awaiting destruction, Russia

environment be thoroughly addressed and publicly discussed. All destruction technologies produce waste, some more toxic than others. Credible, independent risk and health assessments must be undertaken to help officials and the public decide what choices to make. For example, the proposed release of neutralized nerve agent effluents into the Delaware River in the United States has raised many unaddressed questions, as has the long-term storage of toxic bitumen waste in Russia's Kurgan region.[6]

Increased transparency is also essential in order to promote consensus and progress. Secrecy has long dominated the field of nuclear, chemical, and biological weapons, for obvious reasons. With the rise of terrorism and the threat of weapons theft and stockpile attack, officials are once again tempted to limit public information and discussion. In most cases, this is a mistake. Local, regional, national, and international stakeholders need to have confidence that their interests are being protected, a goal that can only be accomplished through openness and verification.[7]

Along with better engagement with civil society, there is a need to expand the technology choices available for dismantling chemical weapons. The physical destruction of these weapons is neither simple nor cheap. It involves a wide variety of hazardous materials and processes, with widely differing risks to the environment and public health. The U.S. Army's preferred destruction technology for chemical weapons has long been incineration, but it was the development and demonstration of non-incineration alternatives such as neutralization that allowed the United States to open facilities at each chemical weapons stockpile site.

Because weapons demilitarization is in the interest of all countries, not just the one possessing the weapons, the international community must share the responsibility and burden for demilitarization, particularly in poorer countries. Russia, in light of its transitioning economy, made this clear when it signed and ratified the Chemical Weapons Convention. Russian chemical weapons destruction will cost at least $5–10 billion; U.S. program costs have now surpassed $25 billion and continue to climb. The U.S. Cooperative Threat Reduction program, along with the 2002 Group of Eight (G-8) Global Partnership pledge of some $20 billion for destroying weapons of mass destruction in Russia, are critical elements in securing and destroying these dangerous stockpiles in a timely fashion.[8]

But donor nations must recognize that providing financial and technical support does not give them the right to dictate priorities. And true partnership requires that recipient nations, such as Russia, help facilitate the demilitarization process with site access, transparency, visas, and liability issues, as required under most bilateral agreements. With cooperation on all sides, the era of chemical weapons can be brought to a close.

—*Paul F. Walker, Global Green USA*

CHAPTER 8

Building Peace Through Environmental Cooperation

Ken Conca, Alexander Carius,
and Geoffrey D. Dabelko

Running along the border separating Peru and Ecuador, the Cordillera del Condor's spectacular cloud forests host a raft of rare and endangered species. Sparsely populated and minimally developed, the mountain range's wealth of biodiversity is rivaled only by the richness of its gold, uranium, and oil deposits. Instead of benefiting from these, however, the people of the Cordillera del Condor have suffered decades of hostility, border conflicts, and government neglect. During the summer months, when the weather let them reach the remote region more easily, military forces from both countries lobbed artillery shells at each other in a low-grade conflict that endangered residents and destabilized the border region. Finally, after decades of simmering conflict and heated border disputes, Peru and Ecuador ceased hostilities under a 1998 peace agreement facilitated by Brazil,

Argentina, Chile, and the United States.[1]

Redrawing the contested border required an innovative arrangement. The governments of Peru and Ecuador agreed to establish conservation zones along the border that would be managed by their national agencies but headed by a binational steering committee. This joint management follows both an ecological and a political logic. The countries' ecosystems are fundamentally interdependent; the Cordillera del Condor conservation zone (or "peace park") uses that interdependence to remove a particularly thorny obstacle to peace.

Yet the people of Cordillera del Condor still face certain enduring challenges: acute poverty, social tensions, and even violence, some of which is prompted by the peace park itself. In the protected forest around Canton Nagaritza, reports of violence between settlers

Ken Conca is Associate Professor of Government and Politics and Director of the Harrison Program on the Future Global Agenda at the University of Maryland. Alexander Carius is Director of Adelphi Research in Berlin. Geoffrey D. Dabelko is Director of the Environmental Change and Security Project at the Woodrow Wilson International Center for Scholars in Washington, D.C.

and conservation agencies suggest that while the governments may have made peace, some people are still fighting—but this time they are fighting the park's architects. The peace park initiative may jump-start conflict transformation between the two governments, but the human struggle for peace and sustainable development remains a daily battle.[2]

Eleven thousand kilometers and an ocean away, environmental cooperation is also helping southern Africa recover from devastating conflicts and prevent new violence from emerging. Following nearly three decades of civil war in Angola, it is peace that now threatens the tranquility of the Okavango River, which drops from its headwaters in Angola down to the wide, flat delta in Botswana, crossing Namibia on a 1,100-kilometer journey south to the Kalahari Desert. This pristine environment in one of the world's few remaining unindustrialized river basins is home to myriad species of animals and plants that have escaped the impact of modern development.[3]

The three basin states' pressing developmental needs are placing demands on the fragile river environment, thereby raising the specter of a different sort of conflict. Angola hopes to resettle citizens displaced by the war, who would need more of the river's water. And as the upstream state, Angola has the power to shake up arrangements that currently favor its downstream neighbors. Newly independent Namibia also has plans for the Okavango's water: it wants to build a pipeline to its arid interior and occasionally threatens to revive its long-standing proposal to build a dam on the short section of the river that crosses through Namibia at the Caprivi Strip. Botswana, on the other hand, favors the status quo, which draws a lucrative stream of tourists to explore the unique ecosystem of the largest inland delta in sub-Saharan Africa and an internationally recognized wetlands area of great ecological significance.[4]

Although these mixed and often contradictory objectives could lead to conflict to gain greater control of the shared water resources, there is hope that cooperative institutions—if strong and vital—could manage competing demands without violence. In 1994, the three countries created the Permanent Okavango River Basin Water Commission (known as OKACOM) to manage the river basin. There is considerable rapport among the OKACOM commissioners and a growing recognition that cooperation can bring greater benefits to all than would fighting over or merely dividing the water.[5]

Environmental cooperation is helping southern Africa recover from devastating conflicts and prevent new violence from emerging.

Unfortunately, the commission has struggled to find the financial resources and political formula to catalyze proactive cooperation. Recently, OKACOM asked nongovernmental organizations (NGOs) and civil society to play a more active role than is commonly found in other shared river basins, acknowledging that the three countries cannot implement effective basin management strategies in isolation. They have also pursued opportunities to collaborate with international donors and conservation groups promoting environmentally sustainable development. Thus far these institutional mechanisms have been sufficiently effective, equitable, and participatory to tip the balance toward confidence building and cooperation rather than tension and violence.[6]

Blending ecology and politics in the service of peace, the Cordillera del Condor and the

Okavango River basin institutions are two examples of a growing array of initiatives—including peace parks, shared river basin management plans, regional seas agreements, and joint environmental monitoring programs—that seek to promote environmental peacemaking. This involves using cooperative efforts to manage environmental resources as a way to transform insecurities and create more peaceful relations between parties in dispute. As such initiatives become more frequent and gain momentum, they may provide a way to transform both how people approach conflict and how they view the environment. Surprisingly, however, relatively little is known about the best designs for these initiatives or the conditions under which they are likely to succeed. While a large body of research examines the contribution of environmental degradation to violent conflict, little in the way of systematic scholarship evaluates an equally important possibility: that environmental cooperation may bring peace.

Environment and Conflict: A History

Over the past 15 years, many scholars have considered whether environmental problems cause or exacerbate violent conflict. Although scarce nonrenewable resources such as oil have long been viewed as a potential source of conflict, this new research shifted the focus to renewable resources such as forests, fisheries, fresh water, and arable land. Most of this work, including projects by Canadian and Swiss researchers in the mid-1990s, found little evidence that environmental degradation contributed significantly to war between countries. Yet the studies found some evidence that environmental problems can trigger or exacerbate local conflicts that emerge from existing social cleavages such as ethnicity, class, or religion. (See Table 8–1.)[7]

As the environment-conflict debate progressed within the scholarly community, the concept of "environmental security" began to attract attention from security institutions and policymakers throughout the industrial world. (As the term is commonly used, environmental security encompasses a diverse set of concerns beyond the narrower question of environment-conflict linkages, including understanding environmental impacts of the preparation for and conduct of war, redefining security to focus on environmental and health threats to human well-being, and using security institutions to aid in the study and management of the environment.) Most recently, U.N. Secretary-General Kofi Annan called for integrating environmental contributions to conflict and instability into the U.N.'s conflict prevention strategy and the deliberations of his High-Level Panel on Threats, Challenges, and Change.[8]

Several national governments and intergovernmental organizations have commissioned state-of-the-art reviews of the concept of environmental security in recent years, with an eye toward developing policy guidelines and implementation procedures. The European Union has discussed ways of integrating the concept into its emerging foreign and security policy and promoted environmental security as a theme for the 2002 World Summit on Sustainable Development in Johannesburg. In the United States, several government agencies—including the Department of State, the Department of Defense, the Environmental Protection Agency, the Agency for International Development, and various intelligence agencies—developed mandates and policies in the 1990s to grapple with environment, conflict, and security connections. Although the events of September 11, 2001, pushed these ideas into the background, many U.S. federal agencies and NGOs continue to search for ways to translate these ideas into tangible programs.[9]

Claims that environmental degradation induces violent conflict remain controversial. Skeptics point out that the causal chain in most environment-conflict models is long and tenuous, with a myriad of social, economic, and political factors lying between environmental change and conflict. Others have questioned the implications of this concept, fearing that casting environmental problems as conflict triggers will "securitize" environmental policy, injecting militarized "us-versus-them" thinking into a realm that demands interdependent, cooperative responses.[10]

These reactions are not surprising in light of the national security framework that is

Table 8–1. Selected National and International Initiatives on the Environment, Conflict, Peace, and Security

Group or Country	Year	Initiative
Club of Rome/ U.S. Department of State	1972 1981	The Club of Rome's *The Limits to Growth* and the U.S. government's *Global 2000 Report to the President* called attention to environmental risks and an array of associated socioeconomic changes (population growth, urbanization, migration) that could lead to social conflict.
Independent Commission on Disarmament and Security Issues	1982	In its first report, *Common Security*, the Commission stressed the connection between security and environment.
World Commission on Environment and Development	1987	The Commission expanded the concept of security in *Our Common Future*: "The whole notion of security as traditionally understood—in terms of political and military threats to national sovereignty must be expanded to include the growing impacts of environmental stress—locally, nationally, regionally, and globally." The Commission concluded that "environmental stress can thus be an important part of the web of causality associated with any conflict and can in some cases be catalytic."
U.N. Environment Programme (UNEP)/ Peace Research Institute, Oslo (PRIO)	1988	A joint program between UNEP and the Peace Research Institute, Oslo on "Military Activities and the Human Environment" included empirical research projects that were largely conceived and implemented by PRIO. From this initiative, PRIO developed a strong research focus on environment and security.
Soviet Union	1989	Proposals for creating an Ecological Security Council at the United Nations have emerged repeatedly over the past 15 years, beginning when Soviet Foreign Minister Eduard Shevardnadze and President Mikhail Gorbachev suggested to the 46th General Assembly that environmental issues be elevated to such a lofty status.
Norwegian Government	1989	In 1989, Defense Minister Johan Jørgen Holst pointed out that environmental problems can become important factors in the development of violent conflicts.
U.N. Development Programme (UNDP)	1994	The U.N. Development Programme explicitly included environmental security as one of the components of "human security," a frame that continues to find favor among UNDP and some prominent national governments, such as that of Canada.

Table 8–1. (continued)

Group or Country	Year	Initiative
German Government	1996	The Federal Ministry for Environment commissioned a state-of-the-art report on environment and conflict in order to explore opportunities to strengthen international environmental policy and law.
Organisation for Economic Co-operation and Development	1998	The Development Assistance Committee of the Organisation for Economic Co-operation and Development commissioned a state-of-the-art report on environment and conflict.
North Atlantic Treaty Organization	1999	In March 1999, the North Atlantic Treaty Organization's Committee on the Challenges of Modern Society published a comprehensive report, *Environment and Security in an International Context*, following a three-year consultation among security, environmental, and foreign policymakers and experts.
European Union (EU)	2001	In April 2001, the General Affairs Council of the EU presented its environmental integration strategy on the issue of environment and security and the contribution of sustainable development to regional security (adopted March 2002).
	2002	The EU discussed how to integrate environmental security into its emerging common foreign and security policy and promoted it as a theme for the 2002 World Summit on Sustainable Development.
Swiss Agency for Development Cooperation	2002	The Swiss Agency for Development Cooperation explored ways to adapt peace and conflict impact assessments to selected projects of their environment program.
United Nations	2002	U.N. Secretary-General Kofi Annan called for better integration of environmental contributions to conflict and instability in the organization's strategy on conflict prevention and the deliberations of his High-Level Panel on Threats, Challenges, and Change.
German Government	2004	The Federal Action Plan on Civilian Crisis Prevention, Conflict Resolution, and Post-Conflict Peace-Building (published in May 2004 after receiving Cabinet approval) identified sustainable development and transboundary environmental cooperation as key ways to foster peace and stability.

SOURCE: See endnote 7.

often attached to the environmental security debate. Consider this 1996 statement by the Director of the U.S. Central Intelligence Agency, John Deutsch: "National reconnaissance systems that track the movement of tanks through the desert, can, at the same time, track the movement of the desert itself…. Adding this environmental dimension to traditional political, economic, and military analysis enhances our ability to alert policymakers to potential instability, conflict, or human disaster and to identify situations which may draw in American involvement." [11]

Many observers have read statements such as this as evidence of ulterior motives. Attention to environment-conflict linkages is suspected to reflect not genuine concern but a desire to predict and isolate troublesome hotspots. Environmental concerns might even be used as a rationale for intervention—as in the U.S. government's rather sudden interest in the long-standing plight of Iraq's "marsh Arabs," which emerged in tandem with the military intervention against Saddam Hussein's regime. Seen in this light, the U.S. military's interest in, say, Haiti's devastatingly denuded countryside could be grounded in a desire to forestall waves of Haitian refugees rather than to find ways to address systemic poverty or reverse the degradation of vital natural resources.

Despite its momentum in many parts of the industrial world, the idea of environmental security has not played particularly well on the global stage. Governments in the global South have long been wary that the North's increased interest in international environmental protection might hamper their own quest for economic development. In the context of an already contentious North-South environmental dialogue, poor countries often view the concept of environmental security as a rich-country agenda serving rich-country interests to control natural resources and development strategies. Seen in this light, northern emphasis on southern security threats shifts the burden of responsibility for global ills, suggests a rationale for intervention in southern resource use, and underscores the tenuous sovereignty of poor countries in the face of unequally distributed economic, military, and institutional power. Many Brazilians have long viewed the North's characterization of the Amazon as the "lungs of the Earth" with suspicion, for example, seeing it as part of a campaign to "internationalize" the rainforest and inhibit development.[12]

Given these concerns, recasting environmental debates in security terms has not been an effective catalyst for global environmental cooperation. Thus a conundrum: posing a problem as one of "environmental security" may inhibit cooperation in the very places where the ecological insecurities of people and communities are most stark.

Why the Environment?

A growing number of voices have suggested that focusing on peace—not security—may provide a way to break this impasse. As a peacemaking tool, the environment offers some useful, perhaps even unique qualities that lend themselves to building peace and transforming conflict: environmental challenges ignore political boundaries, require a long-term perspective, encourage local and nongovernmental participation, and extend community building beyond polarizing economic linkages. These properties sometimes make cross-border environmental cooperation difficult to achieve. But where cooperation does take root, it might help enhance trust, establish cooperative habits, create shared regional identities around shared resources, and establish mutually recognized rights and expectations.[13]

Ecosystem interdependencies present opportunities for mutual gain. When viewed in isolation, environmental problems often create severe upstream/downstream dichotomies, greatly complicating cooperation. For example, most international water law is based on the premise that upstream and downstream states have fundamentally different interests in water use and environmental protection. But communities typically are joined by many simultaneously overlapping ecological interdependencies; places that are upstream from a neighbor in one ecological relationship may well be downstream in another. (See Chapter 5.) Japan is downwind of China's smokestack industries, for instance, but the two countries share a regional marine ecosystem. The United States is upstream of Mexico on the Colorado River, but downstream (at least in the physical sense) of toxic industries flourishing on the U.S.-Mexican border. These complex interdependencies create opportunities to bundle different environmental problems into more robust forms of environmental cooperation.

By their very nature, environmental problems demand anticipatory action, entail longer time horizons, and require an appreciation for sudden, surprising, and dramatic changes. Given these characteristics, environmental cooperation could push decisionmakers to embrace a longer time horizon, such that future gains weigh more heavily in current calculations. For example, it has become more common in recent years for states signing accords on shared river basins to create a permanent basin commission as a platform for information exchange, joint knowledge initiatives, and a longer-term perspective on shared basin management.[14]

Environmental issues encourage people to work at the society-to-society level as well as the interstate level. Domestic constituencies can link up across borders around ecological interdependencies, at times taking the first steps at dialogue that is difficult to pursue through official channels. Over time, regular interaction among scientists and NGOs may help to build a foundation of trust and implicit cooperation. Despite daily battles in the streets of the West Bank and the Gaza Strip, to cite just one example, Palestinians and Israelis continue to meet informally as a means to manage aspects of their shared water resources.

It is almost an article of faith among liberal internationalists that growing interdependence is a force for peace in world politics. Yet interdependence based predominantly on trade and investment linkages can have deeply polarizing effects, as seen in the backlash against economic globalization. Environmental cooperation provides an important opportunity to extend cross-border community building beyond the narrow and often polarizing sphere of economic linkages. For example, many citizens' organizations and

grassroots groups in Mexico and the United States that opposed the North American Free Trade Agreement (NAFTA) are involved in joint environmental protection efforts along and across the border.[15]

More ambitiously, and also more speculatively, it may be that cross-border environmental cooperation can also help to build a more broadly shared conception of place and community. One result may be to loosen the traditional moorings of exclusionary political identities in favor of a broader sense of ecological community.

Using Environmental Cooperation to Build Peace

Most environmental peacemaking initiatives fall into one of three partially overlapping categories: efforts to prevent conflicts related directly to the environment, attempts to initiate and maintain dialogue between parties in conflict, and initiatives to create a sustainable basis for peace. If the minimum requirement for peace is the absence of violent conflict, then environmental cooperation may have a role to play in forestalling the sort of violence that can be triggered by resource overexploitation, ecosystem degradation, or the destruction of people's resource-based livelihoods. Not surprisingly, most of the scholarship linking environmental degradation with violent outcomes has pointed to the need to relieve pressures on people's livelihood resources and to enhance the ability of institutions to respond to environmental challenges. In other words, the most direct form of environmental peacemaking may be action to forestall environmentally induced conflict.[16]

Environmental cooperation may also soften group grievances that form around or are worsened by ecological injustices. Festering environmental problems can create a dangerous link between material insecurity and people's identification as a marginalized group. In settings where ethnicity affects political and economic opportunity, environmental effects often play out unevenly along ethnic lines as well. Thus, many of the most industrially polluted areas in the post-Soviet Baltic states are home mainly to ethnic Russians—creating a potentially combustible mix of reinforced ethno-national identity, heightened social inequality, and environmental grievances. Proactive environmental cooperation could help dampen an important source of grievance that is aggravating these types of social divisions.

A second approach to environmental peacemaking moves beyond conflicts with a specifically environmental component, seeking to build peace through cooperative responses to shared environmental challenges. Initiatives that target shared environmental problems may be used to establish a direct line of dialogue when other attempts at diplomacy have failed. In many instances, governments locked into relationships marked by suspicion and hostility—if not outright violence—have found environmental issues to be one of the few topics around which ongoing dialogue can be maintained.

One of the most serious unresolved conflicts in the politically unstable Caucasus region is the struggle between Armenia and Azerbaijan for control of Nagorno-Karabakh. In autumn 2000, Georgia, which has mediated a dialogue on conservation issues, persuaded Armenia and Azerbaijan to establish a trilateral biosphere reserve in the Southern Caucasus region. The organizers hope that regional environmental cooperation will enhance nature conservation, sustainable development, and, above all, political stability. This long-term project will first collect data, build capacity, and raise awareness. Although Armenia and Azerbaijan are cur-

rently unwilling to cooperate directly, the agreement anticipates that natural biosphere reserves will be established and eventually merged. The two governments have also asked for an independent international environmental assessment of Nagorno-Karabakh; objective data acceptable to both parties could at least lay the groundwork for cooperation.[17]

Despite daily battles in the streets of the West Bank, Palestinians and Israelis continue to meet informally to manage aspects of their shared water resources.

A similar attempt is being made in Kashmir, which has been bitterly contested by India and Pakistan since British decolonization at the end of World War II. Some international conservationists argue that establishing a peace park in the Karakoram mountains between India and Pakistan, which mark the western end of the greater Himalayan mountain chain, would help to manage the border conflict by promoting joint management of the unique glacier environment, where many military casualties are caused by the elements rather than enemy fire. The idea of joint management is also rooted in the recognition that pollution is the greatest threat to this unique environment. To be sure, a joint conservation program in a remote, unpopulated area, where the cost of mounting sustained military operations is prohibitive, seems unlikely to transform the structural dynamics of the India-Pakistan conflict. Yet given the current ceasefire and the recent thaw in relations, there is a growing sense that enhanced cross-border engagement of this sort has a useful role to play in conflict transformation.[18]

Shared environmental challenges may be useful not only for initiating dialogue but also for actually transforming conflict-based relations by breaking down the barriers to cooperation—transforming mistrust, suspicion, and divergent interests into a shared knowledge base and shared goals. Technically complex issues, in which parties work from rival bases of fragmentary knowledge, can heighten distrust. To overcome this, the technical complexity surrounding many environmental issues could be used to create jointly held cooperative knowledge. For instance, OKACOM identified joint assessments of the Okavango's water flow and the potential impacts of hydropower and irrigation diversions as a key step toward developing agreed-upon baselines for successful and peaceful management of water resources.[19]

Narrow government-to-government initiatives risk creating the conditions for more-efficient resource plunder, promoting neither peace nor sustainability.

Skeptics might be tempted to dismiss such initiatives as marginal matters, unrelated to the core of hardened conflicts—akin perhaps to superpower cooperation in outer space during the cold war. But the political and economic stakes in environmental cooperation are high; in the examples provided in this chapter, that fact is clearly understood by the actors involved. Problems surrounding shared river basins, regional biodiversity, forest ecosystems, or patterns of land and water use are controversial, high-stakes questions that engage the state at the highest levels.

A third strand of environmental peacemaking recognizes that a robust peace will require a foundation in sustainability. A narrow focus on whether water shortages "cause" violence between Israelis and Palestinians, for example, misses the larger point:

as a high-stakes issue, the resolution of shared water problems becomes a necessary condition for a broader peace. While water-related tensions between Israelis and Palestinians may not have precipitated the larger conflict, the management of water resources is not only a potential lifeline for continued dialogue during the conflict, it is also a key issue in the negotiations for ending the conflict. In the Oslo Peace Accords between the Palestinians and the Israelis, water warranted its own negotiating group, just as it does in the Indian-Pakistani negotiations initiated in 2004. Whether water is a root cause of conflict or merely exacerbates existing differences, there will be no lasting peace without finding a sustainable water footing for the region.[20]

Remaining Challenges

Despite environmental peacemaking's potential, a skeptical eye is warranted when such initiatives remain the narrow purview of governments and political-economic elites. Initiatives that improve trust and reciprocity among governments without promoting a broader, society-to-society foundation for peace run the risk of reinforcing the zero-sum, state-based logic of national security. They are also prone to short-term mitigation efforts that fail to address the full scope of the problem. A side agreement to NAFTA created an innovative mechanism for funding community projects on the U.S.-Mexican border, for instance. But investment in cooperative initiatives during the first several years of operation was only a fraction of what had been projected, and many citizens' groups on both sides of the border complained of being shut out of the process.[21]

Narrow government-to-government initiatives also risk creating the conditions for more-efficient resource plunder, promoting neither peace nor sustainability. Many international river agreements pay lip service to principles of cooperative watershed management while focusing primarily on capital-intensive schemes for water resources development and interbasin transfers.

Similarly, peace parks in southern Africa serve as a means for reconciliation among apartheid-era enemies while achieving conservation gains by pulling down political fences that arbitrarily break up habitats. But there is the danger that governments are simply deciding things over the heads of people most affected by the projects. Ecotourism may benefit wealthy hotel owners and foreign investors far more than locals living in the shadows of cross-border peace parks and transfrontier conservation areas. Within the Southern African Development Community, the establishment of transboundary conservation areas provided a strong impetus to regional cooperation. Yet the projects were most successful when, after a hurried and largely top-down process of establishing the first peace parks, greater control over land and resource use was ceded to local communities.[22]

Transboundary nature conservation has significant potential to contribute to conflict prevention, mainly by facilitating communication, improving local livelihoods, and promoting the ecological, social, economic, and political benefits of protected areas. Nevertheless, tensions remain between the imperatives of state-managed nature conservation on the one hand and the economic activities of indigenous populations on the other.

A healthy dose of realism is also warranted with regard to the crucial question of commitment. Even where initiatives have been designed with peace and confidence-building in mind, there has often been little follow-through. The Aral Sea offers a cautionary tale about the challenges of effective environmental peacemaking. By the time the Soviet Union collapsed in 1991, what had been the fourth largest inland body of water in 1960 was a shadow of its former self. With its feeder rivers dammed and diverted for irrigation schemes, the sea's level fell by about 15 meters, its surface area was cut in half, its salinity level tripled, and its volume diminished by two thirds. The newly independent states of Central Asia faced a mounting socioeconomic crisis, sowing potential for water-related conflict to break out along ethno-national lines.[23]

With the help of the World Bank and other western aid agencies, the riparian states on the Aral Sea's feeder rivers, the Amu and Syr Darya, crafted a cooperative framework for responding to the crisis. By doing so, they stabilized interstate relations during a time of regional political turmoil. According to researcher Erika Weinthal, the initiation of water-related cooperation by the newly independent post-Soviet states may have helped prevent water-related violence.[24]

Their shared interdependence during the post-Soviet tumult was enough to draw Uzbekistan, Kazakhstan, Turkmenistan, Tajikistan, and Kyrgyzstan to the bargaining table, but it has not changed the fundamental problem: the slow death of the Aral Sea has heightened insecurities for the region's people. The basic problem—unsustainable agricultural practices—has barely been addressed, and the new cooperative framework creates little or no democratic space for stakeholders and civil society. The World Bank and the bilateral aid agencies may have played a catalytic role in brokering the interstate agreement on the crisis, but they have largely failed to create more robust forms of regional environmental governance. In fact, the common syndrome of flagging commitment known as "donor fatigue" has set in, and cynicism about the motives of the region's governments and international actors runs deep. Transforming this situation will require a long-term commitment of resources to put the region's economy on a sustainable footing and renewed initiatives to increase civil society's engagement in the process.[25]

Making Environmental Peacemaking a Reality

It has long been apparent, if not always acted upon, that cross-border environmental cooperation can yield tangible environmental, economic, and political gains. If properly designed, environmental initiatives can also reduce tensions and the likelihood of violent conflict between countries and communities. Environmental peacemaking strategies offer the chance to craft a positive, practical policy framework for cooperation that can engage a broad community of stakeholders by combining environment, development, and peace-related concerns.

Environmental peacemaking strategies offer the chance to craft a positive, practical policy framework for cooperation.

Obviously, environmental cooperation does not occur easily or automatically, nor will it automatically enhance peace. It all depends on the specific institutional form of cooperation. Yet knowledge of environmental initiatives designed specifically to address violence and insecurity is limited. Simply put, governments and other actors have not pursued enough peace-oriented cooperative activity on environmental problems to allow firm conclusions. Where they have started programs, they have just begun to share experience and knowledge about environmental

peacemaking through peace-and-conflict assessments of environmental projects and programs. Without such knowledge, the international community may be missing powerful peacemaking opportunities in the environmental domain.

The challenge, therefore, is to amass evidence—however partial or indirect—that more-aggressive environmental peacemaking strategies could create opportunities. Such evidence might be used to nudge governments, intergovernmental organizations, social movements, and other actors to be more aggressive about environmental cooperation. Identifying credible peacemaking spin-offs may make people more willing to invest in these projects.

The Environment and Security Initiative (ENVSEC), a partnership among the Organization for Security and Co-operation in Europe (OSCE), the U.N. Environment Programme, and the U.N. Development Programme that was launched in fall 2002, is an important attempt to test environmental peacemaking arguments. Its objective is to identify, map, and respond to situations where environmental problems threaten to generate tensions or offer opportunities for cooperative synergies among communities, countries, or regions.[26]

The effort is noteworthy not only for its application of an environmental peacemaking approach. It is also the first formal cooperation among these three organizations, which specialize individually in security, environment, and development. As a result, ENVSEC greatly benefits from their distinct but complementary expertise as well as a network of field presences in its regions of operation: Southeastern Europe, Central Asia, and the Southern Caucasus countries. (See Box 8–1, on pages 156–57.)[27]

As a rough division of labor, the OSCE takes the lead on policy development and political issues, UNEP contributes experi-ence in assessment, visual communication, and presentation, and UNDP is most closely involved in institutional development and project implementation. To be sure, chal-lenges remain: the three partners have very different organizational cultures and opera-tional modes, which are not designed for for-mally cooperating with other international organizations or jointly managing projects.

ENVSEC illustrates the hurdles commonly faced by attempts to put environmental peace-making ideas into operation. The concept of environment-security linkages is sometimes contested by host governments or at least deemed less significant than other problems in the regions. At the same time, the initia-tive faces financial, political, or other devel-opment-support expectations that are beyond its reach. Various stakeholders have different expectations, and political sensitivities must always be considered. Despite these prob-lems, the value of ENVSEC lies precisely in its practical application, which serves to reveal the complexity of environmental peacemak-ing on a daily basis.

World regions as different as post-Soviet Eastern Europe, post-apartheid southern Africa, post–cold war Northeast Asia, and North America under NAFTA are sorting out new security relationships in the wake of a particularly turbulent period of interna-tional change. In each region, the transfor-mations of the past decade have created the political space between states and across soci-eties to seek a more peaceful, cooperative future, even as daunting new challenges to peace and security have emerged.

Another new wrinkle is globalization. Its effects are complex and by no means entirely healthy for ecological sustainability. But glob-alization's ability to move political dynamics out of narrow interstate settings and into a broader society-to-society context is an important and healthy sign. This new social space holds much of the potential for envi-ronmental peacemaking. It is well worth finding out whether these changes create opportunities to build peace, lessen envi-ronmental insecurity, and break out of the zero-sum logic that so often plagues inter-national relations.[28]

BOX 8–1. ADDRESSING ENVIRONMENT AND SECURITY RISKS AND OPPORTUNITIES IN THE SOUTHERN CAUCASUS

The Southern Caucasus—composed of Armenia, Georgia, and Azerbaijan—has long been a focal point for change, a bridge between Asia and Europe. Today, social, political, and economic transformations are altering centuries-old relationships between countries and communities and affecting the natural environment. The region is marked by instability that can broadly be divided into two categories. First, there is continuous danger of identity-based violence related to conflicts inherited from the collapse of the Soviet Union, including the Armenian-Azerbaijani conflict over the Nagorno-Karabakh region, the Georgian-Ossetian and the Georgian-Abkhaz conflicts, and possible spillover from the Northern Caucasus. Second, other conflicts (generally less violent ones) can arise from the decline in living standards and the shifts in the political landscape that occur due to clashes between dominant groups and rival elite groups or between the "winners" and "losers" of post-Soviet socioeconomic development.

The Southern Caucasus countries are also facing immense environmental problems as a legacy of the Soviet period. Some of the main ones pressuring interstate relationships and human security include the lack of up-to-date, precise data; water quality and wastewater treatment; irrigation and drainage system degradation; deforestation; and land and soil degradation, including landslides and desertification. Oil pollution, earthquakes, the condition of the Black and Caspian Seas, and radioactive contamination do not affect all three countries equally, but they still carry the risk of negative transboundary environmental effects.

In May 2004, an Environment and Security Initiative assessment took place in Armenia, Azerbaijan, and Georgia involving representatives from ministries of the environment, foreign affairs, agriculture, defense, and health, as well as civil society and the scientific community. A central question for the Initiative was how environmental cooperation can be fostered in transboundary priority areas where security concerns and environmental or natural resource pressures coincide.

Three sets of environment-security linkages were identified. Environmental degradation in zones of conflict and the lack of information about the state of the environment are points of contention regarding Nagorno-Karabakh and Abkhazia. In addition, rising economic productivity could increase tension over access to natural resources such as clean water, soil, and living space and could increase pollution. Last, failure by a government to manage natural resources and environmental conditions appropriately could add to public frustration and lead governments to lose legitimacy during this fragile post-Soviet period.

Despite conflicting interests, the governments of the Southern Caucasus recognize that some environmental challenges require joint action, as a number of Kura-Araks River basin management projects reveal. Even while serious disputes between parties continue to hamper cooperation efforts, different groups stated clearly during the assessment that they wanted to cooperate with international bodies to increase the amount of environmental information and data on pollution in order to address common concerns and reduce tension over natural resources.

To address environment and security priorities, ENVSEC has compiled a Preliminary Work Program of activities that the partner organizations suggest implementing within the framework of the Initiative. The activities will be developed in close cooperation with stakeholders in the regions and will form part of a "three pillar" approach: in-depth vulnerability assessment, early warning, and monitoring of areas "at risk"; policy development and implementation; and institutional development, capacity building, and advocacy.

—*Gianluca Rampolla, Organization for Security and Co-operation in Europe, and Moira Feil, Adelphi Research*

SOURCE: See endnote 27.

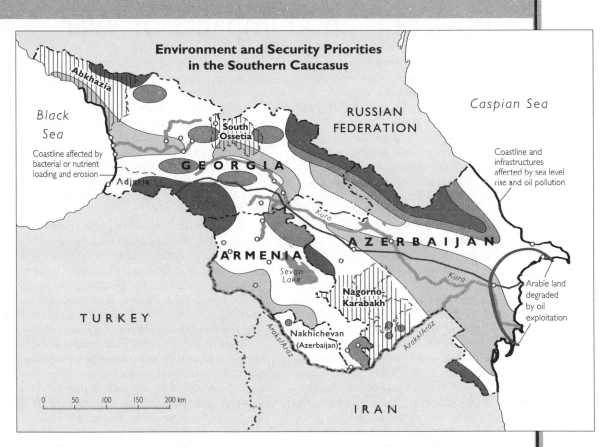

Environment and Security Priorities in the Southern Caucasus

Black Sea

Coastline affected by bacterial or nutrient loading and erosion

RUSSIAN FEDERATION

Caspian Sea

Coastline and infrastructures affected by sea level rise and oil pollution

Abkhazia

South Ossetia

GEORGIA

Adjaria

Kura

AZERBAIJAN

Kura

Arable land degraded by oil exploitation

ARMENIA

Sevan Lake

Nagorno-Karabakh

TURKEY

Araks/Araz

Nakhichevan (Azerbaijan)

Araks/Araz

IRAN

0   50   100   150   200 km

Soil degradation: contamination due to high levels of pesticides or heavy metals (mainly inherited from the Soviet period); salinization and erosion due to poorly maintained irrigation systems and rise of the water table

Areas affected by deforestation

Pastures degraded by overgrazing

Water pollution

Transboundary and domestic polluted waters

Affected coastlines

o   Aging Soviet industrial complex, mining center, or processing plant (oil terminal, refinery): major source of air, soil, and water pollution

Baku-Tbilisi-Ceyhan oil pipeline route

Security issues

– – –   Former Soviet Union administrative boundaries of autonomous regions (South Ossetia and Nagorno-Karabakh have been officially disbanded)

Areas of conflict out of control of central authorities

The map does not imply the expression of any opinion on the part of the three agencies concerning the legal status of any country, territory, city, or area of its authority, or delineation of its frontiers and boundaries. Adapted and simplified from a color map by Philippe Rekacewicz, UNEP/Grid-Arendal, July 2004.

Environmental Impacts of War

Military conflicts always bring human suffering. They also bring longer-term security threats, such as environmental degradation and new risks to human health. For the past seven years the U.N. Environment Programme (UNEP) has been working in areas of the world where natural and human environments have been damaged as a consequence of conflict. In 1999, as the ruins of targeted industrial facilities in Kosovo, Serbia, and Montenegro were still smoldering, UNEP teams conducted the first "post-conflict environmental assessment."[1]

The work in the Balkans concluded that there were several environmental hot spots where immediate cleanup action was needed to avoid further threats to human health, such as targeted oil refineries in Pancevo and Novi Sad and industrial facilities in Kragujevac and Bor. The Danube River was at risk due to the leakage of more than 60 different chemicals, including mercury, from Pancevo. These findings led the international community for the first time to include environmental cleanup in their post-conflict humanitarian aid.[2]

After the Balkans, this new environmental tool has been used in Liberia, the Occupied Palestinian Territories, Afghanistan, and most recently in Iraq. Each situation is unique due to the particular nature of the conflict, the society, and the ecology.

In Afghanistan, two decades of warfare have degraded the environment to the extent that it now presents a major stumbling block for the country's reconstruction efforts. Conflict has put previous environmental management and conservation strategies on hold, brought about a collapse of local and national governance, destroyed infrastructure, hindered agricultural activity, and driven people into cities already lacking the most basic public amenities.[3]

Over 80 percent of Afghanistan's people live in rural areas, where they have seen many of their basic resources—water for irrigation, trees for food and fuel—lost in just a generation. In urban areas, safe water—the most basic necessity for human well-being—may be reaching as few as 12 percent of the people. Badly managed solid waste sites have contaminated groundwater and spread air pollution, and illegal logging has caused widespread loss of forest cover.[4]

In Iraq, a similar picture can be painted. There, UNEP's assessment concluded that the conflict in 2003 and looting after the war have added to chronic environmental stresses already in place from the Iran-Iraq war of the 1980s, the 1991 Gulf War, environmental mismanagement by the former Iraqi regime, and the unintended effects of the sanctions.[5]

A major threat to the Iraqi people is the accumulation of physical damage to the country's environmental infrastructure. In particular, the destruction and lack of investment in water and sanitation systems have led to higher levels of pollution and health risks. When power shortages stop pumping stations, both freshwater supply and wastewater treatment are threatened.[6]

The destruction of military and industrial infrastructure during Iraq's various conflicts has released heavy metals and other hazardous substances into the air, soil, and water. Smoke from oil-well fires and burning oil trenches during the war, looting, and sabotage have caused local air pollution and soil contamination. Lack of investment in the oil industry has reduced maintenance and raised the risk of leaks and spills.[7]

One of the main projects of Saddam Hussein's regime—draining the Mesopotamian Marshes and building artificial waterways—has ruined some of the most valuable areas

UNEP

Unexploded ordnance, Bosnia

of biodiversity in Iraq. The water pollution is affecting not only the Euphrates and Tigris Rivers, but also the wider Persian Gulf region.[8]

In Iraq, as in many post-conflict situations, environmental issues are closely linked to humanitarian and reconstruction needs. Priorities include restoring the water supply and sanitation systems, cleaning up pollution hot spots, and cleaning up waste sites to reduce the risk of disease epidemics from municipal and medical wastes. During the 1991 Gulf War and the 2003 Iraq War, weapons with depleted uranium were used in several places in Iraq. To protect the local populace, sites with these remnants of war need to be assessed and cleaned up.[9]

In all conflict areas there are both chronic, long-term environmental problems and problems directly related to military action. Furthermore, UNEP post-conflict environmental assessments clearly demonstrate that military crises are almost always followed by an environmental crisis.

Consequently, a key lesson is the need to minimize the risks for human health and environment during conflict through preparedness and civil protection. And as soon as the conflict is over, proper assessment and cleanup should take place. Support and capacity building of the existing or newly established environmental administration is crucial for long-term sustainability. When considering how to revive the environment after the guns fall silent, a region's entire environmental history must be addressed.

In addition, after conflict ends efforts must be made to reengage the country in regional and international environmental cooperation— especially when dealing with shared resources like water. In spring 2004, for the first time in 29 years Iraqi and Iranian water and environmental authorities together discussed the issue of the shared Mesopotamian Marshes. Old enemies are once again negotiating on environmental matters. Along with improving the state of these resources, the management of shared resources can serve as an important way to build confidence between formerly hostile countries.[10]

One important way to minimize environmental and health risks is through stricter regulations of warfare by limiting possible targets and weapons. A good example of the legal tools that can be used is the ENMOD convention, which prevents the use of artificial changes in the environment—like human-caused floods—as weapons of war. Since the negative environmental impacts of different types of weapons are known, and since there is enough evidence of the risks that targeting chemical facilities can bring to a population, new international regulations are needed.[11]

Adding environmental costs to the long list of the negative consequences of conflict— human casualties, refugees, economic losses— should make nonviolent solutions even more attractive.

—*Pekka Haavisto,*
UNEP Post-Conflict Assessment Unit

CHAPTER 9

Laying the Foundations for Peace

Hilary French, Gary Gardner, and Erik Assadourian

As people around the world watched in horror when the twin towers of the World Trade Center crumbled on September 11, 2001, it was the immediate human toll that was uppermost in their minds. But it soon became clear that the events of that day had a larger significance, ushering in a new era in world history. Just as the Japanese attack on Pearl Harbor on December 7, 1941, caused the United States to declare war on Japan the following day, the events of September 11th led to President George W. Bush's assertion of a war on terrorism before the day was over. And just as the postwar period came to define a historical epoch, the post–9/11 years will long be recognized as fundamentally different from the time before.[1]

Yet today's global security problems differ significantly from those of the World War II era. Unlike the territorial expansionism of that time, most contemporary flashpoints involve new kinds of challenges, such as internal civil conflicts and international terrorism. These problems are rooted in societal instabilities that are paired with a complex array of phenomena—from poverty and disease to population growth and environmental degradation to religious fundamentalism and ethnic hatred. (See Chapter 1.) Traditional military techniques are of limited use in responding to these underlying forces.[2]

The stance taken by the United States toward the larger world community was also markedly different in the aftermath of September 11th than it was during World War II. President Bush initially spoke of the importance of international cooperation in combating global terrorism. But his subsequent decision to invade Iraq in early 2003 without securing backing from the U.N. Security Council shattered initial hopes that the struggle against terrorism would be a uniting rather than a divisive effort. During World War II, in contrast, the United States worked with its allies before it even entered the war to begin laying the foundations for a lasting postwar peace by developing a detailed blueprint for creating the United Nations. This effort culminated in the signing of the U.N. Charter in San Francisco in June 1945, as the war entered its final months.[3]

Still another way in which the current security environment differs from that after World War II is the growing influence of global civil society. Citizens' organizations have long been powerful advocates of a more peaceful world, including pushing hard for the creation of the United Nations. But recent decades have seen a pronounced surge in civil society's role, power, and global reach.[4]

Despite the many differences between 1945 and today, a central insight of that era still holds true: laying the foundations for lasting global peace will require international cooperation on a broad range of fronts—from resisting aggression to combating terrorism, mediating peace settlements, and addressing the underlying causes of conflict and instability. At the same time, the experience of recent decades has made it clear that building a secure world will require extensive interactions among a broad range of actors, including visionary and committed national and local politicians and government officials as well as engaged, globally minded citizens.

Reinventing Global Governance

The international divide over the wisdom of the Iraq war plunged the United Nations into an identity crisis. As U.N. Secretary-General Kofi Annan put it in fall 2003 when he addressed world leaders at the U.N. General Assembly: "Three years ago, when you came here for the Millennium Summit, we shared a vision, a vision of global solidarity and collective security.... Recent events have called that consensus in question.... We have come to a fork in the road. This may be a moment no less decisive than 1945, when the United Nations was founded.... Now we must decide whether it is possible to continue on the basis agreed then, or whether radical

changes are needed." The crisis created by the controversy over the Iraq war thus had the silver lining of creating a moment of opportunity to lay the foundations for peace by redesigning the United Nations for the security challenges of today and tomorrow.[5]

As the world sets about this task, it is important to consider how well the original structures of 1945 have withstood the test of time. The first purpose of the United Nations, as defined by its charter, is "to maintain international peace and security." Toward that end, the U.N. charter stipulates a set of mechanisms for the Security Council that are designed to galvanize a collective response from U.N. members when confronted with a compelling threat to global peace and stability.[6]

Contrary to expectations, cross-border military incursions have been relatively rare since the United Nations was created. But there has been no shortage of civil strife, and the organization has often played an important role in helping to negotiate and then maintain the peace. The United Nations has helped to bring about over 170 peace settlements, including those that ended the Iran Iraq war in 1988, led to the withdrawal of Soviet troops from Afghanistan in 1988, and brought the El Salvador civil war to a close in 1992. The 59 U.N. peacekeeping missions since 1948 have helped countries maintain ceasefires, conduct free and fair elections, and monitor troop withdrawals in countries as diverse as Cambodia, Cyprus, and East Timor.[7]

But from the very beginning the United Nations was intended to be about much more than military matters. The U.N. Charter states that one of the organization's central purposes is "to achieve international cooperation in solving international problems of an economic, social, cultural, or humanitarian character." These provisions came about in part in response to a widely shared belief that the disastrous world economic conditions of the 1930s had indirectly helped precipitate World War II by creating a climate ripe for the rise of Nazism.[8]

This same conviction also underlay a major international conference held in Bretton Woods, New Hampshire, in 1944 that led to the creation of the World Bank, the International Monetary Fund (IMF), and the General Agreement on Tariffs and Trade (which has since been transformed into the WTO—the World Trade Organization). Technically speaking, the World Bank and the IMF are specialized agencies of the United Nations, but from the beginning they have shown little inclination to associate themselves closely with the rest of the organization. In fact, a 1947 agreement between the World Bank and the United Nations has been described as being "as much, or more, a declaration of independence from the U.N., as an agreement to work together." Similar problems have plagued the relationship with the WTO, with U.N. agencies such as the International Labour Organization and the U.N. Environment Programme (UNEP) forced to battle for the right to even observe WTO deliberations.[9]

In the half-century since the United Nations and the Bretton Woods institutions were created, poverty and destitution around the world have proved to be formidable foes. Nonetheless, the U.N. system has seen its share of successes on a range of social issues. In the field of global health, for instance, the World Health Organization (WHO), a U.N. specialized agency, initiated a global campaign to eradicate smallpox in 1967. At that time, the disease afflicted up to 15 million people annually, leading to some 2 million deaths. In 1980, WHO certified that the disease had been conquered globally. (See Chapter 3.) It is now nearing similar successes

with leprosy, guinea worm, polio, and Chagas disease. Eradication is unfortunately nowhere in sight for a number of other deadly diseases, including HIV/AIDS, tuberculosis, and malaria, but WHO is working with other international institutions and partners to reduce the number of people stricken by these diseases and to expand access to treatment for those who need it.[10]

The United Nations has also proved adaptable in the face of new problems and challenges. Neither rapid population growth nor environmental degradation, for instance, was recognized as a significant global problem in 1945. As a result, neither of them is even mentioned in the U.N. Charter. But as the seriousness of both problems gradually became apparent, new institutions were set up to address them: the U.N. Fund for Population Activities in 1962; UNEP in 1972; and in the early 1990s the Global Environment Facility, a joint undertaking of the World Bank, the U.N. Development Programme, and UNEP that funds projects in developing countries that address global environmental threats such as climate change and the loss of biological diversity.[11]

Similarly, the spread of terrorism and weapons of mass destruction are relatively new preoccupations of the world community, and the United Nations is being called on to play a growing role in combating them. As Secretary-General Annan argued before the U.N. General Assembly within weeks of the September 11th attacks: "The legitimacy that the United Nations conveys can ensure that the greatest number of States are able and willing to take the necessary and difficult steps—diplomatic, legal, and political—that are needed to defeat terrorism." He went on to discuss the importance of governments moving forward to adopt and ratify the 12 international conventions and protocols on international terrorism that already exist and to implement and enforce key international treaties designed to minimize the spread of weapons of mass destruction, such as those that ban chemical and biological weapons and nuclear proliferation.[12]

Through a series of high-profile international conferences over the last few decades, the United Nations has shone the spotlight on emerging issues of global concern and helped to propel action to address them globally and nationally. The 1994 U.N. Conference on Population and Development in Cairo, for example, forged a new global consensus on the relationship between population stabilization, reproductive health care, and women's empowerment, including agreement on a series of goals on access to universal education and reproductive health services.[13]

The new understandings on the range of issues addressed by the global conferences of the 1990s ultimately found expression in the Millennium Development Goals (MDGs), adopted unanimously in preliminary form at the 2000 U.N. Millennium Assembly. (See Box 9–1.) And the 2002 World Summit on Sustainable Development in Johannesburg, South Africa, brought renewed political attention to sustainable development challenges, including the adoption or reaffirmation by governments of a broad range of targets on water, energy, health, agriculture, and biological diversity. (See Box 9–2.) The United Nations is currently finding a growing role for itself in encouraging governments to implement the policy reforms needed to achieve these goals and targets and in tracking their progress along the way.[14]

Despite all the achievements to date, there can be little question that bold reforms are needed to lay the foundations for peace by better equipping the United Nations for the security challenges of today and tomorrow. The need for periodic renovations to U.N. structures was in fact foreseen from the beginning, with U.S. President Harry Truman noting in his speech to the 1945 San Francisco conference that "this charter, like our own Constitution, will be expanded and improved upon as time goes on. No one claims that it is now a final or a perfect instrument.... Changing world conditions will require readjustments." Toward this end, in September 2003 Secretary-General Annan announced the appointment of a panel of eminent world leaders charged with examining current threats and challenges to global peace and security and considering far-reaching changes to address them. The panel's report will form the basis for Annan's recommendations to the U.N. General Assembly in fall 2005.[15]

Despite achievements to date, bold reforms are needed to lay the foundations for peace by better equipping the United Nations for the security challenges of today and tomorrow.

One particularly high priority in preparing the United Nations for the future is to rethink the composition of the Security Council. In 1945, China, France, the Soviet Union, the United States, and the United Kingdom were given a special status as permanent Council members, with the right to veto resolutions. Without these provisions, it is unlikely that either the United States or the Soviet Union would have joined the new organization. But these arrangements had a price: heavy resort to the veto has at times hamstrung the effectiveness of the Security Council, particularly during the cold war, and the council's limited permanent membership is now widely viewed as anachronistic and undemocratic.[16]

Although proposals for altering the status quo are bound to bump up against formidable opposition, a consensus is nonetheless

---

BOX 9–1. MILLENNIUM DEVELOPMENT GOALS AND TARGETS

Eradicate extreme poverty and hunger
By 2015, reduce by half both the proportion of people living on less than $1 a day and the share suffering from hunger.

Achieve universal primary education
Ensure that by 2015 all boys and girls complete a full course of primary schooling.

Promote gender equality and empower women
Eliminate gender disparity in primary and secondary education, preferably by 2005, and at all levels by 2015.

Reduce child mortality
By 2015, reduce by two thirds the mortality rate among children under five.

Improve maternal health
By 2015, reduce by three quarters the maternal mortality rate.

Combat HIV/AIDS, malaria, and other diseases
Halt and begin to reverse the spread of HIV/AIDS, malaria, and other major diseases by 2015.

Ensure environmental sustainability
Integrate the principles of sustainable development into country policies and programs and reverse the loss of environmental resources. By 2015, cut in half the proportion of people without sustainable access to safe drinking water and sanitation. By 2020, improve significantly the lives of 100 million slum dwellers.

Develop a global partnership for development
Develop an open trading and financial system that is rule-based, nondiscriminatory, and includes a commitment to good governance, development, and poverty reduction. Address the special needs of least developed countries, small island developing states, and landlocked countries. Make debt sustainable, increase youth employment, and provide access to essential drugs and new technologies.

SOURCE: See endnote 14.

---

building that changes are needed in order to make the Security Council more representative of today's world. In September 2004, the governments of Brazil, Germany, Japan, and India issued a joint statement noting that "the Security Council must reflect the realities of the international community in the 21st Century." In addition to pushing their own cause as strong candidates for permanent membership, the four countries underscored that similar status should also be granted to an African nation.[17]

It is also important to bolster the United Nations' ability to address underlying threats to international peace and security, including poverty, disease, environmental decline, and rapid population growth. The Security Council could be given a broadened mandate to address nontraditional security issues, as happened in 2000 on HIV/AIDS. Unlike other U.N. organs, the Security Council has significant enforcement capabilities at its disposal, so addressing new security threats there offers important practical as well as symbolic benefits. Other possible approaches include strengthening and streamlining current economic and social organs, such as the Economic and Social Council, or creating a new

Economic Security Council or a similar high-level body that is dedicated to preventing conflict by reducing poverty and addressing other underlying causes of insecurity.[18]

There have also been a number of calls over the years to give environmental issues a more central home within the U.N. system. Among the ideas put forward have been proposals to create an Environmental Security Council, use the now-disbanded U.N. Trusteeship Council for this purpose, create a U.N. High Commissioner for Environment or Sustainable Development, or create a new Global Environmental Organization. The most politically salient proposal is a variation on the last idea: led by President Jacques Chirac, the government of France is promoting the transformation of the Nairobi-based UNEP into a full-fledged U.N. specialized agency, like WHO and UNESCO. This proposal is currently being actively considered at a range of international meetings, although it remains unclear if it will garner sufficient support to be acted on in the near term.[19]

In addition to improving the social, economic, and environmental machinery of the United Nations, it will also be important to reform the World Bank, the IMF, and the WTO, each of which has become both increasingly powerful and increasingly controversial over the years. These institutions are widely seen to disproportionately represent the interests of major industrial countries, either as a result of their formal voting procedures or through less formal but no less influential entrenched ways of doing business. Each organization has also been criticized in recent years for promoting orthodox economic globalization strategies that in some cases have harmed rather than helped poor people and the environment.[20]

One way to address these deficiencies would be for the global economic institutions to work more closely with the United Nations. This collaboration would help ensure that the new development consensus expressed in the Millennium Development Goals and in the broad range of U.N. environmental, social, and human rights accords is more clearly reflected on the ground, including in post-conflict situations. Creating a new high-level oversight board with some measure of authority over both the United Nations and global economic institutions would be one strategy for promoting the needed collaboration.[21]

Another high priority for a peaceful and secure future is redesigning global governance structures so that they do more to harness the energy and insights of a broad array of actors, including civil society organizations (CSOs) and the private sector. In part spurred by pressure from the globalization protest movement, both the United Nations and the international economic institutions have recently taken steps to make their operations more transparent to civil society. But many hurdles remain in bringing about full and meaningful public participation.[22]

Shifting Government Priorities

Reshaping international institutions is only a first step. The United Nations and affiliated organizations, acting through their member governments, lay out visions, enumerate goals for the global community, and help guide implementation efforts. But national governments have the tough tasks of marshaling the domestic political will and resources needed to make that vision a reality and of ensuring that their priorities are in line with today's burgeoning new global security threats.

One of the first things governments can do is recognize how misdirected security spending is today. Nearly $1 trillion is spent annually on the world's militaries, most of which is targeted at traditional security threats. As political leaders recognize poverty, rapidly growing populations, disease, and environmental degradation to be legitimate security issues, these concerns could assume greater importance in government budgets. At the same time, a tabulation of military programs that are outdated, ineffective, or otherwise wasteful will likely highlight rich sources of funding that could be redirected to addressing social and environmental threats. In this new framework, social and environmental programs long deemed too expensive could suddenly be viewed as affordable—in fact,

even indispensable.[23]

Fortunately, the international framework to address this complex array of threats already exists—the Millennium Development Goals and the World Summit on Sustainable Development targets. At the 2000 Millennium Assembly, the members of the United Nations agreed to reduce global poverty, disease, and societal inequities significantly by 2015. The World Summit targets, adopted two years later, rounded out the picture by addressing how countries can further improve social conditions by protecting critical natural systems. These goals were primarily adopted in order to address growing global inequities in a sustainable manner. In the post–9/11 world, however, where security threats have become the dominant concern, the MDGs can equally be seen as a means to strengthen national and global security.[24]

While the commitment on paper to achieving the MDGs is strong, progress for the most part has been excruciatingly slow. In 2004, the World Economic Forum asked some of the world's leading development experts to analyze the progress made during the first three years of working toward the Millennium Development Goals. The results were discouraging: the world had only put in a third of the effort needed to achieve these goals.[25]

While some countries have made notable progress in reaching a number of the MDG targets (see Table 9–1), few nations are on track to achieve the majority of the goals (see Table 9–2). According to the World Bank, less than one fifth of all countries are currently on target to reduce child and maternal mortality and provide access to water and sanitation, for example, while even fewer are on course to contain HIV, malaria, and other major diseases. The World Economic Forum analysis makes it clear that the primary

Table 9–1. Progress in Increasing Access to Food and Water in Selected Countries

Country	MDG Target: Reduce Hunger by Half				MDG Target: Reduce by Half Those Lacking Access to Water			
	1990–92	1999–2001	2015 Objective	On Track?	1990	2000	2015 Objective	On Track?
	(percent of population undernourished)				(percent of population without access to improved water source)			
Bangladesh	35	32	18		6	3	3	Yes
Brazil	12	9	6	Yes	17	13	8	Yes
China	17	11	9	Yes	29	25	14	
Egypt	5	3	3	Yes	6	3	3	Yes
India	25	21	13		32	16	16	Yes
Kenya	44	37	22		55	43	27	Yes
Mexico	5	5	3		20	12	10	Yes
Peru	40	11	20	Yes	26	20	13	Yes
Thailand	28	19	14	Yes	20	16	10	Yes
Uganda	23	19	12	Yes	55	48	27	

SOURCE: See endnote 26.

reason for failure is a lack of focus on basic development priorities.[26]

When governments do set the achievement of certain goals as a priority, however, they can rapidly register great success—success that is often multiplied because of the strong connection between different societal problems. By investing in AIDS prevention, for example, governments not only curtail the spread of the disease, they also reduce health care costs, the number of orphaned children, the loss of economic productivity, and the loss of much-needed professionals such as teachers and doctors.

Thailand saw the wisdom of preventive investments early on. In 1990, after receiving a study stating that if HIV were left unchecked it would infect 4 million Thais by 2000 and cost 20 percent of gross domestic product (GDP) per year, Minister to the Prime Minister's Office Mechai Viravaidya recognized that AIDS was not just a health issue but "a major threat to national security." After encouragement from Mechai, as he is known throughout the country, Prime Min-

ister Anand Panyarachun personally led an AIDS prevention campaign. With this level of commitment, all government ministries were empowered to tackle AIDS. Funding sky-rocketed from $684,000 in 1988 to $82 million in 1997, and Thailand was able to reduce new infections from a high of 143,000 in 1991 to 19,000 in 2003.[27]

Other countries have come up with creative ways to tackle many goals simultaneously. In Mexico, for instance, almost 20 million people in 1995 could not afford to eat enough to meet their minimum daily nutritional needs, 10 million lacked basic health care, and at least 1.5 million children were not in school. The government created a "conditional cash transfer" welfare program that provided payments based on a family's commitment to specific health and education requirements. Recipients had to show that their children were enrolled in school, that mothers received monthly nutrition and hygiene lessons, and that families got routine health checkups. The results were striking. Illness fell 25 percent among infants

Table 9–2. Regional Progress in Achieving Selected Millennium Development Goals

Region	Poverty	Hunger	Primary Education	Child Mortality	Access to Water	Access to Sanitation
Arab States	achieved	reversal	on track	lagging	n. a.	n. a.
Central/Eastern Europe and CIS	reversal	n. a.	achieved	lagging	achieved	n. a.
East Asia/ Pacific	achieved	on track	achieved	lagging	lagging	lagging
Latin America/ Caribbean	lagging	on track	achieved	on track	on track	lagging
South Asia	on track	lagging	lagging	lagging	on track	lagging
Sub-Saharan Africa	reversal	reversal	lagging	lagging	lagging	reversal
WORLD	on track	lagging	lagging	lagging	on track	lagging

SOURCE: See endnote 26.

and 20 percent among children under five. Children's height and weight increased significantly, while rates of anemia fell 19 percent. School enrollment rates also increased since families felt less financial pressure to have their children go to work. By 2004, the program was providing benefits to more than 25 million people, at a cost of just 0.3 percent of Mexico's GDP.[28]

Although national governments are the natural leaders in pursuing the MDGs, a great deal can be done at the regional and local level as well when policymakers are determined to address societal problems. One of the most famous examples is the state of Kerala in India. Compared with the whole country, Kerala's development statistics are impressive: infant mortality is one quarter the national rate, immunization rates are almost double, and the fertility rate is two thirds that of India's. (In fact, at 1.96 births per woman, Kerala has a lower fertility rate than

the United States does.) In conjunction with strong civic engagement, a large measure of Kerala's success derives from dedication by government officials that made the broad provision of health care, education, and other basic services a priority.[29]

The city of Porto Alegre in Brazil has also made huge gains in improving health and social conditions. In just a decade the percentage of the population with access to water and sanitation jumped from 75 to 98 percent and the number of schools quadrupled. This happened mainly because the municipal government gave local people the power to set government funding priorities. People decided to devote resources to ensuring their basic needs were met, which meant increasing the health and education budget from 13 percent in 1985 to almost 40 percent in 1996.[30]

Yet even as governments work to reach basic development goals, they will need to pursue them in an ecologically sustainable manner to avoid making short-term gains at the expense of long-term well-being and security. One example of how not to develop is provided by the Aral Sea basin in Central Asia. In 1960, government planners started an aggressive economic development program to transform an arid region into the cotton belt of the Soviet Union. For a time, they succeeded: irrigated land grew to 7 million hectares (twice the irrigated area of California), farmers consistently exceeded production quotas, and the area became a leading supplier of cotton and produce for the Soviet Union. But water was drained too rapidly from the rivers that fed the Aral Sea, and the rivers started to run dry.[31]

Today, the Aral Sea is less than half the area it once was, with less than a fifth as much volume. The fishery that originally supplied 45,000 tons of marketable fish a year is dead. And salt from the dried seabed, carried throughout the region on the wind, now contaminates the area and poisons remaining agricultural lands. Worse, without the sea to regulate the climate, the growing season has shortened and rainfall has shrunk, straining agriculture even further. Overall, this environmental disaster has affected 3.5–7 million people.[32]

Although not always as dramatic, similar tragedies due to unsustainable development initiatives are unfolding around the world. Southeast Asia's mangrove forests have been decimated by shrimp farms that themselves have short productive lives; tropical rainforests have been cleared across the Amazon, erasing traditional lifestyles and countless undiscovered species; and 15,000 square kilometers of the Gulf of Mexico—an area nearly the size of Kuwait—is now dead from the spilling of farm wastes into the Mississippi River.[33]

Overburdening the ecological systems people depend on is thus creating grave new threats. Some of the strategies called for in the MDGs will naturally help counter these— for example, providing basic education to women tends to reduce fertility rates and, subsequently, population pressures. But they may also exacerbate the threats—education may provide the means or incentive to join the global consumer class, which could greatly increase resource use. Incorporating principles of sustainability directly into development strategies would help governments prevent further ecological stresses.[34]

China is working to simultaneously reduce poverty and alleviate environmental problems with its ambitious rural electrification program. Ninety percent of the poorest people in China live in rural areas. The government has recognized that electricity is an effective means to alleviate poverty as it lowers dependence on biomass fuel (the burning of which often contributes to respiratory disease) and leaves more time for education by reducing the hours spent collecting water and fuel. Starting in late 2001, over a period of 20 months the government installed wind turbines, solar photovoltaics, and small hydroelectric arrays in more than a thousand townships, providing electricity to almost a million people. By using renewable energy resources, the government not only helped raise living standards in rural areas, it also reduced local environmental problems such as deforestation and desertification and lowered China's overall contribution to climate change.[35]

As important as national development plans and policy changes are, however, a new definition of economic success is needed if nations are to set their economies on a sustainable path. Current understandings of success focus mainly on whether national economies, often measured in terms of gross domestic product, grow or shrink. Yet GDP hides the fact that some growth is destructive;

an alternative that provides a better measure of success is needed.

While many nongovernmental organizations (NGOs) have created alternatives over the past three decades that have incorporated environmental and social costs into the GDP measure, 2004 may mark a turning point in this new approach. China announced that within the next three to five years it would adopt a Green GDP measure that would subtract resource depletion and pollution costs from GDP. Already this is being field-tested in the city of Chongqing and the province of Hainan. Early work suggests that China's average GDP growth would have been 1.2 percent lower between 1985 and 2000 had environmental costs been subtracted from the calculation. If fully implemented, not only would this lead China to pursue a more sustainable development path, it could push the world's other major economies to follow suit—which could set in motion a powerful transformation in the types of economic development the world values.[36]

Achieving the Millennium Development Goals will require greater investment. Some countries are already recognizing this and acting accordingly. In 2003, for example, Brazil delayed the purchase of $760 million worth of jet fighters and cut its military budget by 4 percent in order to finance an ambitious anti-hunger program. Costa Rica, by having no military for the past 50 years, has been able to devote a much larger portion of its budget to social spending—with impressive results. With a similar GDP per capita as Latin America as a whole, Costa Rica has the highest life expectancy and one of the highest literacy rates in the entire region. Even if developing countries redirect just a small portion of their estimated military expenditures of over $220 billion to achieving the MDGs, significant additional funding could be available.[37]

But most of these countries will need more funding than they can provide themselves. Indeed, for the poorest countries it will be nearly impossible to find enough funds within their own budgets to provide basic services. WHO estimates, for example, that to sustain a public health system, a minimum of $35–40 per person each year is necessary. For the poorest countries, where GDP per capita is in the low hundreds, this will be impossible without outside aid. As the eighth MDG makes clear, a concerted effort from industrial countries and global institutions will be essential—both in providing additional development aid and in "leveling the playing field" through initiatives like increased debt relief and fairer trade.[38]

Too little aid is currently provided to achieve the MDGs. In 2003, donor countries gave $68 billion in official development assistance (ODA), or just 0.25 percent of their gross national incomes (GNI). At the Johannesburg summit, governments reconfirmed the need to provide 0.7 percent of GNI in aid. But only five countries have done this—Denmark, Luxembourg, the Netherlands, Norway, and Sweden. If all donors actually met this readily attainable goal, annual development aid would increase by over $110 billion—more than twice the estimated $50 billion in additional annual funds needed to achieve the MDGs. So far only Belgium and Ireland have announced plans to increase their ODA to 0.7 percent.[39]

In addition, donor countries will have to do better at targeting the aid they provide. In 2001, more than a fifth of the aid was conditioned on purchasing goods and services from the donor country, while less than a third went to improving basic health, sanitation, and education services. To address nontraditional security threats successfully, more aid will have to go directly toward achieving the MDGs.[40]

Donor countries must also do more to reduce the unpayable burdens of highly indebted poor countries, many of which spend a significant percentage of their annual GDP servicing outstanding debts—often at the expense of providing basic social services. After a long campaign for debt relief in the 1990s, the benefits are starting to accrue. The 26 countries that have received some relief have reduced their debt service by 42 percent, from $3.8 billion in 1998 to $2.2 billion in 2001. Some 65 percent of these savings have been redirected to health and education programs. This has helped Uganda, for instance, achieve nearly universal primary school enrolment. Yet sub-Saharan Africa—the region furthest behind in achieving the MDGs—continues to pay creditor nations $13 billion a year in debt service.[41]

While aid and debt relief will help significantly, these gains are often overshadowed by the disparities created by the trade subsidies and tariffs of industrial countries. For example, while the European Union gives about $8 in aid per person in sub-Saharan Africa each year, it gives $913 in subsidies per cow in Europe. In total, more than $300 billion in annual subsidies and agricultural tariffs weaken the ability of farmers in developing countries to compete with farmers elsewhere. According to a 2004 study by the Institute for International Economics and the Center for Global Development, removing these tariffs and subsidies could pull 200 million people out of poverty by 2020.[42]

Another potential source of significant ODA could be money from redirected military funding. (See Figure 9–1.) In fact, redirecting just 7.4 percent of donor governments' military budgets to development aid would provide all the additional funds—$50 billion a year—needed to pay for the MDGs. According to a 2004 report by the Center for Defense Information and Foreign Policy in Focus,

$51 billion—or 13 percent—could be cut from the U.S. military budget just by removing outdated, unnecessary programs. This alone could provide the additional funds needed to attain the MDGs.[43]

In 2003, Brazil delayed the purchase of $760 million worth of jet fighters and cut its military budget by 4 percent in order to finance an ambitious anti-hunger program.

One of the most promising and comprehensive commitments to development comes from Sweden. At the end of 2003, the Swedish government passed a bill entitled Shared Responsibility—Sweden's Policy for Global Development. This commits the government to facilitate development not just through aid, which it also plans to increase to 1.0 percent of GDP, but by aligning all government policies—trade, agriculture, environment, defense—around a guiding principle of equitable and sustainable global development. In September 2004, the Swedish government released its first annual progress report. Used as a way to provide an overview of the current policy climate, the report documented the many inconsistencies within current policies and provided a starting point to engage government ministries and civil society in reorienting Swedish policy around a global sustainable development plan.[44]

Even if the Millennium Development Goals were achieved by 2015, however, there would still be 400 million people who are undernourished, 600 million who live on less than $1 per day, and 1.2 billion without access to improved sanitation. And the world is not even close to meeting these modest goals. To do so, governments will have to make strong commitments—and then live up to them.[45]

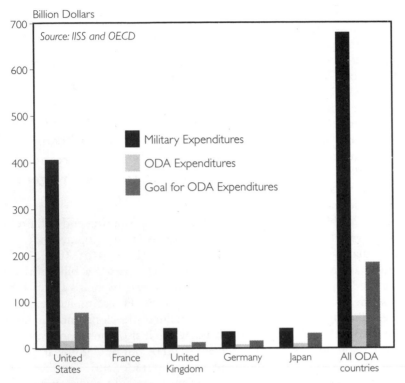

Billion Dollars

Source: IISS and OECD

■ Military Expenditures

ODA Expenditures

Goal for ODA Expenditures

United States · France · United Kingdom · Germany · Japan · All ODA countries

Figure 9–1. Military Expenditures versus Development Assistance, Selected Countries and All Donors, 2003

Engaging Civil Society

Success in creating a more secure and more peaceful world is likelier if civil society is involved in the effort. Fortunately, the record of the past 15 years suggests that actors from the civil sector—especially NGOs, a subset of civil society organizations—have emerged as skilled players in global politics and even as leaders on the broad range of issues relevant to security. (See Box 9–3.) The selection of Wangari Maathai, leader of Kenya's Green Belt movement, to receive the 2004 Nobel Prize for Peace is an encouraging example of the acceptance of such leaders on the international stage and of the environment's link to concerns about peace and security. The

growing effectiveness of civil society can be credited to a diverse set of assets that strengthen groups' capacity to "network"—perhaps the emblematic verb of this globalizing age. Civil society may best be able to help lay the foundations for peace by further developing this capacity to be effective partners and applying these skills to security issues.[46]

A powerful illustration of the civil sector's skill in reaching across national borders on a security issue came in the run-up to the 2003 Iraq war, when a global antiwar movement emerged that generated the largest demonstrations in history: millions of people gathered in hundreds of cities worldwide during the weekend of February 15, 2003, to protest the looming hostilities in Iraq. Although the movement failed to stop the war, it posted some noteworthy successes. Mobilizing a global public at a single moment on a critical issue was itself a considerable advance for civil society. And for the first time since the founding of the United Nations, public opinion helped prevent the United States from gaining a majority of Security Council votes on an issue it considered of vital importance—aided, of course, by concern among member states that weapons inspectors had not been allowed

---

BOX 9–3. THE RISE OF CIVIL SOCIETY

The enhanced stature of civil society is the product of several societal trends that have emerged in the last two decades. Setting the stage was the advance of democracy in scores of countries, which opened up greater operating space for citizens and civic organizations. Since the 1980s, and especially since the fall of the Berlin Wall, dozens of countries in Eastern Europe, Asia, and Latin America have abandoned totalitarian or authoritarian governments in favor of political systems that offered a greater degree of freedom of speech and of the press—the lifeblood of a vibrant civil sector. Meanwhile, some long-established democracies in Europe and the Americas began to turn to civil society organizations to take on responsibilities shunned by government and business, everything from running soup kitchens to implementing overseas development projects.

As the operating latitude of citizens' groups expanded, powerful and inexpensive communications technologies helped them organize and share information, enhancing their status as political players. By the 1980s, computers had become relatively cheap, portable, decentralized, and interconnected—a combination of attributes that has multiplied networking opportunities for organizations and individuals. In particular, the rapid advance of the Internet greatly enhanced opportunities for participatory democracy and for direct appeals to decisionmakers.

At the same time, international issues such as climate change and competition for water and other resources were gradually being recognized as too difficult for a single government, or even a group of governments, to address. Governments and businesses began to realize that partnerships with a liberated and empowered civil society could be an effective way of tackling some of today's more intractable issues.

Into this energized political space stepped the diverse set of civil entities known as NGOs. These generally work for a public purpose, typically on issues such as human rights, environmental protection, women's issues, and health care—and often from a wide range of political perspectives. NGOs are commonly regarded as being flexible, efficient, small, closely connected to citizens, and able to marry the operating efficiency of a business with the public purpose of government. Their growth has been notable even at the international level: between 1975 and 2000 the number of international NGOs has grown from fewer than 5,000 to roughly 25,000.

The newly empowered civil sector produced by this combination of historical trends led a *New York Times* reporter in 2003 to brand global public opinion a "second superpower"—a power whose activities political leaders ignore at considerable political risk.

SOURCE: See endnote 46.

---

to complete their work. Emboldened by the public protests and by polls showing that majorities opposed the war in nearly all nations surveyed on the question, the Security Council resisted U.S. pressure for an authorization of war. The Council's reluctance to give its blessing in turn energized antiwar organizers to continue their efforts.[47]

The protests differed from peace marches of the twentieth century in ways that highlight the collaborative thread that runs through today's civil society initiatives. Most obviously, the new demonstrations were coordinated by NGOs globally, although they were organized primarily at the local level. In the United States, for example, a new NGO known as United for Peace and Justice emerged to help coordinate more than 70 demonstrations across the country—and to publicize the demonstrations held in other countries. No previous cross-border peace demonstrations—neither the ones against the Viet Nam War in the 1960s nor those in opposition to nuclear weapons in the 1980s—had such extensive international coordination.[48]

In addition, the February 2003 protests were distinctive because they were embedded in a larger web of civil society activity on issues that extend well beyond war. The genesis of the protests that day, in fact, was an organizing call made at a meeting of the European Social Forum in November 2002 and seconded at the World Social Forum (WSF) in January 2003, gatherings of CSOs and other civic actors that focus primarily on social and economic issues. And some of the organizing groups for the February 15 marches were veterans of the 1999 protests that shut down the World Trade Organization meeting in Seattle. The linkages to a broader and globally active civil society movement suggests that the February 15 mobilization was not a passing moment of public pique.[49]

Indeed, there is evidence that civil society's capacity to form the networks that give birth to events like regional and global Social Forums has been developing steadily over more than a decade. The Centre for the Study of Global Governance (CSGG) in London reports that CSOs have stepped up their convening activities markedly in recent years: nearly a third of the major international meetings on peace, environment, and development issues organized by such groups since 1988 were held in just a 15-month period in 2002 and 2003. And these meetings are increasingly sophisticated. Many are large—some 55 percent had more than 10,000 participants—and are increasingly likely to be independent ventures rather than "parallel" events to official governmental meetings. Beyond offering a global communications platform, the meetings are excellent opportunities for face-to-face networking: CSOs

surveyed for the CSGG report listed networking and partnering as primary objectives for attendance.[50]

At the same time, some of the assets associated with CSO mobilizations and meetings cut two ways, suggesting a need for caution as these groups build on their successes to date. For starters, the energies of a broadly mobilized citizenry may have limited staying power and may need to be tapped sparingly. Perhaps tellingly, a call for global antiwar demonstrations in March 2004, on the first anniversary of the start of the Iraq war, produced only a fraction of the turnout of a year earlier and had little if any evident impact on the U.S. occupation of Iraq. Large-scale mobilizations may be difficult to organize with great frequency and may need to be used strategically for maximum effect. This reality will challenge civil society leaders globally to work together to determine when global mobilizations are warranted.[51]

In addition, CSO success in organizing large meetings may ironically create its own challenges. The World Social Forum has grown impressively—from 10,000 participants at the first gathering in 2001 to 100,000 or more in 2004, numbers that could easily strain the capacity for effective participation and could lead to the gatherings becoming little more than gabfests. This is a particular danger for the WSF, which was designed not to push a particular action agenda but to offer a space in which diverse views could be articulated under the rubric "another world is possible." Now WSF veterans such as Arundhati Roy are suggesting that action opportunities should become a regular part of the meetings.[52]

Finally, as public mobilizations achieve greater success, civil society will need to be alert for countermeasures that dilute its effectiveness. Citing security concerns, the city of New York, for example, went to great lengths to minimize the impact of the February 15 marches by diverting protestors from planned routes and refusing to let the demonstration pass in front of the United Nations. Similar efforts were evident 18 months later when the city rerouted demonstrations planned for the Republican Convention in the summer of 2004 and arrested thousands of demonstrators on weak legal grounds. Challenges such as these in a country with a long history of legal protections of public protest suggest that civil actors cannot take their operating space—which in many countries is newly conquered terrain—for granted.[53]

CSO networking is also facilitated through the use of new communications technologies. The International Campaign to Ban Landmines (ICBL), for example, was a coordinated effort in the 1990s of hundreds of CSOs tied together through e-mail and the Internet. The campaign conceived, drafted, and gained government support for a Treaty to Ban Landmines that by October 2004 had 143 signatories—the first time a treaty had been drafted and brought to fruition with leadership primarily from civil society. This achievement earned the ICBL the Nobel Peace Prize in 1997. The group arguably was doubly deserving of the Peace Prize: for the treaty itself, which shows real promise of eliminating one of the great scourges afflicting postwar civilian populations, and for the innovative way in which the group worked, which strengthened civil society as a force for peace.[54]

Other CSOs may be learning from the networking success of the ICBL. Research and advocacy on biological weapons, for example, were until recently spearheaded largely by pockets of specialists in the West, including small groups of academics and scientists who targeted policymakers rather than the public with information. But since 2001 a few NGOs like the Sunshine Project in

Germany and the United States have worked to broaden interest by reframing the topic to include issues CSOs are already active on, such as biodiversity and biosafety. Another group, the BioWeapons Prevention Project, has borrowed from the toolbox of grassroots activities to ramp up action on biological warfare issues. It has established networks of citizen groups in Europe, North America, and Africa, along with an annual *BioWeapons Monitor*, to help the public track compliance with the Biological Weapons Convention. Using Web pages, e-mail, and other modern communications technologies, these two groups are broadening the constituency interested in biological and chemical issues beyond scientists, beyond western industrial countries, and beyond the traditional security community.[55]

> **The International Campaign to Ban Landmines was a coordinated effort in the 1990s of hundreds of CSOs tied together through e-mail and the Internet.**

Another impressive example of the use of technology is the citizen mobilization that forced Philippine President Joseph Estrada to resign in January 2001. Alerted that his impeachment trial for corruption had been suspended indefinitely, outraged citizens used text messaging on cell phones and computers to organize a protest that drew 150,000 people to downtown Manila within two hours. Protesters kept vigil for four days in numbers large enough that the president felt compelled to step down.[56]

Such successes are possible, of course, only where the technology is available. CSOs in wealthier countries could help ensure that less prosperous organizations are as effective as possible by getting them the technologies they need. An inspiring example of such collaboration is the work of Witness, a U.S. nonprofit established in 1992 to provide camcorders, technical training, and coaching in message development to CSOs around the world. Capitalizing on the increased power and reduced price of handheld movie cameras and video editing equipment in the past two decades, Witness set out to help civil actors document abuses of people and the environment. By 2004, the group had collaborated with more than 200 CSO partners on projects in 50 countries and had scored several impressive successes, including the closing of a notorious Mexican mental health hospital following public broadcast of footage taken by a Witness-supported CSO. Witness videos are also credited with prompting the Philippine government to investigate the murder of indigenous activists who had been pursuing ancestral land claims.[57]

Civil society, government, and businesses are forming partnerships to tackle issues of common interest, including problems of peace and security.

Beyond their work with other actors in civil society, CSOs are also gaining valuable experience in collaborating with government and industry to address some of society's most intractable problems. The traditional pattern of international diplomacy, in which crossborder policy initiatives were largely undertaken by governments and international organizations (with pressure, at times, from business and occasionally from civil society) is giving way to a new dynamic. Civil society, government, and businesses are forming partnerships—often temporary and nonhierarchical in character—to tackle issues of common interest, including problems of

peace and security. These "global public policy networks" offer a seat at the policymaking table for NGOs and other civil society organizations in unprecedented ways. (See Table 9–3.)[58]

One example of the new collaboration is the Kimberley Process Certification Scheme (KPCS), a cooperative arrangement of diamond companies, governments, and CSOs that certifies that exported diamonds are not "conflict diamonds"—rough gems whose sale generated revenues that were used to fund civil conflict in Angola, Sierra Leone, Liberia, and other countries. Begun in early 2003 following a U.N. General Assembly call for diamond certification in 2000, Kimberley process certification now covers some 98 percent of the world's diamond exports. Industry, CSOs, and governments sit together on working groups that administer the scheme and monitor its functioning.[59]

How successful the Kimberley Process will be remains in question. Critics charge that diamond retailers have been slow to back the process by ensuring that diamonds are conflict-free. On the other hand, the KPCS has proved itself willing to get tough with governments, as in its July 2004 decision to evict the government of the Democratic Republic of the Congo from the organization after it was unable to document the origin of Congolese diamonds and guarantee that they were clean. The action prevents Congo from exporting diamonds to any other of the 43 KPCS members that engage in diamond trade.[60]

Collaborative NGO, government, and business networks hold great promise for addressing assorted security issues and deserve the support of governments and international institutions. U.N. promotion of this kind of partnership at the World Summit on Sustainable Development in 2002 is an example of the kind of institutional support

Table 9–3. Selected Global Public Policy Networks

Network Name	Selected Partners	Details
Roll Back Malaria	Bayer Environmental Science, CORE, UNDP, UNICEF, World Bank, WHO, governments of Ghana, India, and Italy	Launched in 1998, the goal is to halve the burden of malaria by 2010 through a coordinated international approach.
World Commission on Dams	FAO, International Energy Agency, IUCN, Transparency International, UNEP, World Bank, WHO	In 1998, the commission undertook two years of consultations and case studies on the role of big dams in development. The final report was released in 2000, and in 2001 the UNEP Dams and Development Project was created to disseminate the report's findings.
Global Water Partnership	European Union, IFPRI, Peking University, Swedish International Development Agency, UNDP, World Bank	The partnership was established after the Dublin and Rio de Janeiro conferences of 1992 to support countries in the sustainable management of their water resources.
Africa Stockpiles Programme	African Union, CropLife International, GEF, Pesticide Action Network–Africa, UNEP, WHO, WWF	This program began in 2000 as a multistakeholder effort to clean up stockpiles of obsolete pesticides in Africa, to dispose of persistant organic chemicals according to international guidelines, and to prevent future pesticide accumulation.
Global Village Energy Partnership	BP Solar, USAID, UNDP, Winrock International, World Bank	Launched in 2002 at the World Summit on Sustainable Development, this partnership aims to increase communication between energy investors, entrepreneurs, and users; to develop village energy policies; and to provide 400 million more people with access to modern energy services such as heating, cooling, and cooking.

SOURCE: See endnote 58.

these initiatives need. At the meeting in Johannesburg, more than 100 major partnerships of governments, businesses, and NGOs were established that address issues from water management to promotion of renewable energy.[61]

International institutions can also support cross-sectoral networks indirectly by working with CSOs and giving them legitimacy as potential partners for government and business. The World Bank has increasingly consulted with CSOs in its work over the past decade—it claims that some 70 percent of its projects involved collaboration with CSOs in 2002, up from 50 percent five years earlier, a promising development that raises the stature of civil society.[62]

Meanwhile, the United Nations is also currently taking steps to promote greater inclusion of NGOs. Civil society has long been active in U.N. economic and social work, particularly through major U.N. conferences and, following the Earth Summit in Rio, through the Commission on Sustainable Development. But the Security Council has traditionally been off-limits to any but official U.N. delegations. This is slowly starting to change, with the Council now allowing closed-door, off-the-record sharing of views between NGOs and official government delegates. In addition, U.N. Secretary-General Kofi Annan is considering reforms that could lead to still more dialogue between civil society and the Security Council, involve civil society groups more closely in U.N. field work, and establish a special fund to help CSOs in developing countries increase their capacity to work effectively with the United Nations.[63]

Education, the media, and religion are in strong positions to shape public understanding of how to make societies more peaceful and just.

Many issues remain to be tackled regarding the place of civil society in these policy networks. Of the diverse actors in the civil sector, which get access to a policy network, and who makes this decision? How representative are CSOs, whose leadership is seldom elected by and accountable to the public? What kinds of checks are needed to ensure that CSOs are not co-opted by their government or business partners? These and other complex issues remain to be settled as the public policy network movement matures. But the same spirit of collaboration that characterizes the operation of these networks can presumably help resolve process questions as well.

The efforts of relatively new and fluid networks—whether transient collaborations among NGOs or the more institutionalized efforts of policy networks—can themselves be buttressed by the values-shaping work of long-established centers of influence in civil society. In particular, education, the media, and religion are in strong positions to shape public understanding of global

political processes and of how to make societies more peaceful and just. Each of these institutions has a checkered history, of course, in wielding power. Schools, the media, and centers of worship are sometimes as effective at calling citizens to arms as in leading them in peacemaking.

Twentieth-century education, for example—despite all its success—has been criticized for turning out the citizens and leaders who engineered the most violent and most environmentally destructive century in human history. It is also worth noting that some of the most durable civilizations on record were led by people with no formal schooling as we know it today. Yet schools could just as conceivably be institutions that turn out "global citizens": those who understand their connectedness to the people and problems of other lands, who wrestle with fundamental questions of global justice, and who feel deeply that the natural environment is an integral part of their well-being and therefore deserves protection. Creating such an educational system is a major challenge for the twenty-first century.

Meanwhile, the world's media—television, radio, newspapers, books, music, and the Internet, among other outlets—might be thought of as a parallel education system, so widespread is its reach and so powerful its capacity to shape worldviews. A Pew Research Center poll in March 2003 found that 41 percent of Americans identified the media as the primary influence in shaping their views on the Iraq war. A media that broadens citizens' visions, that offers a diversity of perspectives on great societal issues, and that is retooled to depend far less for its sustenance on advertising would powerfully influence societal values in a direction more consistent with the needs of a globalized, environmentally and socially stressed world.[64]

Finally, religious influence over worldviews

is considerable, often operating at the deepest levels of the human psyche and expressed through ritual, scriptural teachings, and moral exhortation. Sometimes wielded violently and for repressive ends, this power has nevertheless been used impressively in constructive ways as well. Gandhi's movement for Indian independence, the U.S. civil rights struggle, the global boycott of infant formula in the 1970s, the anti-nuclear movement of the 1980s, and the campaign to restructure developing-country debt in the 1990s were all led or heavily influenced by religious people and organizations. And collaborative efforts to end conflict, such as the initiatives of Sri Lanka's Interreligious Peace Foundation—a group of Buddhists, Christians, Muslims, Hindus, and Baha'is working for peace in the island nation—offer hope that religious groups can combine their influence in the cause of peace.[65]

Drawing on the power of the world's diverse religious traditions to shape perspectives on the suite of crises facing the global community today—especially war, inequity, and environmental degradation—could profoundly affect the course of events in this new century.

Such a new focus among these three centers of influence would contribute greatly to strengthening an invigorated and empowered civil sector. It would also facilitate reforming international institutions and achieving the social, economic, and environmental visions endorsed by the Millennium Assembly and the World Summit on Sustainable Development. A globally oriented citizenry that embraced a sense of solidarity with the world's poorest and responsibility for the planet that sustains us all would likely not only support new policy initiatives, it would insist on them.

Notes

Preface

1. Politician quoted in Patrick E. Tyler, "In Wartime, Critics Question Peace Prize for Environmentalism," *New York Times*, 10 October 2004.

2. Alex Kirby, "AID Agencies' Warning on Climate," *BBC News*, 20 October 2004.

State of the World: A Year in Review

October 2003. "US Traffic Delays Cost $8 Billion in Wasted Fuel," *Reuters*, 1 October 2004; UN-Habitat, "The Challenge of Slums—Global Report on Human Settlements 2003," press release (Nairobi: 1 October 2003); "Iranian Human Rights Activist Wins Nobel Peace Prize," *UN Wire*, 10 October 2003; "ChevronTexaco on Trial in Ecuador," *Frontiers in Ecology*, December 2003.

November 2003. "Thousands of Children Serving Armed Groups, U.N. Says," *UN Wire*, 10 November 2003; Environmental Investigation Investigation Agency, "Lost in Transit: Global CFC Smuggling Trends and the Need for Faster Phase Out" (Nairobi: 10 November 2003); International Telecommunication Union, "ITU Digital Access Index: World's First Global ICT Ranking," press release (Geneva: 19 November 2003); CSIRO, "Global Halt to Major Greenhouse Gas Growth," press release (Australia: 25 November 2003); "Wild Animal Slaughter Surges for Fashion's Sake," *Reuters*, 28 November 2003.

December 2003. "Illegal Coltan Mining Threatens Endangered D.R.C. Gorillas," *Knight Ridder News Service*, 9 December 2003; "2003 Is Third-Hottest Year on Record, WMO Says," *UN Wire*, 17 December 2003; U.S. Department of Agriculture, "Case of BSE in the United States Chronology of Events," press release (Washington DC: 22 December 2003).

January 2004. Fundacion Ambiente y Recursos Naturales, "Informe Expecial sobre la Cuenca Matanza-Riachuelo," press release (Buenos Aires: 5 January 2004); Chris Thomas et al., "Extinction Risk from Climate Change," *Nature*, 8 January 2004, pp. 145–48; "Farmed Salmon Loaded with Chemicals, Study Confirms," *Reuters*, 9 January 2004; "Shell Pares Reserves A Third Time, Institutes Policy Changes," *Petroleum Finance Week*, 26 April 2004; Keith Bradsher and Lawrence K. Altman, "A War and a Mystery: Confronting Avian Flu," *New York Times*, 12 October 2004; "Polio Spreads to Seventh African Country from Nigeria," *UN Wire*, 28 January 2004.

February 2004. "Shark Species Declining Precipitously in Gulf of Mexico," *UN Wire*, 5 February 2004; European Environment Agency, "Commission and EEA Make Public Extensive Information about Industrial Pollution in Your Neighbourhood," press release (Brussels: 23 February 2004); Munich Re, *TOPICS Geo—Annual Review of Natural Catastrophes 2003* (Munich: 2004).

March 2004. "Obesity Catching Up to Smoking as No. 1 U.S. Killer," *UN Wire*, 11 March 2004; "China Faces Huge Shortages of Girls, U.N. Says," *UN Wire*, 17 March 2004; "Carbon Dioxide Levels Sky High," *Associated Press*, 20 March 2004; U.N. Environment Programme, *Global Environ-*

mental Outlook 3 (London: Earthscan, 2004); "Dust Storm Blankets Chinese Capital," *Reuters*, 30 March 2004.

April 2004. Center for International Forestry Research, "Hamburger Connection Fuels Amazon Destruction," press release (Bogor, Indonesia: 2 April 2004); "Global Energy Demand to Rise 54 Percent by 2025, U.S. Says," *UN Wire*, 15 April 2004; "Report Finds World Bank Dam Projects Harmful," *UN Wire*, 23 April 2004; Basel Action Network, *Mobile Toxic Waste: Recent Findings on the Toxicity of End-of-Life Cell Phones* (Seattle: 2004); "Patagonian Glacier Melting Fast, Scientists Warn," *UN Wire*, 29 April 2004.

May 2004. "India Bans Public Smoking, Tobacco Sales to Minors," *UN Wire*, 3 May 2004; "Convention Banning 12 Hazardous Chemicals Enters Into Force," *UN Wire*, 17 May 2004; U.N. Food and Agriculture Organization, "Locust Situation in Northwest Africa is Very Worrying," press release (Rome: 26 May 2004); "Rabbits Back to Public Enemy No. 3 as Cats and Foxes Reign," *Sydney Morning Herald*, 28 May 2004.

June 2004. "German Green Energy Conference Draws 100 Nations," *UN Wire*, 1 June 2004; World Health Organization, *The Economic Dimensions of Interpersonal Violence* (Geneva: 2004); U.N. Economic Commission for Europe, "Water and Energy in Central Asia: Preventing the Slow Death of the Aral Sea," press release (Geneva: 18 June 2004); National Center for Atmospheric Research, "NCAR Releases New Version of Premier Global Climate Model," press release (Boulder, CO: 22 June 2004); "Africa's Black Rhino Numbers Rising, Conservationists Say," *UN Wire*, 25 June 2004; "Global Treaty on Sharing Plant Genes Becomes Law," *UN Wire*, 29 June 2004.

July 2004. "Australian Reef Becomes World's Biggest Reef Sanctuary," *UN Wire*, 2 July 2004; "New Japan Car Recycling Law to Fan Competition," *Reuters*, 14 July 2004; Sea Shepherd Conservation Society, "Victory in the Galapagos," press release (Galapagos: 27 July 2004); "China Faces Summer Energy Shortage," *Environmental News Network*, 30 July 2004.

August 2004. "Danube Delta Terns Victims of Canal Construction," *Environment News Service*, 2 August 2004; "Plutonium Particles from 1954 Bikini Atoll Nuclear Test Accumulating in Japanese Bay," *Planetsave.com*, 3 August 2004; "South Asia Monsoon Death Toll Spirals Beyond 1,800," *Environmental News Network*, 5 August 2004; "Occidental Signs Controversial $50 Million Peru Oil Deal," *Reuters*, 16 August 2004.

September 2004. "60% of World Will Live in Cities by 2030," *FT.com*, 10 September 2004; WWF Arctic Programme, "More Scientific Evidence that Polar Bears are Affected by Toxic Chemicals," press release (Gland, Switzerland: 13 September 2004); World Resources Institute, *Reefs at Risk in the Caribbean* (Washington DC: 2004); "Haitian Storm Deaths Blamed on Deforestation," *Environment News Service*, 27 September 2004; "Russian Cabinet Approves Kyoto Protocol, Will Send It to Parliament for Ratification," *Environmental News Network*, 30 September 2004.

Chapter 1. Security Redefined

1. Canadian diplomat Rob McRae uses the term "age of anxiety" in "Human Security in a Globalized World," in Rob McRae and Don Hubert, eds., *Human Security and the New Diplomacy: Protecting People, Promoting Peace* (Montreal, PQ, Canada: McGill–Queen's University Press, 2001), p. 18.

2. Gallup International and World Economic Forum, "Voice of the People 2003" (Geneva: 5 January 2004); World Bank, "Fighting Poverty Key to Global Peace and Stability, Says Global Poll," press release (Washington, DC: 5 June 2003); Deepa Narayan et al., *Global Synthesis: Consultations with the Poor* (Washington, DC: World Bank Poverty Group, 1999), pp. 2, 7, 11.

3. Risks and benefits of globalization from World Commission on the Social Dimension of Globalization, *A Fair Globalization: Creating Opportunities for All* (Geneva: International Labour Office, 2004); International Labour Organization (ILO), "World Commission Says Globalization Can and Must Change, Calls for Urgent

Rethink of Global Governance," press release (Geneva: 24 February 2004).

4. Fred Halliday, "The Crisis of Universalism: America and Radical Islam after 9/11," *open-Democracy*, 16 September 2004.

5. Several independent commissions addressed these questions, including the Independent Commission on International Development Issues (1980), the Independent Commission on Disarmament and Security Issues (1982), the World Commission on Environment and Development (1987), and the Commission on Global Governance (1995). The 1994 edition of the *Human Development Report* defined seven distinct categories of human security: economic security, food security, health security, environmental security, personal security, community security, and political security; U.N. Development Programme, *Human Development Report 1994* (New York: Oxford University Press, 1994), p. 24.

6. War deaths from World Health Organization, *World Report on Violence and Health: Summary* (Geneva, 2002), p. 7; deaths from unsafe water and inadequate sanitation from "Annan Opens U.N. Expert Panel on Water, Sanitation," *UN Wire*, 23 July 2004.

7. Alyson J. K. Bailes, "Introduction: Trends and Challenges in International Security," in Stockholm International Peace Research Institute (SIPRI), *SIPRI Yearbook 2003: Armaments, Disarmament and International Security* (New York: Oxford University Press, 2003), p. 8.

8. Michael Renner, *The Anatomy of Resource Wars*, Worldwatch Paper 162 (Washington, DC: Worldwatch Institute, 2002).

9. Somini Sengupta, "Land Quarrels Unsettle Ivory Coast's Cocoa Belt," *New York Times*, 26 May 2004; Ed Stoddard, "African Conflict is Seen as Rooted in Environment," *Reuters*, 5 October 2004; Michael Renner, *Fighting for Survival* (New York: W.W. Norton & Company, 1996); Thomas Homer-Dixon and Jessica Blitt, eds., *Ecoviolence: Links Among Environment, Population, and Security* (New York: Rowman & Lit-

tlefield Publishers, 1998).

10. Sandra Postel, "Water Stress Driving Grain Trade," in Worldwatch Institute, *Vital Signs 2002* (New York: W.W. Norton & Company, 2002), p. 134.

11. World Bank, *World Development Report 2003* (New York: Oxford University Press, 2003), pp. 60–61.

12. U.N. Food and Agriculture Organization, *The State of Food Insecurity in the World 2003* (Rome: 2003), pp. 6, 10–11; "AIDS Crisis Could Fuel African Famine—U.N.," *Reuters AlertNet*, 12 October 2004.

13. Jonathan Ban, *Health, Security, and U.S. Global Leadership* (Washington, DC: Chemical and Biological Arms Control Institute, 2001), p. 14.

14. Rising impoverishment and inequality from "AIDS Causing Widening Inequality in Africa, FAO Study Shows," *UN Wire*, 1 December 2003, and from Asian Development Bank and Joint United Nations Program on HIV/AIDS (UNAIDS), *Asia Pacific's Opportunity: Investing to Avert an HIV/AIDS Crisis*, ADB/UNAIDS Study Series (Geneva: July 2004), p. 1; "AIDS Plunges Life Expectancy Below 35 in African Countries," *UN Wire*, 14 July 2004; Zambia from "AIDS Crisis Could Fuel African Famine—U.N.," op. cit. note 12; UNAIDS, UNICEF, and U.S. Agency for International Development, *Children on the Brink 2004: A Joint Report of New Orphan Estimates and a Framework for Action* (New York: 2004); Ban, op. cit. note 13, pp. 11, 13–14.

15. Toll in 2003 from United Nations, "Press Briefing on Disaster-Reduction Initiatives," press release (New York: 14 July 2004); "Flood Risk to Double by 2050, U.N. Report Says," *UN Wire*, 14 June 2004; Haiti from David Suzuki, "Peace Prize is a Well-Deserved Honor," *Environmental News Network*, 15 October 2004; "UNDP Study Shows Poor Suffer Most from Natural Disasters," *UN Wire*, 2 February 2004.

16. People at risk due to desertification from

"Creeping Desertification: The Cause and Consequence of Poverty," *Environmental News Service*, 18 June 2004; Anderson quoted in David Ljunggren, "Global Warming Bigger Threat Than Terrorism, Says Canada," *Reuters*, 6 February 2004.

17. Richard P. Cincotta, Robert Engelman, and Daniele Anastasion, *The Security Demographic: Population and Civil Conflict After the Cold War* (Washington, DC: Population Action International, 2003); 2004 ILO report from "Economic Insecurity Fosters World 'Full of Anxiety and Anger'—UN Report," *UN News Service*, 1 September 2004; youth unemployment and working poor from "Global Youth Unemployment Skyrockets to All-Time High, Action Needed—UN," *UN News Service*, 11 August 2004, and from ILO, *World Employment Report 1998–99* (Geneva: 1998).

18. "Gewalt als Ventil," *Der Spiegel Online*, 29 May 2004; Tibaijuka from "Spreading Slums May Boost Extremism—UN Agency," *Reuters AlertNet*, 13 September 2004.

19. ILO, *International Labour Conference, 92nd Session, 2004, Report of the Director-General. Appendix: The Situation of Workers of the Occupied Arab Territories* (Geneva: 2004), pp. 22–23; poverty rate from James Bennet, "In Chaos, Palestinians Struggle for a Way Out," *New York Times*, 15 July 2004; Iraq from Coalition Provisional Authority, "Administrator's Weekly Report," 14–20 February 2004; higher estimates from Institute for Policy Studies (IPS) and Foreign Policy In Focus (FPIF), *Paying the Price: The Mounting Costs of the Iraq War*, FPIF Special Report (Washington, DC: 2004), and from Ahmed Janabi, "Iraqi Unemployment Reaches 70%," *AlJazeera.net*, 1 August 2004.

20. Robert Picciotto, "Towards a Comprehensive Security and Development Framework," paper prepared for Global Policy Project Workshop on Security and Development, New Delhi, 25–26 January 2004, pp. 1, 6; Thomas L. Friedman, "Ask Not What..." (op-ed), *New York Times*, 9 December 2001; Liberia and Sierra Leone from Bryan Bender, "Former Liberian Leader Allegedly Aided Al Qaeda," *Boston Globe*, 4 August 2004, from Douglas Farah, "Al-Qaida Tied to Africa Diamonds Trade," *Washington Post*, 30 December 2002, and from Douglas Farah and Richard Shultz, "Al Qaeda's Growing Sanctuary," *Washington Post*, 14 July 2004.

21. William Reno, "Shadow States and the Political Economy of Civil Wars," in Mats Berdal and David M. Malone, eds., *Greed and Grievance: Economic Agendas in Civil Wars* (Boulder, CO: Lynne Rienner Publishers, 2000).

22. Ahmed S. Hashim, "Iraq's Chaos: Why the Insurgency Won't Go Away," *Boston Review*, October/November 2004.

23. John Tirman, "The New Humanitarianism: How Military Intervention Became the Norm," *Boston Review*, December 2003/January 2004.

24. Ahmed Rashid, *Taliban: Militant Islam, Oil and Fundamentalism in Central Asia* (New Haven: Yale University Press, 2000); John K. Cooley, *Unholy Wars: Afghanistan, America and International Terrorism* (London: Pluto Press, second edition 2000).

25. Christopher Smith, "Light Weapons and the International Arms Trade," in Christopher Smith, Peter Batchelor, and Jakkie Potgieter, *Small Arms Management and Peacekeeping in Southern Africa*, UNIDIR Disarmament and Conflict Resolution Project (New York: United Nations, 1996).

26. Sebastian Mallaby, "The Reluctant Imperialist: Terrorism, Failed States, and the Case for American Empire," *Foreign Affairs*, March/April 2002; Max Boot, "The Case for American Empire," *Weekly Standard*, 15 October 2001.

27. United Nations, "Secretary-General's Address to the General Assembly," New York, 23 September 2003; "Annan Stresses Importance of Striking Global Consensus on Major Threats," *UN News Service*, 16 March 2004.

28. All percentage data based on arms export change calculated from International Institute for Strategic Studies, *The Military Balance* (London: Oxford University Press, annual), on nuclear war-

head changes calculated from Natural Resources Defense Council (NRDC), "Table of Global Nuclear Weapons Stockpiles, 1945–2002," at www.nrdc.org/nuclear/nudb/datab19.asp, viewed 26 October 2003, and on other data calculated from Bonn International Center for Conversion (BICC), *Conversion Survey* (Baden-Baden, Germany: Nomos Verlagsgesellschaft, annual), Data Appendix. Figure 1–1 based on BICC, op. cit. this note, and on NRDC, op. cit. this note.

29. Owen Greene et al., *Implementing the Programme of Action: Action by States and Civil Society* (London: International Action Network on Small Arms, 2003); International Campaign to Ban Landmines, *Landmine Monitor Report 2003* (Washington, DC: 2003), and earlier editions.

30. Coalition for the International Criminal Court, "Signatures and Ratifications Chart," at www.iccnow.org/countryinfo/worldsigsandratifications.html, viewed 22 October 2004.

31. Patrick E. Tyler, "Threats and Responses: News Analysis; A New Power in the Streets," *New York Times*, 17 February 2003; Thomas Fues and Dirk Messner, "Die Beziehungen zwischen Nord und Süd im Schatten der Irak-Krise: Perspektiven Kooperativer Weltpolitik nach der Johannesburg-Konferenz," in Corinna Hauswedell et al., eds., *Friedensgutachten 2003* (Münster, Germany: Lit Verlag, 2003), p. 51.

32. BICC, *Conversion Survey 2004* (Baden-Baden, Germany: Nomos Verlagsgesellschaft, 2004), pp. 94–95.

33. David Albright and Kimberly Kramer, "Fissile Material: Stockpiles Still Growing," *Bulletin of the Atomic Scientists*, November/December 2004, pp. 14–16.

34. David Krieger, *US Policy and the Quest for Nuclear Disarmament* (Santa Barbara, CA: Nuclear Age Peace Foundation, 2004); "A Deeply Flawed Review," Testimony of Joseph Cirincione, Carnegie Non-Proliferation Project Director, before the Senate Foreign Relations Committee, 16 May 2002; excerpts from the Nuclear Posture Review available at www.globalsecurity.org/wmd/library/policy/dod/npr.htm.

35. Figure 1–2 from Nils Petter Gleditsch et al., "Armed Conflict 1946–2001: A New Dataset," *Journal of Peace Research*, September 2002, pp. 615–37, and from International Peace Research Institute, Oslo (PRIO), at www.prio.no/cwp/armedconflict; emergence of private military companies from P. W. Singer, *Corporate Warriors: The Rise of the Privatized Military Industry* (Ithaca, NY: Cornell University Press, 2003); privatization of violence from Michael T. Klare, "The Global Trade in Light Weapons and the International System in the Post-Cold War Era," in Jeffrey Boutwell, Michael T. Klare, and Laura W. Reed, eds., *Lethal Commerce: The Global Trade in Small Arms and Light Weapons* (Cambridge, MA: American Academy of Arts and Sciences, 1995).

36. Peacekeeping trends from U.N. Department of Public Information, "United Nations Peacekeeping Operations: Background Note," New York, 1 August 2004, and earlier editions.

37. Number of vetoes and hidden veto use from Céline Nahory, "The Hidden Veto," Global Policy Forum, New York, 19 May 2004, and from Global Policy Forum, "Subjects of UN Security Council Vetoes," at www.globalpolicy.org/security/membship/veto/vetosubj.htm, viewed 21 September 2004.

38. Nahory, op. cit. note 37; Morton Abramowitz and Samantha Power, "A Broken System" (op-ed), *Washington Post*, 13 September 2004.

39. "Honeymoon" reference from Bailes, op. cit. note 7, p. 12.

40. Ibid., p. 14; Jeffrey Record, *Bounding the Global War on Terrorism*, Strategic Studies Institute (Carlisle, PA: U.S. Army War College, 2003), p. 2; John V. Whitbeck, "'Terrorism': A World Ensnared by a Word," *International Herald Tribune*, 18 February 2004.

41. European Union Institute for Security Studies, *A Secure Europe in a Better World: European Security Strategy* (Paris: 2004); critique of European strategy document from "Summing Up Dis-

armament and Conversion Events: Getting Priorities Refocused," in BICC, op. cit. note 32, p. 17, and from Corinna Hauswedell and Herbert Wulf, "Die EU als Friedensmacht? Neue Sicherheitsstrategie und Rüstungskontrolle," in Christoph Weller et al., *Friedensgutachten 2004* (Münster, Germany: LIT Verlag, 2004), pp. 122–30.

42. Whitbeck, op. cit. note 40; Jason Burke, "It's Too Easy to Blame Bin Laden," *The Observer* (London), 5 September 2004; Human Rights Watch, "Opportunism in the Face of Tragedy: Repression in the Name of Anti-Terrorism," www.hrw.org/campaigns/september11/opportunismwatch.htm, viewed 20 July 2004; idem, "U.S. Torture and Abuse of Detainees," at www.hrw.org/campaigns/torture.htm, viewed 14 July 2004; arms supplies from Debbie Hillier and Brian Wood, *Shattered Lives: The Case for Tough International Arms Control* (London and Oxford: Amnesty International and Oxfam International, 2003), pp. 41–42, and from Tamar Gabelnick, "Security Assistance After September 11," *Foreign Policy In Focus*, May 2002; "Amnesty Says U.S.-Led War on Terror Undermines Human Rights," *UN Wire*, 26 May 2004.

43. Nicole Deller, Arjun Makhijani, and John Burroughs, eds., *Rule of Power or Rule of Law? An Assessment of U.S. Policies and Actions Regarding Security-Related Treaties* (New York: Apex Press, 2003); Patricia Jurewicz and Kristin Dawkins, *The Treaty Database: A Monitor of U.S. Participation In Global Affairs* (Minneapolis, MN: Institute for Agriculture and Trade Policy, 2004); Dafna Linzer, "U.S. Shifts Stance on Nuclear Treaty," *Washington Post*, 31 July 2004.

44. White House, *The National Security Strategy of the United States* (Washington, DC: September 2002), p. 15; "Russia Plans Preemptive Anti-Terror Strikes," *The Russia Journal Daily*, 8 September 2004; Jonathan Clarke, "An Ominous U.S. Model" (op-ed), *Los Angeles Times*, 22 September 2004; Steve Weissman, "How Soon Will the U.S. or Israel Bomb Iran?" (editorial), *Truthout.org*, 2 September 2004.

45. Record, op. cit. note 40.

46. James Glanz, "Iraqis Warn That U.S. Plan to Divert Billions to Security Could Cut Off Crucial Services," *New York Times*, 21 September 2004; Paul Richter, "Costs Whittle Funds to Iraqis," *Los Angeles Times*, 26 September 2004; National Intelligence Estimate from Douglas Jehl, "U.S. Intelligence Shows Pessimism on Iraq's Future," *New York Times*, 16 September 2004; sources of Iraqi fragmentation from Chatham House, *Iraq in Transition: Vortex or Catalyst?* Briefing Paper (London: Royal Institute for International Studies, 2004).

47. Anonymous, *Imperial Hubris: Why the West Is Losing the War on Terror* (Washington, DC: Brassey's, Inc., 2004); Jim Lobe, "U.S.: Three Years On, War on Terrorism Looks Like a Loser," *Inter Press Service*, 11 September 2004; Institute for International Security Studies findings from Roger Hardy, "Al-Qaeda 'Spurred on' by Iraq War," *BBC News Online*, 25 May 2004, and from Kim Sengupta, "Occupation Made World Less Safe, Pro-War Institute Says," *The Independent* (London), 26 May 2004; U.S. Department of State, *Patterns of Global Terrorism 2003* (Washington, DC: corrected version, June 2004).

48. Hans Greimel, "Iraq War Diverts Key Aid from Devastated Afghanistan, Minister Says," *Associated Press*, 31 March 2004; attention diverted from Afghan demobilization efforts from BICC, op. cit. note 32, p. 73; per capita data from Jeff Madrick, "Skimpy Aid to Afghanistan," *New York Times*, 13 May 2004.

49. World military expenditure from SIPRI, "Recent Trends in Military Expenditure," at web.sipri.org/contents/milap/milex/mex_trends.html, viewed 1 October 2004.

50. These are principally United Nations–generated estimates, as collected by the World Game Institute, "Global Priorities," wall chart, 2000, available at www.worldgame.org, and by Phyllis Bennis et al., *A Failed "Transition": The Mounting Costs of the Iraq War* (Washington, DC: IPS and FPIF, 2004), p. 54; malaria spending from "War on Terror Drains Funds from Poverty Fight," *Agence France Presse*, 2 December 2003; Iraq costs from Bennis et al., op. cit. this note, p. 15

(based on congressional appropriations of $151 billion in 2003 and 2004, plus a widely expected request for an additional $60 billion after the November 2004 U.S. elections).

51. Development aid, converted to 2002 dollars, from Development Co-operation Directorate, Organisation for Economic Co-operation and Development, Paris; United Nations General Assembly, "International Financial System and Development: Report of the Secretary-General," New York, 16 September 2003; military spending surpassing health and education budgets from Elisabeth Sköns et al., "Military Expenditure," in SIPRI, *SIPRI Yearbook 2002: Armaments, Disarmament and International Security* (New York: Oxford University Press, 2002), p. 238.

52. Risk of missing Millennium Development Goals from "World Is Lagging in Fight to Eliminate Poverty and Hunger, Annan Warns," *UN News Service*, 23 September 2004, and from "More Needs to be Done Quickly to Achieve Millennium Goals, UN Official Says," *UN News Service*, 1 October 2004.

53. "U.N. Report Highlights Growing Poverty in Sub-Saharan Africa," *UN Wire*, 20 July 2004; quote from "Panel Should Help Refashion Global Notion of Collective Security—Annan," *UN News Service*, 29 September 2004.

54. Anatol Lieven, "The Bush Blip?" *Prospect*, August 2004.

55. Donors tend to spend $100 in post-disaster relief for every $1 in pre-disaster preparedness, for instance; United Nations University, "Two Billion People Vulnerable to Floods by 2050," news release (Tokyo: 13 June 2004).

56. Robert Picciotto, "Introduction to Global Policy Project Workshop on Security and Development," paper prepared for Global Policy Project Workshop on Security and Development, New Delhi, 25–26 January 2004, p. 2.

57. "Summing Up Disarmament and Conversion Events," op. cit. note 41, p. 14; Picciotto, op. cit. note 20, p. 4; "Synthesis of Workshop Delibera-

tions," Global Policy Project Workshop on Security and Development, New Delhi, 25–26 January 2004, p. 11.

58. Picciotto, op. cit. note 20, p. 4.

59. "10 Years After Rwanda Genocide, Annan Unveils Plan to Stop Future Massacres," *UN News Service*, 7 April 2004.

60. Tirman, op. cit. note 23; International Commission on Intervention and State Sovereignty, *The Responsibility to Protect* (Ottawa, ON, Canada: December 2001).

61. George Monbiot, "A Charter to Intervene: Human Rights Interventions Can Only Be Divorced from Imperialism with New UN Rules," *Guardian* (London), 23 March 2004.

62. Lee Feinstein and Anne-Marie Slaughter, "A Duty to Prevent," *Foreign Affairs*, January/February 2004.

63. Hauswedell and Wulf, op. cit. note 41, pp. 125–26; "Proliferation Security Initiative," at www.globalsecurity.org/military/ops/psi.htm, viewed 29 September 2004.

64. Tirman, op. cit. note 23.

Transnational Crime

1. Jerome C. Glenn and Theodore J. Gordon, *State of the Future 2004* (Washington, DC: American Council for the United Nations University, 2004), p. 34; World Bank, *World Development Report 2004* (New York: Oxford University Press, 2004).

2. Brian Halweil, "Harvesting of Illegal Drugs Remains High," in Worldwatch Institute, *Vital Signs 2003* (New York: W.W. Norton & Company, 2003), p. 98; death toll from World Health Organization, *The World Health Report 2002* (Geneva: 2002), p. 226; HIV from U.N. Office on Drugs and Crime (UNODC), *World Drug Report 2004* (New York: United Nations, 2004), pp. 47–51; illicit drug sales from Frank Cilluffo, "The Threat Posed from the Convergence of Orga-

nized Crime, Drug Trafficking, and Terrorism," Testimony Before the U.S. House Committee on the Judiciary Subcommittee on Crime, Washington, DC, 13 December 2000.

3. U.S. National Security Council, *International Crime Threat Assessment* (Washington, DC: 2000).

4. Patrick Tracey, "UNEP Launches Initiative to Combat Illegal Trade in Toxics, Ozone Depleters," *International Environment Reporter*, 4 June 2003; Ezra Clark, *Lost In Transit: Global CFC Smuggling Trends and the Need for a Faster Phase-Out* (London: Environmental Investigation Agency, 2003).

5. Sales of $6–10 billion from U.S. National Security Council, op. cit. note 3; 350 million specimens and convention from "Trade in Lions, Eagles, Turtles, Elephants Under Scrutiny," *Environmental News Service*, 14 May 2004.

6. U.S. Department of State, *Trafficking in Persons Report 2004* (Washington, DC: 2004).

7. Revenue less than $1 billion from Graduate Institute of International Studies, *Small Arms Survey 2003* (New York: Oxford University Press: 2003), p. 98; deaths from idem, *Small Arms Survey 2004* (New York: Oxford University Press: 2004), pp. 175–77; Jim Wurst, "Black Market Small Arms Readily Available Despite Global Efforts," *UNWire*, 30 June 2004; no convictions from Peter Landesman, "Arms and the Man," *New York Times Magazine*, 17 August 2003, p. 33.

8. UNODC, "The United Nations Convention Against Transnational Organized Crime and Its Protocols," at www.unodc.org/unodc/en/crime_cicp_convention.html, viewed 10 October 2004.

9. Ibid.; Wurst, op. cit. note 7.

10. Convention on International Trade in Endangered Species of Wild Fauna and Flora Secretariat, *2002 Annual Report* (Geneva: 2002).

11. UNODC, "Sweden Introduces New Anti-Drug Strategy to Strengthen Its Partnership with the United Nations Office on Drugs and Crime," press release (Vienna: 15 March 2004).

Chapter 2. Examining the Connections Between Population and Security

1. Daniel C. Esty et al., "State Failure Task Force Report: Phase II Findings," *Environmental Change and Security Project, Report* 5 (Washington, DC: Woodrow Wilson Center for International Scholars, Environmental Change and Security Project, 1999), pp. 49–72; Chester A. Crocker, "Engaging Failing States," *Foreign Affairs*, September/October 2003, pp. 32–44; U.S. Department of State, *Patterns of Global Terrorism* (Washington, DC: 2000).

2. Though the group found no direct connection between infant deaths and political crisis, infant mortality was a good proxy for quality of material life, per Esty et al., op. cit. note 1.

3. Early studies and reports on the nexus of population and armed conflict include Herbert Moller, "Youth as a Force in the Modern World," *Comparative Studies in Society and History*, vol. 10 (1967/68), pp. 237–60; Nazli Choucri, *Population Dynamics and International Violence: Propositions, Insights and Evidence* (Cambridge, MA: The MIT Press, 1973); idem, *Population Dynamics and International Violence* (Lexington, MA: Lexington Press, 1974); idem, ed., *Multidisciplinary Perspectives on Population and Conflict* (Syracuse, NY: Syracuse University Press, 1984); National Security Council, *Implications of Worldwide Population Growth for U.S. Security and Overseas Interests*, National Security Study Memorandum 200 (Washington, DC: 1974); M. Green and Robert Fearey, "World Population: The Silent Explosion," *Department of State Bulletin*, fall 1978, pp. 1–32; and Population Crisis Committee (PCC), "World Population Growth and Global Security," *Population*, September 1983, pp. 1–8. More recent studies include P. G. Barnett, "Population Pressures—Threat to Democracy," wall chart (Washington, DC: PCC, 1989); Brian Nichiporuk, *The Security Dynamics of Demographic Factors* (Santa Monica, CA: RAND, 2000); Thomas Homer-Dixon and Valerie Percival, "Environmental Scarcity and Violent Conflict: Briefing Book," in Brian Smith, ed., *Project on Environment, Population and Security* (Washington, DC: American Association for the Advancement of Science, 1996); and International Crisis Group, *HIV/AIDS as a Security Issue*, Issue Report (Brussels: 19 June 2001).

4. U.N. Population Division, *World Population Prospects: 1990* (New York: United Nations, 1991); idem, *World Population Prospects: The 2002 Revision* (New York: United Nations, 2003).

5. U.N. Population Division, *The 2002 Revision*, op. cit. note 4.

6. Population Action International (PAI) found that, on average, a decline in the annual birth rate of five births per thousand people corresponded to a decline of about 5 percent in the likelihood of conflict during the 1970s, 1980s, and 1990s, per Richard P. Cincotta, Robert Engelman, and Daniele Anastasion, *The Security Demographic: Population and Civil Conflict After the Cold War* (Washington, DC: PAI, 2003), p. 12.

7. "Thai Forces Overreacted in Mosque Attack—Thaksin," *Reuters*, 31 July 2004.

8. "Muslim Ideology Fuels Thailand's Southern Violence," *Associated Press*, 4 May 2004; "Thai Leader Tours Scarred South," at CNN.com, 8 May 2004.

9. Youth bulge countries and Table 2–1 based on projections from U.N. Population Division, *The 2002 Revision*, op. cit. note 4, as cited in Cincotta, Engelman, and Anastasion, op. cit. note 6, pp. 96–101.

10. U.N. Population Division, *The 2002 Revision*, op. cit. note 4.

11. John G. Bauer, "Demographic Change, Development, and the Economic Status of Women in East Asia," in A. Mason, ed., *Population Change and Economic Development in East Asia: Challenges Met, Opportunities Seized* (Stanford, CA: Stanford University Press, 2001), pp. 359–84.

12. International Labour Organization (ILO), *Global Employment Trends for Youth 2004* (Geneva: 2004); idem, "Global Unemployment Remains at Record Levels in 2003 But Annual ILO Jobs Report Sees Signs of Recovery," press release (Geneva: 22 January 2004).

13. Leif Ohlsson, *Livelihood Conflicts—Linking Poverty and Environment as Causes of Conflict* (Stockholm: Swedish International Development Cooperation, Environmental Policy Unit, 2000); Chris Dolan, "Collapsing Masculinities and Weak States—A Case Study of Northern Uganda," in Frances Cleaver, ed., in *Masculinities Matter!* (London: Zed Books, 2003), pp. 57–83.

14. UNICEF, "Armed Conflict," at www.unicef .org/protection/index_armedconflict.html, viewed 27 July 2004; U.N. Development Programme (UNDP), *Human Development Report 2004* (New York: 2004); official cited in Women's Commission for Refugee Women and Children, *Precious Resources: Adolescents in the Reconstruction of Sierra Leone* (New York: 2002).

15. William Ochieng, "Looming Dilemma for Africa," *The Nation* (Kenya), 27 January 2002; Jack A. Goldstone, *Revolution and Rebellion in the Early Modern World* (Berkeley: University of California Press, 1991); idem, "Population and Security: How Demographic Change Can Lead to Violent Conflict," *Journal of International Affairs*, fall 2002, pp. 3–22.

16. Goldstone, "Population and Security," op. cit. note 15, p. 15; Samuel Huntington, "So, Are Civilisations At War?" interview with Michael Steinberger, *The Observer* (London), 21 October 2001.

17. "Muslim Ideology Fuels Thailand's Southern Violence," op. cit. note 8; P. W. Singer, *Pakistan's Madrassahs: Ensuring a System of Education Not Jihad*, Analysis Paper 14 (Washington, DC: Brookings Institution, 2001).

18. Christian G. Mesquida and Neil I. Wiener, "Human Collective Aggression: A Behavioral Ecology Perspective," *Ethology and Sociobiology*, vol. 17 (1996), pp. 247–62; idem, "Male Age Composition and the Severity of Conflicts," *Politics in the Life Sciences*, vol. 18, no. 2 (1999), pp. 181–89; Cincotta, Engelman, and Anastasion, op. cit. note 6, p. 48. The PAI study omitted countries where civil war persists, such as in Afghanistan, India (Kashmir), and Colombia, as in these cases it can be virtually impossible to untangle the relationships between social conditions, economic trends, and conflict.

19. Cincotta, Engelman, and Anastasion, op. cit. note 6, pp. 48–49.

20. U.N. Population Division, *The 2002 Revision*, op. cit. note 4. Box 2–1 from the following: Wolfgang Lutz, Brian C. O'Neill, and Sergei Scherbov, "Europe's Population at a Turning Point," *Science*, 28 March 2003, pp. 1991–92; Paul Demeny, "Population Policy Dilemmas in Europe at the Dawn of the Twenty-first Century," *Population and Development Review*, March 2003, pp. 1–28; workers-per-retiree and U.S. funding for seniors from David J. Rothkopf, "The Coming Battle of the Ages," *Washington Post*, 1 February 2004; military impact from Nichiporuk, op. cit. note 3, p. 27; Naohiro Ogawa and Robert D. Retherford, "Shifting Costs of Caring for the Elderly Back to Families In Japan: Will It Work?" *Population and Development Review*, March 1997, pp. 59–94; John G. Bauer, *How Japan and the Newly Industrialized Economies of Asia Are Responding to Labor Scarcity*, Asia-Pacific Research Report 3 (Honolulu: East-West Center, 1995).

21. UNAIDS, *2004 Report on the Global AIDS Epidemic* (Geneva: 2004); Dean T. Jamison et al., *The Effect of the AIDS Epidemic on Economic Welfare in Sub-Saharan Africa* (Cambridge, MA: Commission on Macroeconomics and Health, 2001).

22. UNAIDS, op. cit. note 21; death rates in industrial and war-torn countries from Cincotta, Engelman, and Anastasion, op. cit. note 6; sub-Saharan Africa and Table 2–2 derived from U.N. Population Division, op. cit. note 4, and from UNAIDS, *2002 Report on the Global AIDS Epidemic* (Geneva: 2002), as cited in Cincotta, Engelman, and Anastasion, op. cit. note 6, pp. 96–101.

23. Deaths of 74 million from ILO, *HIV/AIDS and Work: Global Estimates, Impact and Response* (Geneva: 2004); Peter Piot, "Global AIDS Epidemic: Time to Turn the Tide," *Science*, 23 June 2000, p. 2176.

24. ILO, op. cit. note 23, p. 20; Robert Greener, "AIDS and Macroeconomic Impact," in Steven Fortsythe, ed., *State of the Art: AIDS and Economics* (Washington, DC: International AIDS-Economics Network, 2002), pp. 49–54.

25. UNAIDS defines an AIDS orphan as someone under 15 years of age who has lost his or her mother or both parents from an AIDS-related cause, per UNAIDS, *Children and Young People in a World of AIDS* (Geneva: 2001); UNICEF, UNAIDS, and U.S. Agency for International Development, *Children on the Brink 2004: A Joint Report of New Orphan Estimates and a Framework for Action* (New York: 2004); UNICEF, "Armed Conflict," at www.unicef.org/protection/index_armedconflict.html, viewed 27 July 2004.

26. Figure of 53 includes 49 countries with HIV prevalence at 1 percent or greater, plus Russia, United States, India, and China, per U.N. Population Division, *The 2002 Revision*, op. cit. note 4; UNAIDS, op. cit. note 21, p. 23; UNDP, "HIV/AIDS Crisis Drives Down Life Expectancy, Human Development Rankings in Sub-Saharan Africa," press release (Bangkok: 14 July 2004); Greener, op. cit. note 24.

27. Colin Powell, speech at Public-Private Partnerships in the Global Fight Against HIV/AIDS conference, U.S. Department of State, Washington, DC, 24 June 2002; William J. Clinton, speech at XIV International AIDS Conference, Barcelona, Spain, 12 July 2002.

28. "Thousands Displaced in Malawi Floods," *BBC News Online*, 3 February 2002, at news.bbc.co.uk/1/hi/world/africa/1798713.stm; Sara Randall, "The Consequences of Drought for the Populations in the Malian Gourma," in John I. Clarke and Daniel Noin, eds., *Population and Environment in Arid Regions* (Paris: UNESCO, Parthenon, 1998), pp. 149–75; "Malawi Is Fastest Urbanizing Country in the World, UN Says,"

press release (New York: U.N. News Centre, 8 July 2004); Brian Wood, *Malawi Security Sector Reform Pilot Project Report: Sept 1999–Aug 2000* (Oslo: Malawi Centre for Human Rights and Rehabilitation and the International Peace Research Institute in cooperation with the Norwegian Initiative on Small Arms Transfers, 2000).

29. Even though fertility rates are consistently lower in cities than in the rural areas, urban population growth rates in most countries are about 50 percent faster than overall national population growth rates, per U.N. Population Division, *World Urbanization Prospects: The 2003 Revision* (New York: United Nations, 2004).

30. Mark Brockerhoff, "Migration and the Fertility Transition in African Cities," in Richard E. Bilsborrow, ed., *Migration, Urbanization, and Development: New Directions and Issues* (Norwell, MA: Kluwer Academic Publishers, 1996).

31. Jakarta and Delhi from U.N. Population Division, op. cit. note 29; fresh water and sanitation from UN-HABITAT, *Water and Sanitation in the World's Cities* (Nairobi: 2003).

32. Ellen Brennan-Galvin, "Crime and Violence in an Urbanizing World," in *Journal of International Affairs*, fall 2002, pp. 123–45; Marc Sommers, *Urbanization, War, and Africa's Youth at Risk* (Washington, DC: CARE, Inc. and Creative Associates International, Inc., 2003), p. 7.

33. Raj Chengappa and Ramesh Menon, "The New Battlefields," *India Today*, 31 January 1993, p. 28; Peter Gizewski and Thomas Homer-Dixon, *Urban Growth and Violence: Will the Future Resemble the Past?* Project on Environment, Population, and Security (Washington, DC: American Association for the Advancement of Science (AAAS), 1995); Human Rights Watch, *We Have No Orders to Save You: State Participation and Complicity in Communal Violence in Gujarat* (New York: 2002).

34. Cincotta, Engelman, and Anastasion, op. cit. note 6, p. 55; Carole Rakodi, "Global Forces, Urban Change, and Urban Management in Africa," in idem, ed., *The Urban Challenge in Africa: Growth and Management of its Large Cities* (New York: United Nations University Press, 1997); Nichiporuk, op. cit. note 3; Gary Fuller and Forrest R. Pitts, "Youth Cohorts and Political Unrest in South Korea," *Political Geography Quarterly*, January 1990, pp. 9–22; Peter Gizewski and Thomas Homer-Dixon, *Environmental Scarcity and Violent Conflict: The Case of Pakistan*, Project on Environment, Population, and Security (Washington, DC: AAAS, 1996).

35. Kempe R. Hope, "Urbanization and Urban Growth in Africa," *Journal of Asian and African Studies*, vol. 33, no. 3 (1998), pp. 345–58; Mark Brockerhoff and Ellen Brennan, "The Poverty of Cities in the Developing World," *Population and Development Review*, March 1998, pp. 75–114; U.N. Population Division, op. cit. note 29.

36. Urbanization appears to be closely linked with a country's overall population growth: in both developing and industrial countries, when population growth rates have slowed, urban growth rates have also declined, per U.N. Population Division, *World Urbanization Prospects: The 2001 Revision, Data Tables and Highlights* (New York: United Nations, 2002).

37. Arun P. Elhance, *Hydropolitics in the Third World* (Washington, DC: United States Institute for Peace Press, 1999), pp. 123–53; Douglas Jehl, "In Race to Tap the Euphrates, The Upper Hand Is Upstream," *New York Times*, 25 August 2002.

38. Robert Engelman and Pamela LeRoy, *Sustaining Water: Population and the Future of Renewable Water Supplies* (Washington, DC: PAI, 1993); idem, *Conserving Land: Population and Sustainable Food Production* (Washington, DC: PAI, 1995). Table 2–3 is from Cincotta, Engelman, and Anastasion, op. cit. note 6, pp. 96–101.

39. U.N. Environment Programme, *United Nations Environment Programme Water Strategy* (Geneva: March 2000); Sandra Postel and Aaron T. Wolf, "Dehydrating Conflict," *Foreign Policy*, September/October 2001, pp. 60–67; Aaron T. Wolf, "Conflict and Cooperation Along International Waterways," *Water Policy*, April 1998, pp. 251–65.

40. Postel and Wolf, op. cit. note 39; Ken Conca and Geoffrey D. Dabelko, "Problems and Possibilities of Environmental Peacemaking," in idem, eds., *Environmental Peacemaking* (Washington, DC: Woodrow Wilson International Center for Scholars, 2002), pp. 220–33; Thomas F. Homer-Dixon and Jessica Blitt, eds., *Ecoviolence: Links Among Environment, Population and Security* (New York: Rowman & Littlefield, 1998); Thomas F. Homer-Dixon, *The Environment, Scarcity and Violence* (Princeton: Princeton University Press, 1999).

41. Mark W. Rosegrant, Ximing Cai, and Sarah A. Cline, *World Water and Food to 2025: Dealing with Scarcity* (Washington, DC, and Battaramulla, Sri Lanka: International Food Policy Research Institute and International Water Management Institute, 2002).

42. Miriam R. Lowi, "Water and Conflict in the Middle East and South Asia: Are Environmental Issues and Security Issues Linked?" *Journal of Environment and Development*, December 1999, pp. 376–96; Homer-Dixon, op. cit. note 40; Kimberley Kelly and Thomas Homer-Dixon, *Environmental Scarcity and Violent Conflict: The Case of Gaza*, Project on Environment, Population, and Security (Washington, DC: AAAS, 1996).

43. Cincotta, Engelman, and Anastasion, op. cit. note 6, p. 60.

44. James D. Nations, "The Ecology of the Zapatista Revolt," *Cultural Survival Quarterly*, vol. 18.1 (1994), pp. 31–33; Philip Howard and Thomas Homer-Dixon, *Environmental Scarcity and Violent Conflict: The Case of Chiapas, Mexico*, Project on Environment, Population and Security (Washington, DC: AAAS, 1995).

45. Cincotta, Engelman, and Anastasion, op. cit. note 6, p. 60; Esty et al., op. cit. note 1; Science Applications International Corporation (SAIC), *Final Report of the State Failure Task Force* (Washington, DC: SAIC and U.S. Agency for International Development, 1995); Jack A. Goldstone, "Demography, Environment, and Security: An Overview," in Myron Weiner and Sharon Stanton Russell, eds., *Demography and National Security*

(New York: Berghahn Books, 2001), pp. 38–61.

46. Rosegrant, Cai, and Cline, op. cit. note 41. Box 2–2 from the following: Republic of Rwanda, *Draft Land Policy* (Kigali: MINITERE, 2004); Moses Kazoora, "Ombudsman Meets District Officials," *The New Times* (Kigali, Rwanda), 16–18 February 2004; authors' interviews in Kigali City and Butare Province, February 2004; Catherine Andrea, "Custom, Contracts and Cadastres in North-West Rwanda," in Tor A. Benjaminsen and Christian Lund, eds., *Securing Land Rights in Africa* (London: Frank Cass, 2004); Nelson Marongwe and Robin Palmer, "Struggling With Land Reform Issues In Eastern Africa Today," *Independent Land Newsletter–Eastern Africa*, August 2004.

47. PAI identified a total of 66 countries displaying a combination of two or more of the three factors, per Cincotta, Engelman, and Anastasion, op. cit. note 6, p. 73. HIV/AIDS was omitted from the analysis because in the mid-1990s the pandemic had not yet progressed to the alarming rates of prevalence and premature death now experienced in southern African countries.

48. Cincotta, Engelman, and Anastastion, op. cit. note 6, p. 30.

49. U.N. Economic and Social Council, *World Youth Report 2003* (New York: UN, 2003); Cincotta, Engelman, and Anastastion, op. cit. note 6, p. 30.

50. U.N. Population Division, *World Population Prospects: 1990*, op. cit. note 4; idem, *The 2002 Revision*, op. cit. note 4; Shanti R. Conly and Shyami de Silva, *Paying Their Fair Share? Donor Countries and International Population Assistance* (Washington, DC: PAI, 1998).

51. J. D. Shelton and B. Johnston, "Condom Gap in Africa: Evidence from Donor Agencies and Key Informants," *British Medical Journal*, 21 July 2001, p. 139; U.N. Population Fund, *Donor Support for Contraceptives and Logistics, 2000* (New York: 2001).

52. Harry Cross et al., *Completing the Demo-*

graphic Transition in Developing Countries, Policy Occasional Paper 8 (Washington, DC: The Futures Group, 2002); U.S. share calculation from S. Radloff, U.S. Agency for International Development, Population, Heath and Nutrition Center, personal communication with PAI, June 2001; Caroline Preston, "U.S. Blocks Funding to UNFPA for Third Consecutive Year," *U.N. Wire*, 19 July 2004.

53. April 2002 response cited in "CIA Concerned US War on Terror is Missing Root Causes," *Agence France Presse*, 29 October 2002.

54. Amy L. Chua, *World on Fire: How Exporting Free Market Democracy Breeds Hatred and Global Instability* (New York: Doubleday, 2002); Edward D. Mansfield and Jack Snyder, "Democratization and the Danger of War," *International Security*, summer 1995, pp. 5–38.

55. Cincotta, Engelman, and Anastasion, op. cit. note 6, pp. 15–16.

56. Ibid.

57. See U.K. Department for International Development, "Conflict Reduction and Humanitarian Assistance," at www2.dfid.gov.uk/about dfid/organisation/conflicthumanitarianassis tance.aspDFID; Seyoum cited in International Alert, *Women, Violent Conflict and Peacebuilding: Global Perspectives*, Proceedings from International Conference, London, 5–7 May 1999.

58. The chances for successful peace negotiations appear to have improved in Northern Ireland and Sri Lanka, in part due to the effects of transitions in age structure and the growth of the labor force; see Cincotta, Engelman, and Anastasion, op. cit. note 6, p. 12.

59. Paul Collier and Anke Hoeffler, *Greed and Grievance in Civil War* (Washington, DC: World Bank, 2001); on the national and international conditions facilitating conflict mediation, see Ted Robert Gurr, Monty G. Marshall, and Deepa Khosla, *Peace and Conflict 2001: A Global Survey of Armed Conflicts, Self-Determination Movements, and Democracy* (College Park, MD: Center for

International Development and Conflict Management, University of Maryland, 2001), Carnegie Commission on Preventing Deadly Conflict, *Preventing Deadly Conflict* (New York: Carnegie Corporation of New York, 1997), and Ted Robert Gurr, "The Challenge of Resolving Ethnonational Conflicts," in idem, ed., *Peoples Versus States: Minorities at Risk in the New Century* (Washington, DC: United States Institute of Peace, 2000), pp. 195–211; on the role of democratization, see Fareed Zakaria, *The Future of Freedom* (New York: W.W. Norton & Company, 2003), and Mansfield and Snyder, op. cit. note 54; on the role of economic development and poverty alleviation, see Paul Collier, *Breaking the Conflict Trap: Civil War and Development Policy* (Washington, DC: World Bank and Oxford University Press, 2003); on openness to trade and other facets of globalization, see Niall Ferguson, *The Pity of War: Explaining World War I* (New York: Basic Books, 1999), and T. P. M. Barnett, "The Pentagon's New Map," *Esquire*, March 2003, pp. 174–79.

Environmental Refugees

1. U.N. High Commissioner for Refugees (UNHCR), "World Refugee Population Lowest in a Decade, UNHCR Says," *UNHCR Release*, 17 June 2004. A refugee is legally defined as a person fleeing his or her country because of persecution or "owing to the well-founded fear of being persecuted for reasons of race, religion, nationality, membership of a particular social group or political opinion, is outside of the country of his nationality and is unable, or owing to such fear, is unwilling to avail himself of the protection of that country" (*Text of the 1951 United Nations Convention Relating to the Status of Refugees*, Chapter 1, Article 1 (A) 2). The 1967 protocol incorporates the measures included in the 1951 convention, but imposes no time or geographical limits. Internally displaced persons do not have the same protection as refugees. According to experts, many other individuals seeking refuge are not counted by UNHCR because they fall outside its mandate.

2. Many experts attribute the term "environmental refugee" to Essam El-Hinnawi from a report he prepared for the U.N. Environment Programme in 1985, although others attribute it

to earlier sources. Gaim Kbreab, "Environmental Causes and Impact of Refugee Movements: A Critique of the Current Debate," *Disasters*, vol. 21, no. 1 (1997), pp. 20–38; El-Hinnawi cited in Stefania Milan, "Refugee Day: Searching for a Place Under the Sun," *IPS News*, 19 June 2004.

3.   International Federation of the Red Cross and Red Crescent Societies, *World Disasters Report 2002*, at www.ifrc.org/publicat/wdr2002/chapter8.asp; Lester R. Brown, "Troubling New Flows of Environmental Refugees," *Eco-Economy Update* (Washington, DC: Earth Policy Institute, 28 January 2004); Intergovernmental Panel on Climate Change cited in Milan, op. cit. note 2, and in Norman Myers, "Total Environment Refugees Forseen," *BioScience*, December 1993; "Nature: World Day to Combat Desertification on June 17," *Europe Information Service–Europe Environment*, 10 June 2004.

4.   On occasion, UNHCR will contribute to the overall U.N. aid response to a natural disaster; see UNHCR, *Refugees*, July 2002.

5.   Dana Zartner Falstrom, "2001 Yearbook: Perspective: Stemming the Flow of Environmental Displacement: Creating a Convention to Protect Persons and Preserve the Environment," *Colorado Journal of International Environmental Law and Policy*, 2001; Select Committee on International Development, U.K. Parliament, "Memorandum Submitted by the New Economics Foundation" (London: 8 July 2004).

6.   See www.unhdr.ch/environ/enviro.htm; ratio of expenditures from Janet Abramovitz, *Unnatural Disasters*, Worldwatch Paper 158 (Washington, DC: Worldwatch Institute, 2001), p. 7; Rustem Ertegun, "Balancing Upon a Fine Line—Humanitarian Action and Environmental Sustainability," *UN Chronicle*, online edition, December 2002.

Chapter 3. Containing Infectious Disease

1.   Henry Hobhouse, *Forces of Change* (London: Sidgwick & Jackson, 1989), p. 11.

2.   SARS data from World Health Organization (WHO), at www.who.int/csr/sars/country/table2004_04_21/en/print/html, viewed 1 October 2004.

3.   WHO, *The World Health Report 2004* (Geneva: 2004), Annex Table 2.

4.   Historical deaths from conflict estimates are from "Millennium of Wars," *Washington Post*, 13 March 1999; Lynette Iezzoni, *Influenza 1918* (New York: TV Books, 1999), p. 204.

5.   Harvard Working Group on New and Resurgent Diseases, "New and Resurgent Diseases: The Failure of Attempted Eradication," *The Ecologist*, January/February 1995.

6.   David Brown, "WHO Calls for Rise in Health Spending," *Washington Post*, 21 December 2001.

7.   Clinton cited in Barton Gellman, "AIDS Is Declared Threat to Security," *Washington Post*, 30 April 2000.

8.   WHO, "Avian Influenza—Situation in Thailand," *Communicable Disease Surveillance & Response*, 28 September 2004; Working Group II, *Climate Change 2001: Impacts, Adaptation, and Vulnerability*, Contribution to the Third Assessment Report of the Intergovernmental Panel on Climate Change (New York: Cambridge University Press, 2001), Chapter 9.7.

9.   More detailed information on these waves of change is found in Tony McMichael, *Human Frontiers, Environments and Disease* (Cambridge, U.K.: Cambridge University Press, 2001), pp. 100–14.

10.   Disease outbreaks in the Roman Empire are discussed in William McNeill, *Plagues and Peoples* (Garden City, NY: Anchor Press, 1976), pp. 115–17.

11.   Diseases brought to the Americas from ibid., pp. 199–209.

12.   Gautam Naik, "Forget SARS: WHO Expert Fears the Flu," *Wall Street Journal*, 29 May 2003;

UNAIDS, *2004 Report on the Global AIDS Epidemic* (Geneva: 6 July 2004).

13. Population projections from Population Reference Bureau, *2004 World Population Data Sheet* (Washington, DC: 2004); percentage of people living in cities from Eugene Linden, "The Exploding Cities of the Developing World," *Foreign Affairs*, January/February 1996; Asian and African figures from Terry McGee, "Urbanization Takes on New Dimensions in Asia's Population Giants," *Population Today*, October 2001.

14. Ann Gibbons, "Where Are 'New' Diseases Born?" *Science*, 6 August 1993; U.N. High Commissioner for Refugees, "World Refugee Population Lowest in a Decade, UNHCR Says," press release (Geneva: 17 June 2004); Darfur outbreak reported by Marianne Hopp, ProMED-mail, 8 September 2004.

15. Growing gap figures from Bruce R. Scott, "The Great Divide in the Global Village," *Foreign Affairs*, January/February 2001, pp. 162–63; poverty figures from World Bank, *World Development Report 2000/2001* (New York: Oxford University Press, 2000), p. vi; per capita health expenditures from WHO, op. cit. note 3, Annex Table 6.

16. Overconsumption figure from Gary Gardner and Brian Halweil, *Underfed and Overfed: The Global Epidemic of Malnutrition*, Worldwatch Paper 150 (Washington, DC: Worldwatch Institute, 2000), p. 7; deaths from preventable diseases from Dara Carr, *Improving the Health of the World's Poorest People*, Health Bulletin Number 1 (Washington, DC: Population Reference Bureau, 2004), p. 2.

17. Tropical disease data from Carr, op. cit. note 16, p. 3.

18. Hilary French, *Vanishing Borders: Protecting the Planet in the Age of Globalization* (New York: W.W. Norton & Company, 2000), pp. 6–7.

19. Jeff Gerth and Tim Weiner, "U.S. Food-Safety System Swamped by Booming Global Imports," *New York Times*, 29 September 1997.

20. Historical perspective from Charles C. Mann, "1491," *The Atlantic Monthly*, March 2002; Ebola outbreak from Gretchen Vogel, "Can Great Apes Be Saved from Ebola?" *Science*, 13 June 2003; West Nile from Rick Weiss, "West Nile's Widening Toll," *Washington Post*, 28 December 2002; plant information from Anne Simon Moffat, "Finding New Ways to Fight Plant Diseases," *Science*, 22 June 2001.

21. Dennis Normile and Martin Enserink, "Avian Influenza Makes a Comeback, Reviving Pandemic Worries," *Science*, 16 July 2004.

22. For more perspective on possible flu pandemic, see Martin Enserink, "Tiptoeing Around Pandora's Box," *Science*, 30 July 2004; WHO, op. cit. note 8; Richard J. Webby and Robert G. Webster, "Are We Ready for Pandemic Influenza," *Science*, 28 November 2003; genetic changes in swine flu from Bernice Wuethrich, "Chasing the Fickle Swine Flu," *Science*, 7 March 2003.

23. Waterborne disease summarized by David Pimental et al., "Ecology of Increasing Disease," *Bioscience*, October 1998, p. 3.

24. For effects of global warming, see Nathan Y. Chan et al., "An Integrated Assessment Framework for Climate Change and Infectious Diseases," *Environmental Health Perspectives*, May 1999; Paul R. Epstein, "Climate Change and Health," *Science*, 16 July 1999; P. Martens et al., "Climate Change and Future Populations at Risk of Malaria," *Global Environmental Change 9*, 1999.

25. Stuart B. Levy, "The Challenge of Antibiotic Resistance," *Scientific American*, March 1998, p. 50; Shannon Brownlee, "Antibiotics in the Food Chain," *Washington Post*, 21 May 2000.

26. Annual deaths from drug-resistant microbes and disease drug resistance from WHO, *Overcoming Microbial Resistance* (Geneva: 2000), Chapter 4; HIV drug resistance from David Brown, "Study Finds Drug-Resistant HIV in Half of Infected Patients," *Washington Post*, 19 December 2001; pneumonia, chest infections, and cholera from WHO, op. cit. this note.

27. Tuberculosis costs from WHO, op. cit. note 26, and from Erik Stokstad, "Drug Resistant TB on Rise," *Science*, 31 March 2000.

28. New and resurgent diseases discussed in National Intelligence Council, *The Global Infectious Disease Threat and Its Implications for the United States* (Washington, DC: 2000); death registry information from WHO, op. cit. note 3, p. 95.

29. Disease statistics from WHO, op. cit. note 3, Annex Table 2.

30. Disease statistics and Table 3–1 from ibid., and from WHO, *The World Health Report 2001* (Geneva: 2001), Annex Table 2; latest AIDS deaths from UNAIDS, op. cit. note 12.

31. WHO, op. cit. note 3, Annex Table 4.

32. Table 3–2 from ibid.

33. HIV/AIDS victim data from UNAIDS/WHO, *AIDS Epidemic Update December 2003* (Geneva: 2003), p. 36; children data from WHO, op. cit. note 3, p. 1.

34. National Intelligence Council, *The Next Wave of HIV/AIDS: Nigeria, Ethiopia, Russia, India, and China* (Washington, DC: 2002).

35. Naik, op. cit. note 12.

36. Peter Daszak, Andrew A. Cunningham, and Alex D. Hyatt, "Emerging Infectious Diseases of Wildlife—Threats to Biodiversity and Human Health," *Science*, 21 January 2000; Rob Stein, "Infections Now More Widespread: Animals Passing Them to Humans," *Washington Post*, 15 June 2003; Dan Ferber, "Human Diseases Threaten Great Apes," *Science*, 25 August 2000.

37. Disease outbreak costs from Jonathan Ban, *Health, Security, and U.S. Global Leadership* (Washington, DC: Chemical and Biological Arms Control Institute, 2001), p. 31.

38. Pete Engardio, Mark L. Clifford, and Michael Shari, "Epidemics and Economics," *BusinessWeek*, 28 April 2003.

39. Lost economic growth estimate from ibid.; tourism losses from Chris Prystay, "SARS Squeezes Asia's Travel Sector," *Wall Street Journal*, 16 May 2003.

40. Table 3–3 compiled from Richard P. Cincotta, Robert Engelman, and Daniele Anastasion, *The Security Demographic: Population and Civil Conflict After the Cold War* (Washington, DC: Population Action International, 2003), Appendix 4.

41. HIV-positive data from UNAIDS/WHO, op. cit. note 33, p. 7; cuts in economic growth from economic losses from International Labour Organization (ILO), *HIV/AIDS and Work: Global Estimates, Impact and Response* (Geneva: 2004), p. 13, and from WHO, op. cit. note 3, p. 3. Box 3–1 from the following: infection rates in Zimbabwe and Malawi from International Crisis Group, *HIV/AIDS as a Security Issue* (Washington, DC: 2001), p. 20; prospects for soldiers in Zimbabwe from Poverty Reduction Forum, *Zimbabwe Human Development Report 2003* (Mt. Pleasant, Zimbabwe: Institute of Development Studies, University of Zimbabwe, 2004), p. 17; National Intelligence Council, *The Global Infectious Disease Threat and Its Implications for the United States* (Washington, DC: 2000), p. 53; P. W. Singer, "AIDS and International Security," *Survival*, spring 2002, pp. 148–49; U.S. General Accounting Office, *U.N. Peacekeeping: United Nations Faces Challenges in Responding to the Impact of HIV/AIDS on Peacekeeping Operations* (Washington, DC: 2001).

42. General HIV/AIDS situation from Jennifer Brower and Peter Chalk, *The Global Threat of New and Reemerging Infectious Diseases* (Santa Monica, CA: RAND, 2003), pp. 45–48; economic data from ILO, op. cit. note 41, and from Andrew T. Price-Smith, *Pretoria's Shadow: The HIV/AIDS Pandemic and National Security in South Africa* (Washington, DC: Chemical and Biological Arms Control Institute, 2002), pp. 12–16.

43. For more detail on the need to rebuild the world's public health infrastructure, see Laurie Garrett, *Betrayal of Trust: The Collapse of Global*

Public Health (New York: Hyperion, 2000).

44. ProMED information from www.pro medmail.org, viewed 3 October 2004; Box 3–2 from W. Wayt Gibbs, "An Uncertain Defense," *Scientific American*, October 2004, and from Richard Danzig, *Catastrophic Bioterrorism—What Is To Be Done?* (Washington, DC: Center for Technology and Security Policy, 2003).

45. Alicia Ault, "Shifting Tactics in the Battle Against Influenza," *Science*, 27 February 2004; Martin Enserink, "Crisis Underscores Fragility of Vaccine Production System," *Science*, 15 October 2004.

46. Phil Temples, ProMED-mail, 12 March 2004.

47. Ann Marie Nelson, "The Cost of Disease Eradication: Smallpox and Bovine Tuberculosis," *Annals of the New York Academy of Sciences*, December 1999, pp. 83–91.

48. Ibid.

49. Basic information on campaign from WHO press release circulated in ProMED-mail, 22 June 2004; information on campaign to eliminate polio from ProMED-mail, 27 May 2004; immunization problem from David Brown, "Polio Eradication Faces Setback in Africa," *Washington Post*, 23 June 2004.

50. Leslie Roberts, "Health Workers Scramble to Contain African Epidemic," *Science*, 2 July 2004.

51. Ibid.

52. Donald G. McNeil, Jr., "Plan to Battle AIDS Worldwide is Falling Short," *New York Times*, 28 March 2004.

53. Marilyn Chase, "Gates Foundation Bets It Can Stem India's AIDS Crisis," *Wall Street Journal*, 3 May 2004.

54. David Brown, "Bush's AIDS Program Balks at Foreign Generics," *Washington Post*, 27 March 2004; Tom Hamburger, "U.S. Flip on Patents

Shows Drug Makers' Growing Clout," *Wall Street Journal*, 6 February 2003.

55. ILO, op. cit. note 41; Ellen Nakashima and David Brown, "AIDS Plan Falls Short of Target," *Washington Post*, 11 July 2004.

56. Antonio Regalado, Matt Pottinger, and Betsy McKay, "Across the Globe, A Race to Prepare for SARS Round 2," *Wall Street Journal*, 9 December 2003.

57. John Pomfret, "China's Slow Reaction to Fast-Moving Illness," *Washington Post*, 3 April 2003; idem, "Outbreak Gave China's Hu an Opening," *Washington Post*, 13 May 2003.

Bioinvasions

1. World trade (in 2002 dollars) from IMF, *World Economic Outlook* (Washington, DC: April 2004), p. 216; World Tourism Organization, "World Tourism Barometer," January 2004, p. 3, with percentage increase based on preliminary 2003 figure.

2. Definition of invasive from Executive Order 13112, 3 February 1999, at www.invasive species.gov, and from "Decision VI/23: Alien Species that Threaten Ecosystems, Habitats, or Species," Sixth Conference of the Parties, Convention on Biological Diversity; U.S. food system from *Statistical Abstract of the United States 1996*, cited in David Pimentel et al., "Environmental and Economic Costs Associated with Non-Indigenous Species in the United States," Cornell University, Ithaca, NY, 12 June 1999; Mexican fish from "Pilot Assessments: The Ecological and Socio-Economic Impact of Invasive Alien Species on Inland Water Ecosystems," Convention on Biological Diversity, 5 November 2003.

3. China losses from "Invasive Species Costly to Ecology," *Xinhuanet*, 22 May 2004; U.S. losses from Pimentel et al., op. cit. note 2; H. De Groote et al., "Economic Impact of Biological Control of Water Hyacinth in Southern Benin," *Ecological Economics*, April 2003, pp. 105–17.

4. Robert J. Pratt, "Invasive Threats to the

American Homeland," *Parameters: US Army War College Quarterly*, spring 2004, pp. 44–61.

5. "Alien Invaders in Ballast Water—New Convention to be Adopted at IMO," press release (London: International Maritime Organization, 13 February 2004); Aquatic Plant Control Section, Army Corps of Engineers, *Okeechobee Waterway Zebra Mussel Monitoring Plan* (Jacksonville, FL: 2003).

6. "Decision VI/23," op. cit. note 2; "Resolution VIII.18 on Invasive Species and Wetlands," Ramsar Convention on Wetlands, 2002.

7. Agents' role from "CBP Agricultural Specialist Fact Sheet" (Washington, DC: U.S. Department of Homeland Security, undated); funding concerns from "Homeland Security Department Could Open Floodgates to Biological Invaders," press release (Cambridge, MA: Union of Concerned Scientists, 10 July 2002) and from Eugene Russo, "Cooperation Urged on Invasives," *The Scientist*, 22 March 2004; "Council Members," National Invasive Species Council, at www.invasivespecies.gov.

8. European Commission, *Alien Species and Nature Conservation in the EU: The Role of the LIFE Program* (Brussels: European Communities, 2004), p. 24; "Decision VI/23," op. cit. note 2.

Chapter 4. Cultivating Food Security

1. U.N. Food and Agriculture Organization (FAO), "HIV/AIDS and the Food Crisis in Sub-Saharan Africa," Johannesburg, South Africa, 1–5 March 2004; idem, "Impacts of HIV/AIDS," HIV/AIDS and Food Security, at www.fao.org/hivaids/impacts/knowledge_en.htm, viewed August 2004; Future Harvest, "Will Agriculture Fall Victim to AIDS? Researchers Seek to Protect Harvest From Africa's AIDS Epidemic," feature (Alexandria, VA: 14 November 2001); International Crops Research Institute for the Semi-Arid Tropics, *The Impact of HIV/AIDS on Farmers' Knowledge of Seed: Case Study of Chókwè District, Gaza Province, Mozambique* (Maputo, Mozambique: 2004).

2. Water scarcity from Population Action International, "Why Population Matters to Natural Resources," factsheet (Washington, DC: updated April 2003); soil degradation from Stanley Wood, Kate Sebastian, and Sara J. Scherr, *Pilot Analysis of Global Ecosystems: Agroecosystems* (Washington, DC: International Food Policy Research Institute and World Resources Institute: 2001), and from U.N. Environment Programme (UNEP), *Global Environmental Outlook 2002* (Nairobi: 2002), cited in Erik Millstone and Tim Lang, *The Penguin Atlas of Food: Who Eats What, Where, and Why* (London: Penguin Books, 2003), p. 16; conflict from "Erratic Rains, Civil Strife, and Desert Locusts Severely Threaten Food Security in Sub-Saharan Africa," *FAONewsroom*, 6 July 2004.

3. Number of hungry people from FAO, *State of World Food Insecurity 2003* (Rome: 2003), and from idem, *State of Food and Agriculture 2003–2004* (Rome: 2004), p. 109; nutrient deficiencies from idem, *State of Food and Agriculture 2002* (Rome: 2002).

4. John Sheehy, International Rice Research Institute, Manila, Philippines, e-mail to Brian Halweil, 23 May 2004; UNEP, "Climate Change: Billions Across The Tropics Face Hunger and Starvation as Big Drop in Crop Yields Forecast," press release (Marrakech/Nairobi/Manila: 8 November 2001).

5. Julio Godoy, "Alarm Sounds on Bee-Killing Pesticides," *Inter Press Service News Agency*, 22 March 2004.

6. Ibid.

7. M. E. Watanabe, "Pollination Worries Rise as Honeybees Decline," *Science*, vol. 265, no. 26 (1994), p. 1170.

8. Figure of 75 percent from Hope Shand, *Human Nature: Agricultural Biodiversity and Farm-based Food Security* (Pittsboro, NC: Rural Advancement Fund International, 1997), p. 21; wheat and maize from ibid. and from FAO, *State of the World's Plant Genetic Resources for Food and Agriculture* (Rome: 1997), p. 22; rice from GRAIN, *Hybrid Rice in Asia: An Unfolding Threat*

(Barcelona, Spain: 2000), from Shand, op. cit. this note, p. 21, and from FAO, op. cit. this note, p. 21; Patrick Mulvany, Intermediate Technology Development Group (ITDG), discussion with Molly Norton, Worldwatch Institute, 21 July 2004.

9. Table 4–1 from the following: "Environment Nepal: Indigenous Livestocks on Verge of Extinction," *Himalayan Times*, 1 December 2002; pig, chicken, and sheep from FAO, *World Watch List for Domestic Animal Diversity* (Rome: 2000), pp. 132, 517, 493; William K. Stevens, "Fierce Debate on Future of Ocean Species," *New York Times*, cited by South Coast Today, at www.s-t.com/daily/09-96/09-29-96/m02li060 .htm, viewed 9 August 2004. First seed banks from N. I. Vavilov Research Institute of Plant Industry, at www.vir.nw.ru.

10. FAO, op. cit. note 9; idem, "Loss of Domestic Animal Breeds Alarming," press release (Rome: 31 March 2004); Shand, op. cit. note 8, p. 46.

11. Shand, op. cit. note 8, p. 22.

12. Ibid.

13. FAO, "Biological Diversity in Food and Agriculture," at www.fao.org/biodiversity/index.asp, viewed August 2004.

14. Jose Esquinas Alcazar, Secretary of the Commission on Genetic Resources for Food and Agriculture, Rome, discussion with Danielle Nierenberg, July 2004.

15. Shand, op. cit. note 8, p. 22; FAO, op. cit. note 8, p. 21.

16. "Seeds of Freedom," Navdanya, at www.navdanya.org, viewed 4 July 2004; crop diversity reduces dependence from P. Mader et al., "Soil Fertility and Biodiversity in Organic Farming," *Science*, 31 May 2002, pp. 1694–97; J. P. Reganold et al., "Sustainability of Three Apple Production Systems," *Nature*, 19 April 2001, pp. 926–30; Peter B. Reich et al., "Plant Diversity Enhances Ecosystem Responses to Elevated CO_2 and Nitrogen Deposition," *Nature*, 12 April 2001 pp. 809–10.

17. Box 4–1 from Peter Chalk, RAND Corporation, discussion with Danielle Nierenberg, 14 July 2004, and from idem, *Hitting America's Soft Underbelly: The Potential Threat of Deliberate Biological Attacks Against the U.S. Agricultural and Food Industry* (Arlington, VA: RAND Corporation, 2004), pp. 8–9, 17; Chuck Bassett, Executive Director of the American Livestock Breeds Conservancy, discusson with Molly Norton, Worldwatch Institute, 23 June 2004.

18. For a discussion of the effects of factory farming on human health, see World Society for the Protection of Animals (WSPA), *Industrial Animal Agriculture: The Next Global Health Crisis?* (London: 2004); Table 4–2 from World Health Organization (WHO), "Avian Influenza," "Nipah Virus," and "Bovine Spongiform Encephalopathy" at www.who.int/mediacentre/factsheets, viewed 4 August 2004.

19. WSPA, op. cit. note 18; FAO, Animal Health Health and Production Division, "Avian Influenza—Questions & Answers," at www.fao.org/ag/againfo/subjects/en/health/diseases-cards/avian_qa.html; idem, "High Geographic Concentration May Have Favored the Spread of Avian Flu," *FAONewsroom*, 28 January 2004; WHO, "Avian Influenza—Situation in Thailand," *Communicable Diseuse Surveillance & Response*, 28 September 2004.

20. FAO, Animal Production and Health Division, "Avian Influenza—Background" and "Avian Influenza–Disease Card," at www.fao.org/ag/againfo/subjects/en/health/diseases-cards; WHO, Communicable Disease Surveillance & Response, "Pandemic Preparedness," at www.who.int/csr/disease/influenza/pandemic/en; H. Chen et al., "The Evolution of H5N1 Influenza Viruses in Ducks in Southern China," *Proceedings of the National Academy of Sciences*, 13 July 2004, pp. 10452–57; Keith Bradsher and Lawrence K. Altman, "Health Experts Worry Over Return of Bird Flu in Asia," *New York Times*, 8 July 2004; K. S. Li et al., "Genesis of a Highly Pathogenic and Potentially Pandemic H5NI Influenza Virus in Eastern Asia," *Nature*, 8 July 2004, pp. 209–13.

21. WSPA, op. cit. note 18; FAO, "Avian Flu:

FAO Provides $1.6 Million in Emergency Aid to Four Asian Countries," *FAONewsroom*, 3 February 2004; Emmanuelle Guerne-Bleich, Animal Production Officer, FAO, discussion with Danielle Nierenberg, April 2004.

22. WSPA, op. cit. note 18; Peter Fritsch, "Scientists Search for Human Hand Behind Outbreak of Jungle Virus," *Wall Street Journal*, 19 June 2003; Leslie Bienen, "Bats Suspected in Disease Outbreak," *Frontiers in Ecology*, April 2004, p. 117.

23. WSPA, op. cit. note 18; Fritsch, op. cit. note 22; Bienen, op. cit. note 22; "Peter Daszak's Comments on the 60 Minutes Nipah Virus Report," Consortium for Conservation Medicine, at www.conservationmedicine.com/index.htm; Peter Daszak, Executive Director, Consortium for Conservation Medicine, e-mail to Danielle Nierenberg, September 2004; Wildlife Trust, "Nipah Virus Breaks Out in Bangladesh: Mortality Rates of 60 to 74 Percent, Human-to-Human Transmission May Be Implicated," press release (New York: 28 April 2004).

24. WHO, "Bovine Spongiform Encephalopathy," Fact Sheet No. 113 (Geneva: 2002).

25. WSPA, op. cit. note 18; Paul Brown et al., "Bovine Spongiform Encephalopathy and Variant Creutzfeldt-Jacob Disease, Background, Evolution, and Current Concerns," *Emerging Infectious Diseases*, January-February 2001, pp. 6–17; Millstone and Lang, op. cit. note 2, p. 37.

26. WSPA, op. cit. note 18; Kim Willsher, "French Mad Cow Disease Cases Went Undetected," *The Daily Telegraph* (London), 7 April 2004.

27. WSPA, op. cit. note 18; Donald G. McNeil, "Research in Italy Shows New Form of Mad Cow Disease," *New York Times*, 17 February 2004; James Meikle, "More Illness Linked to BSE, Government Committee Alarmed by the Report," *The Guardian* (London), 18 February 2004.

28. WSPA, op. cit. note 18; WHO, "Foodborne Diseases—Possibly 350 Times More Frequent Than Reported," press release (Geneva: 13 August

1997); FAO database at www.apps/fao/org.uk; 44 million from FAO, *FAOSTAT Statistical Database*, at apps.fao.org, and from Millstone and Lang, op. cit. note 2, p. 62; A. R. Barham et al., "Effects of the Transportation of Beef Cattle from the Feedyard to the Packing Plant on Prevalence Levels of *Escherichia coli* O157 and *Salmonella* spp.," *Journal of Food Protection*, February 2002, pp. 280–83.

29. WHO and FAO, "Antimicrobial Resistance," Fact Sheet No. 194 (Geneva: WHO, 2003).

30. Alexandra Goho, "Fishy Alpha Males," *Science News Online*, 6 March 2004.

31. Geoffrey Lean, "Revealed: Shocking New Evidence of the Dangers of GM Crops," *Independent Digital* (London), 7 March 2004.

32. Ibid.; Margaret Mellon and Jane Rissler, *Gone to Seed: Transgenic Contaminants in the Traditional Seed Supply* (Cambridge, MA: Union of Concerned Scientists, 2004).

33. Hubert Zandstra, "A Colorful Solution to Late Blight," *DG's Quarterly Report* (La Molina, Peru: Centro Internacional de la Papa, March 2004).

34. Jared Diamond, *Guns, Germs, and Steel: The Fates of Human Societies* (New York: W.W. Norton & Company, 1999), p. 110; William F. Ruddiman et al., "The Anthropogenic Greenhouse Era Began Thousands of Years Ago," *Climatic Change*, December 2003, pp. 261–93.

35. Peter Schwartz and Doug Randall, *An Abrupt Climate Change Scenario and Its Implications for United States National Security* (Emeryville, CA: Global Business Network, 2003); see also David Stipp, "The Pentagon's Weather Nightmare," *Fortune*, 9 February 2004.

36. Anderson from David Ljunggren, "Global Warming Bigger Threat than Terrorism," *Reuters*, 6 February 2004; see also David Anderson, "Notes for Remarks by the Honourable David Anderson, P.C., M.P. Minister of the Environment to the Calgary Chamber of Commerce," Calgary, AB,

Canada, 27 February 2004; Doug Randall, Global Business Network, Emeryville, CA, discussion with Brian Halweil, 10 March 2004.

37. Sheehy, op. cit. note 4; UNEP, op. cit. note 4.

38. Tim Lockette, "Global Warming May Cut Production of Key Protein Crop, Peanuts," *UF/IFAS News* (Institute of Food and Agricultural Sciences, Gainesville, FL), 9 February 2004; Hartwell Allen, University of Florida and Agricultural Research Service, U.S. Department of Agriculture, Gainesville, FL, discussion with Brian Halweil, 29 March 2004.

39. "Summary for Policymakers, A Report of Working Group I of the Intergovernmental Panel on Climate Change," in J. T. Houghton et al., (eds.), *Climate Change 2001: The Scientific Basis*, Contribution of Working Group I to the Third Assessment Report of the Intergovernmental Panel on Climate Change (Cambridge, U.K.: Cambridge University Press, 2001), p. 15.

40. David Rhind, Goddard Institute for Space Studies, Columbia University, New York, discussion with Brian Halweil, 10 April 2004.

41. Cynthia Rosenzweig, Goddard Institute for Space Studies, Columbia University, New York, discussion with Brian Halweil, 29 March 2004, and Cynthia Rosenzweig et al., *Climate Change and U.S. Agriculture: The Impacts of Warming and Extreme Weather Events on Productivity, Plant Diseases, and Pests* (Boston, MA: Center for Health and the Global Environment, Harvard Medical School, 2000); D. T. Patterson et al., "Weeds, Insects, and Diseases," *Climatic Change*, December 1999, pp. 711–27; Chi-Chung Chen and Bruce A. McCarl, "An Investigation of the Relationship between Pesticide Usage and Climate Change," *Climatic Change*, September 2001, pp. 475–87.

42. Robert Watson, World Bank, Washington, DC, discussion with Brian Halweil, 1 April 2004; Rosenzweig, op. cit. note 41.

43. Patrick Luganda, "Are Subsistence Farmers Coping with Climate Change?" *New Agricultur-*

alist, 1 November 2003; Bambang Irawan, "El Niño and La Niña: Tendency of Occurrence and Impact on Food Production," *Palawija News* (Regional Coordination Centre for Research and Development of Coarse Grains, Pulses, Roots and Tuber Crops in the Humid Tropics of Asia and the Pacific Newsletter), Bogor, Indonesia, December 2003.

44. Cary Fowler, "Accessing Genetic Resources: International Law Establishes Multilateral System," *Genetic Resources and Crop Evolution*, September 2004, pp. 615–18.

45. Ibid.

46. "Karen Commitment Pastoralist/Indigenous Livestock Keepers Rights," Indigenous Livestock Breeders Workshop, 27–30 October 2003, at www.ukabc.org/karen.htm.

47. Pat Mooney, Executive Director ETC Group, discussion with Danielle Nierenberg, June 2004.

48. ITDG, "Sustaining Agricultural Biodiversity and the Integrity and Free Flow of Genetic Resources for Food and Agriculture," Forum for Food Sovereignty, 8–13 June 2002, Rome, p. 7.

49. Susie Emmett, "Conserving Animal Genetic Resources—The Race is On," *New Agriculturalist*, May 2004; Heritage Foods USA, at www.heritagefoodsusa.com.

50. Jacob Wanyama, ITDG, e-mail to Danielle Nierenberg, 10 April 2004.

51. FAO, "Transboundary Animal Diseases on the Rise," *FAONewsroom*, 25 May 2004.

52. Louis V. Verchot, International Centre for Research in Agroforestry, Nairobi, Kenya, discussion with Brian Halweil, 30 April 2004.

53. Ibid.; UNEP, op. cit. note 4.

54. Verchot, op. cit. note 52.

55. Ibid.; Mary Knapp, Kansas State Univesity, State Climatologist, Manhattan, KS, e-mail to

Brian Halweil, 20 June 2004; mission statement from Martin Bender, "Energy in Agriculture and Society: Insights from the Sunshine Farm," Land Institute, Salina, KS, March 2001; Martin Bender, senior scientist, Land Institute, Salina, KS, discussion with Brian Halweil, 28 May 2004.

56. Bender, discussion with Halweil, op. cit. note 55; Rosenzweig, op. cit. note 41; C. Rosenzweig and D. Hillel, "Soils and Global Climate Change: Challenges and Opportunities," *Soil Science*, January 2000, pp. 47–56.

57. Rosenzweig and Hillel, op. cit. note 56; U.S. Environmental Protection Agency, *Inventory of U.S. Greenhouse Gas Emissions and Sinks: 1990–2002* (Washington, DC: 2004) pp. 24, 27; farmers' potential role from Bender, discussion with Halweil, op. cit. note 55.

58. Annika Carlsson-Kanyama, "Climate Change and Dietary Choices: How Can Emissions of Greenhouse Gases From Food Consumption Be Reduced," *Food Policy*, fall/winter 1998, pp. 288–89.

Toxic Chemicals

1. V. Ramana Dhara et al., "Personal Exposure and Long-Term Health Effects in Survivors of the Union Carbide Disaster at Bhopal," *Environmental Health Perspectives*, May 2002, pp. 487–500; Timothy H. Holtz, "Tragedy Without End: The 1984 Bhopal Gas Disaster," *Dying For Growth: Global Inequality and the Health of the Poor* (Monroe, ME: Common Courage Press, 2000), p. 245; U.S. chemical discharges from National Response Center, "Incident Type Per Year," at www.nrc.uscg.mil/incident97-02.html, viewed 22 July 2004.

2. Lois Ember, "Worst-Case Scenario for Chemical Plant Attack," *Chemical & Engineering News*, 18 March 2002, p. 8; Jeff Johnson, "Lawsuit Filed Over Terrorism Study," *Chemical & Engineering News*, 18 March 2002, p. 8; Soraya Sarhaddi Nelson, "Blasts Kill 10 in Israel Port," *Mercury News*, 15 March 2004.

3. World Wide Fund for Nature–UK, *Compro-*

mising Our Children—Chemical Impacts on Children's Intelligence and Behaviour (Surrey, U.K.: 2004); Inuit from Marla Cone, "Ancestral Diet Gone Toxic," *Los Angeles Times*, 13 January 2004, and from "Eat Less Blubber," *CBC News*, 26 June 2003.

4. Rachel Carson, *Silent Spring* (Boston: Houghton Mifflin Company, reprint ed., 1994); U.S. Environmental Protection Agency (EPA), "DDT Ban Takes Effect," press release (Washington, DC: 31 December 1972).

5. Anne Platt McGinn, "Reducing Our Toxic Burden," in Worldwatch Institute, *State of the World 2002* (New York: W.W. Norton & Company, 2002), pp. 87–88; Brian Halweil, "Sperm Counts Dropping," in Worldwatch Institute, *Vital Signs 1999* (New York: W.W. Norton & Company, 1999), pp. 148–49.

6. EPA, "TRI Public Data Release 2002," at www.epa.gov/newsroom/tri2002/TRI2002_files/frame.htm, viewed 8 October 2004; idem, "What Is the Toxics Release Inventory (TRI) Program," at www.epa.gov/tri/whatis.htm, viewed 8 October 2004.

7. Lowell Center for Sustainable Production, *Update on the European Commission's REACH Proposal* (Lowell, MA: 25 November 2003); Richard Carter, "Verheugen Hints at Reform of Controversial Chemicals Plan," *EU Observer*, 30 September 2004; Lennart Hardell and Mikael Eriksson, "Is the Decline of the Increasing Incidence of Non-Hodgkin Lymphoma in Sweden and Other Countries a Result of Cancer Preventive Measures?" *Environmental Health Perspectives*, November 2003, pp. 1704–05.

8. "Stockholm Convention on Persistant Organic Pollutants," at www.pops.int/documents/signature, viewed 13 September 2004; U.N. Environment Programme, "Stockholm Convention on POPS to Become International Law, Launching a Global Campaign to Eliminate 12 Hazardous Chemicals," press release (Stockholm, Sweden: 14 May 2004).

9. Eric J. Beckman, "Using CO_2 to Produce

Chemical Products Sustainably," *Environmental Science & Technology*, 1 September 2002, pp. 347A–53A.

10. Edward Cohen-Rosenthal and Thomas N. McGalliard, "Eco-Industrial Development: The Case of the United States," *The IPTS Report* (European Commission), September 1998.

Chapter 5.
Managing Water Conflict and Cooperation

1. Stanley Crawford, *Mayordomo: Chronicle of an Acequia in Northern New Mexico* (Albuquerque: University of New Mexico Press, 1988); Asit K. Biswas, "Indus Water Treaty: The Negotiation Process," *Water International*, vol. 17 (1992), p. 202; Dale Whittington, John Waterbury, and Elizabeth McClelland, "Toward a New Nile Waters Agreement," in Ariel Dinar and Edna Tusak Lochman, eds., *Water Quantity/Quality Management and Conflict Resolution: Institutions, Processes, and Economic Analyses* (Westport, CT: Praeger, 1995).

2. Sandra Postel, *Pillar of Sand* (New York: W.W. Norton & Company, 1999).

3. Aaron T. Wolf, Shira B. Yoffe, and Mark Giordano, "International Waters: Identifying Basins at Risk," *Water Policy*, February 2003, pp. 29–60; Alexander Carius, Geoffrey D. Dabelko, and Aaron T. Wolf, "Water, Conflict, and Cooperation," Policy Briefing Paper for United Nations & Global Security Initiative (Washington, DC: United Nations Foundation, 2004).

4. Table 5–1 from the following: Cauvery River from Ramaswamy Iyer, "Water-Related Conflicts: Factors, Aspects, Issues," in Monique Mekenkamp, Paul van Tongeren, and Hans van de Veen, eds., *Searching for Peace in Central and South Asia* (Boulder, CO: Lynne Rienner Publishers, 2002), and from Aaron T. Wolf, "Water and Human Security," *AVISO Bulletin* (Global Environmental Change and Human Security Project, Canada), June 1999; Okavango River basin from Anthony R. Turton, Peter Ashton, and Eugene Cloete, "An Introduction to the Hydropolitical Drivers in the Okavango River Basin," in idem, eds., *Trans-*

boundary Rivers, Sovereignty and Development—Hydropolitical Drivers in the Okavango River Basin (Pretoria, South Africa: African Water Issues Research Unit and Green Cross International, 2003), pp. 9–30; Mekong River basin from "Pak Mun Communities Ripe for Reparations," *World Rivers Review*, December 2000, p. 13; Incomati River from Annika Kramer, *Managing Freshwater Ecosystems of International Water Resources—The Case of the Maputo River in Mozambique*, Working Paper on Management in Environmental Planning (Berlin, Germany: Technische Universität, 2003); Rhine River from Rainer Durth, *Grenzüberschreitende Umweltprobleme und regionale Integration* (Baden Baden, Germany: Nomos Verlagsgesellschaft, 1996), and from Alexander López, *Environmental Conflicts and Regional Cooperation in the Lempa River Basin: The Role of the Trifinio Plan as a Regional Institution* (San José, Costa Rica: Environment, Development, and Sustainable Peace Initiative, 2004); Syr Darya from Victor Dukhovny and Vadim Sokolov, *Lessons on Cooperation Building to Manage Water Conflicts in the Aral Sea Basin* (Paris: UNESCO, International Hydrological Programme (IHP), 2003).

5. Aaron T. Wolf et al., "International River Basins of the World," *International Journal of Water Resources Development*, vol. 15, no. 4 (1999), pp. 387–427; United Nations, *Register of International Rivers* (New York: Pergamon Press, 1978).

6. Data and Table 5–3 from Wolf et al., op. cit. note 5.

7. Serageldin quoted in Barbara Crossette, "Severe Water Crisis Ahead for Poorest Nations in Next Two Decades," *New York Times*, 10 August 1995.

8. Aaron T. Wolf, "'Hydrostrategic' Territory in the Jordan Basin: Water, War, and Arab-Israeli Peace Negotiations," in Hussein A. Amery and Aaron T. Wolf, eds., *Water in the Middle East: A Geography of Peace* (Austin: University of Texas Press, 2000).

9. For a historical list of events related to water and conflict, see Peter H. Gleick, *Water Conflict*

Chronology, at www.worldwater.org/conflict.htm; U.N. Food and Agriculture Organization, *Systematic Index of International Water Resources Treaties, Declarations, Acts and Cases, by Basin, Vol. I* and *Vol. II* (Rome: 1978 and 1984); number of treaties from the Transboundary Freshwater Dispute Database maintained at Oregon State University, which is available at www.transboundary waters.orst.edu.

10. Excluded are events where water is incidental to a dispute, such as those concerning fishing rights or access to ports, transportation, or river boundaries, as well as events where water is not the driver, such as those where water is a tool, target, or victim of armed conflict; Wolf, Yoffe, and Giordano, op. cit. note 3.

11. Wolf, Yoffe, and Giordano, op. cit. note 3.

12. Ibid.

13. Ibid.

14. Mekong Committee from Ti Le-Huu and Lien Nguyen-Duc, *Mekong Case Study* (Paris: UNESCO, IHP, 2003); Indus River Commission from Wolf, op. cit. note 4; Nile Basin talks from Alan Nicol, *The Nile: Moving Beyond Cooperation* (Paris: UNESCO, IHP, 2003). Box 5–1 from the following: Anders Jägerskog, *Why States Cooperate Over Shared Water: The Water Negotiations in the Jordan River Basin* (Linköping, Sweden: Linköping University, 2003); Israeli-Jordanian peace treaty at www.israel-mfa.gov.il/mfa/go.asp?MFAH00pa0; Israeli-Palestinian interim agreement at www.mfa.gov.il/mfa/go.asp?MFA H00qd0#app-40, and at www.nad-plo.org/fact/annex3.pdf; Tony Allan, *The Middle East Water Question: Hydropolitics and the Global Economy* (New York: I. B. Tauris, 2001); Arun P. Elhance, *Hydropolitics in the 3rd World: Conflict and Cooperation in International River Basins* (Washington, DC: United States Institute of Peace Press, 1999); fears of water wars from Joyce R. Starr, "Water Wars," *Foreign Policy*, spring 1991, pp. 17–36, and from John Bulloch and Adel Darwish, *Water Wars: Coming Conflicts in the Middle East* (London: Victor Gollancz, 1993).

15. Anthony R. Turton, "The Evolution of Water Management Institutions in Select Southern African International River Basins," in C. Tortajada, O. Unver, and A. K. Biswas, eds., *Water as a Focus for Regional Development* (London: Oxford University Press, 2004); Ken Conca and Geoffrey D. Dabelko, eds., *Environmental Peacemaking* (Washington, DC, and Baltimore, MD: Woodrow Wilson Center Press and Johns Hopkins University Press, 2002).

16. Aaron T. Wolf, "Water, Conflict, and Cooperation," in Ruth S. Meinzen-Dick and Mark W. Rosegrant, *Overcoming Water Scarcity and Quality Constraints* (Washington, DC: International Food Policy Research Institute, 2001).

17. Zecharya Tagar, Tamar Keinan, and Gidon Bromberg, *A Seeping Time Bomb: Pollution of the Mountain Aquifer by Sewage*, Investigative Report Series on Water Issues No. 1 (Tel Aviv and Amman: Friends of the Earth Middle East, 2004).

18. Meredith A. Giordano, Mark Giordano, and Aaron T. Wolf, "The Geography of Water Conflict and Cooperation: Internal Pressures and International Manifestations," in Meredith A. Giordano, *International River Basin Management: Global Principles and Basin Practice*, dissertation submitted to Oregon State University, 2002.

19. Arizona and Shandong from Sandra L. Postel and Aaron T. Wolf, "Dehydrating Conflict," *Foreign Policy*, September/October 2001, pp. 60–67; Gleick, op. cit. note 9.

20. Box 5–2 from the following: "Bolivia Water Management: A Tale of Three Cities," *Précis* (Operations Evaluation Department, World Bank), spring 2002; Emanuel Lobina, "Cochabamba—Water War," *Focus* (PSI Magazine, Public Services International, France), vol. 7, no. 2 (2000); William Finnegan, "Letter from Bolivia: Leasing the Rain," *The New Yorker*, 8 April 2002; Andrew Nickson and Claudia Vargas, "The Limitations of Water Regulation: The Failure of the Cochabamba Concession in Bolivia," *Bulletin of Latin American Research*, January 2002, pp. 128–49; Postel and Wolf, op. cit. note 19; Jimmy Langmann, "Bechtel Battles Against Dirt-Poor Bolivia," *San*

Francisco Chronicle, 2 February 2002. For more on the case, see Bechtel Corporation, "Cochabamba and the Aguas del Tunari Consortium," fact sheet (San Francisco: Bechtel Corporation, 2002), and Public Citizen, *Water Privatization Case Study: Cochabamba, Bolivia* (Washington, DC: 2001).

21. Postel, op. cit. note 2; Richard Cincotta, Robert Engelman, and Daniele Anastasion, *The Security Demographic: Population and Civil Conflict After the Cold War* (Washington, DC: Population Action International, 2003).

22. Wolf, op. cit. note 4.

23. Asit K. Biswas and Tsuyoshi Hashimoto, *Asian International Waters: From Ganges–Brahmaputra to Mekong* (New Delhi, India: Oxford University Press, 1996).

24. Leif Ohlsson, *Livelihood Conflicts: Linking Poverty and Environment as Causes of Conflict* (Stockholm, Sweden: Swedish International Development Agency, 2000).

25. Wolf, Yoffe, and Giordano, op. cit. note 3.

26. Ibid.

27. Ibid.

28. Nickson and Vargas, op. cit. note 20.

29. Anthony R. Turton, "Water Wars in Southern Africa: Challenging Conventional Wisdom," in Green Cross International, *Water for Peace in the Middle East and Southern Africa* (Geneva: 2000), pp. 112–30; Victoria Percival and Thomas Homer-Dixon, "Environmental Scarcity and Violent Conflict: The Case of South Africa," *Journal of Peace Research,* vol. 35, no. 3 (1998), pp. 279–98.

30. Box 5–3 from the following: environmental and social impact from Patrick McCully, *Silenced Rivers* (London: Zed Books, 2003), and from World Commission on Dams, *Dams and Development* (London: Earthscan, 2001); marginality of indigenous groups from Thayer Scudder, *The Future of Large Dams: Dealing with Social, Environmental, Institutional and Political Costs* (London: Earthscan, 2004), and from Barbara Rose Johnston, ed., *Life and Death Matters: Human Rights and the Environment at the End of the Millennium* (Walnut Creek, CA: AltaMira Press, 1997); Chixoy Dam development history from William L. Partridge, *Comparative Analysis of BID Experience with Resettlement Based on Evaluations of the Arenal and Chixoy Projects,* Consultant Report (Washington, DC: Inter-American Development Bank, 1983), from Barbara Rose Johnston, "Reparations for Dam-Displaced Communities?" *Capitalism Nature Socialism,* March 2004, pp. 113–19, and from Paula Goldman, Casey Kelso, and Monika Parikh, "The Chixoy Dam and the Massacres at Rio Negro, Agua Fria and Los Encuentros: A Report on Multilateral Financial Institution Accountability," submitted to the U.N. High Commissioner for Human Rights by the Working Group on Multilateral Institution Accountability, Graduate Policy Workshop: Human Rights and Non-State Actors, Woodrow Wilson School, Princeton University, December 2000; Chixoy Dam–related violence and its effect on local people from Jesus Tecu Osorio, *The Rio Negro Massacres,* translated from the Spanish original *Memoria de las Masacres de Rio Negro* (Washington, DC: Rights Action, 2003); U.N. report and other findings from American Association for the Advancement of Science, "Guatemala: Memory of Silence," English version of the report of the Historical Clarification Commission, at shr.aaas.org/guatemala/ceh; Human Rights Office, Archdiocese of Guatemala, *Guatemala: Never Again! Recovery of Historical Memory Project* (New York: Maryknoll Press, 1998); Witness for Peace, *A People Dammed* (Washington, DC: 1996); Jaroslave Colajacomo, "The Chixoy Dam: The Maya Achi' Genocide. The Story of Forced Resettlement," Contributing Paper, World Commission on Dams (Cape Town, South Africa: 1999); dam seizure from Frank Jack Daniels, "Mayan Indians Seize Disputed Guatemalan Dam," *Reuters,* 9 September 2004; Pak Mun and Rasi Salai from "Rivers for Life: Inspirations and Insights from the 2nd International Meeting of Dam-Affected People and Their Allies," International Rivers Network and Environmental Leadership Program, 2004; reparations in the context

of development legacy issues from Barbara Rose Johnston, "Reparations and the Right to Remedy," Contributing Paper, World Commission on Dams (Cape Town, South Africa: 2000).

31. B. Colby and T. P. d'Estree, "Economic Evaluation of Mechanisms to Resolve Water Conflicts," *International Journal of Water Resources Development*, vol. 16, no. 2 (2000), pp. 239–51; Anthony R. Turton, "A Southern African Perspective on Transboundary Water Resources Management: Challenging Conventional Wisdom," *Environmental Change and Security Project Report*, Issue 9 (2003), pp. 75–87.

32. Incomati and Maputo Rivers from Turton, op. cit. note 31.

33. This list is derived from Annika Kramer, *Water and Conflict*, Draft Policy Briefing Paper prepared for U.S. Agency for International Development, Office of Conflict Management and Mitigation (Berlin, Bogor, and Washington, DC: Adelphi Research, Center for International Forestry Research, and Woodrow Wilson International Center for Scholars, 2004).

34. Ibid.; for Boran people, see www.both ends.org/encycl/cases/viewcase.php?cat=2&id=1 3&id_language=1; Arvari River from Iyer, op. cit. note 4; James D. Wolfensohn, "A Vision of Peace and Development on the Nile," Keynote Address at launch of the International Consortium for Cooperation on the Nile, Geneva, 26 June 2001.

35. For more information on the Every River Has Its People Project, see www.everyriver.net.

36. World Commission on Dams, op. cit. note 30, p. 177.

37. Fitzroy Nation, "Wajir Peace Initiative— Kenya: Back to Future Dialogue and Cooperation," *Conflict Prevention Newsletter*, January 1999, pp. 4–5; for more information on the Nile Basin Initiative negotiations, see the Secretariat's Web site, at www.nilebasin.org, or the World Bank's Nile Basin Initiative Web site, at www.worldbank.org/afr/nilebasin.

38. Carius, Dabelko, and Wolf, op. cit. note 3.

Resource Wealth and Conflict

1. One-quarter share from Michael Renner, *The Anatomy of Resource Wars*, Worldwatch Paper 162 (Washington, DC: Worldwatch Institute, 2002); number of deaths estimated from data in Milton Leitenberg, *Deaths in Wars and Conflicts Between 1945 and 2000* (College Park, MD: Center for International and Security Studies, University of Maryland, 2001); refugee numbers derived from U.N. High Commissioner for Refugees, at www.unhcr.ch, viewed 25 August 2002; internally displaced persons derived from U.S. Committee for Refugees, at www.refugees .org, viewed 25 August 2002.

2. Renner, op. cit. note 1; U.N. Development Programme, *Human Development Report 2004* (New York: Oxford University Press, 2004).

3. Renner, op. cit. note 1.

4. Ibid.; for an overview of cases in which indigenous peoples confront resource companies, see International Forum on Globalization, "Globalization: Effects on Indigenous Peoples" wall chart, at www.ifg.org/programs/indig.htm#map; Dulue Mbachu, "Villagers Flee Troops, Militia Fighting Near Nigerian Oil City," *Associated Press*, 10 September 2004.

5. Michael T. Klare, *Resource Wars: The New Landscape of Global Conflict* (New York: Metropolitan Books, 2001); Michael Renner, "Post-Saddam Iraq: Linchpin of a New Oil Order," *Foreign Policy in Focus*, January 2003; Chalmers Johnson, "America's Empire of Bases," *TomDispatch.com*, January 2004, at www.tomdispatch .com/index.mhtml?pid=1181; Matthew Yeomans, "Oil, Guns and Money," 2 September 2004, Global Policy Forum, at www.globalpolicy.org/ empire/analysis/2004/0902yeomans.htm.

6. Caspian from Lutz Kleveman, "Oil and the New 'Great Game'," *The Nation*, 16 February 2004, pp. 12–13; Ayako Doi, "Asian Enmities— China and Japan Revert to Hostility, and Hope for Reconciliation Fades," *Washington Post*, 29 August

2004; James Brooke, "The Asian Battle for Russia's Oil and Gas," *New York Times*, 3 January 2004; Julio Godoy, "U.S. and France Begin a Great Game in Africa," *allAfrica.com*, 11 August 2004; Eric Schmitt, "Pentagon Seeking New Access Pacts for African Bases," *New York Times*, 5 July 2003; Gerald Butt, "Thirst for Crude Pulling China Into Sudan," *The Daily Star* (Beirut, Lebanon), 17 August 2004; Felix Onua, "Nigeria and United States Agree on Military Exercises in Oil Delta," *Reuters*, 13 August 2004; Alexandra Guáqueta, "The Colombian Conflict: Political and Economic Dimensions," in Karen Ballentine and Jake Sherman, eds., *The Political Economy of Armed Conflict: Beyond Greed and Grievance* (Boulder, CO: Lynne Rienner Publishers, 2003), pp. 73–106.

7.   Jeffrey D. Sachs and Andrew M. Warner, "Natural Resource Abundance and Economic Growth," *Development Discussion Paper* (Cambridge, MA: Harvard Institute for International Development, 1995).

8.   Marc Lacey, "War Is Still a Way of Life for Congo Rebels," *New York Times*, 21 November 2002; Finbarr O'Reilly, "Rush for Natural Resources Still Fuels War in Congo," *Reuters*, 10 August 2004; illegal networks from United Nations Security Council, *Final Report of the Panel of Experts on the Illegal Exploitation of Natural Resources and Other Forms of Wealth of the Democratic Republic of the Congo* (New York: 2002).

9.   Renner, op. cit. note 1.

10. For transparency codes and other efforts, see Ian Bannon and Paul Collier, eds., *Natural Resources and Violent Conflict: Options and Actions* (Washington, DC: World Bank, 2003).

11. Ibid.; Ian Smillie, *Conflict Diamonds: Unfinished Business* (Ottawa, ON, Canada: International Development Research Centre, 2002); U.S. Government Accounting Office, *Critical Issues Remain in Deterring Conflict Diamond Trade* (Washington, DC: 2002), pp. 17–21; Jeremy Smith, "E.U. Aims to Stem Illegal Rainforest Timber Trade," *Reuters*, 14 October 2003.

12. See, for example, the Campaign to Eliminate Conflict Diamonds at www.phrusa.org/campaigns/sierra_leone/conflict_diamonds.html, the Fatal Transactions Campaign at www.niza.nl/fatal transactions/partner.html, Global Witness at www.globalwitness.org, Christian Aid at www.christian-aid.org.uk, and Project Underground at www.moles.org.

The Private Sector

1.   United Nations, "Role of Business in Armed Conflict Can Be Crucial—'For Good and For Ill' Secretary-General Tells Security Council Open Debate on Issue," press release (New York: 15 April 2004).

2.   Human Rights Watch, *Sudan, Oil, and Human Rights* (New York: 2003); Essential Action/Global Exchange, *Oil for Nothing—Multinational Corporations, Environmental Destruction, Death and Impunity in the Niger Delta* (San Francisco and Washington, DC: 1999); United Nations, *Sierra Leone: Report of the Panel of Experts Appointed Pursuant to UN Security Council Resolution 1306* (New York: 2000); Global Witness, *All the Presidents' Men—The Devastating Story of Oil and Banking in Angola's Privatised War* (London: 2002); U.S. Senate Committee on Banking, Housing, and Urban Affairs, *The Financial War on Terrorism and the Administration's Implementation of Title III of The USA Patriot Act* (Washington, DC: 29 January 2002).

3.   Karl Vick, "Oil Money is Fueling Sudan's War: New Arms Used to Drive Southerners from Land," *Washington Post*, 11 June 2001; Christian Aid, *The Scorched Earth: Oil and War in Sudan* (London: 2001); correlation between oil revenues and defense spending from Dita Smith and Laris Karklis, "Graphic: Fuelling War," *Washington Post*, 11 June 2001; use of landing strip from John Harker, *Human Security in the Sudan: The Report of a Canadian Assessment Mission* (Ottawa, ON, Canada: Department of Foreign Affairs and Trade, 2000), p. 15.

4.   Alison Azer, "Talismanized—Talisman's Exit, Sudan's Legacy," *Corporate Knights*, 25 March 2004.

5. Feizal Samath, "Sri Lankans Make Peace Their Business," *Asia Times*, 6 September 2002; idem, "Sri Lanka: Amid Peace, Business Turns to Education, Health," *Interpress Service*, 5 December 2002; Women Waging Peace, "Spotlight on Neela Marikkar, Sri Lanka," at www.womenwagingpeace.net/content/articles/0222a.html.

6. Rachel Goldwyn and Jason Switzer, *Assessments, Communities and Peace—A Critique of Extractive Sector Assessment Tools from a Conflict Sensitive Perspective* (Winnipeg, MB, Canada: IISD/International Alert, 2003); Jessica Banfield, Virginia Haufler, and Damian Lilly, *Transnational Corporations in Conflict Prone Zones: Public Policy Responses and a Framework for Action* (London: International Alert, 2003); International Finance Corporation, *Safeguard Policy for Environmental Assessment* (currently under review) (Washington, DC: 1998).

7. Publish What You Pay campaign, at www.publishwhatyoupay.org; Extractive Industries Transparency Initiative, at www2.dfid.gov.uk/pubs/files/eitinewslettermarch04.pdf.

8. Georgette Gagnon, Audrey Macklin, and Penelope Simons, *Deconstructing Engagement: Corporate Self-Regulation in Conflict Zones—Implications for Human Rights and Canadian Public Policy* (Ottawa, ON, Canada: SSHRC/Law Commission of Canada, 2003); Philippe Le Billon, *Getting It Done: Instruments of Enforcement*, in Paul Collier and Ian Bannon, eds., *Natural Resources and Violent Conflict: Options and Actions* (Washington, DC: World Bank, 2003).

9. For more on Afghanistan and other pilot efforts at business investment in post-conflict zones, see Business Humanitarian Forum, at www.bhforum.ch; for more on tri-sector partnerships in the extractives sector, see the Natural Resource Cluster of the Business Partners for Development initiative, at www.bpd-natural resources.org; for multistakeholder groups, see, for example, the Chad-Cameroon Independent Assurance Group, at www.gic-aig.org.

Chapter 6. Changing the Oil Economy

1. Products made with petroleum from www.anwr.org/features/oiluses.htm, viewed 1 August 2004.

2. Amory Lovins, *Energy Security Facts: Details and Documentation*, 2nd ed. (Snowmass, CO: Rocky Mountain Institute, 2003), p. 8; Figure 6–1 adapted from U.S. Department of Energy (DOE), Energy Information Administration (EIA), *Energy in the United States, 1635–2000*, at www.eia.doe.gov/emeu/aer/eh/frame.html.

3. Vaclav Smil, *Energy in World History* (Boulder, CO: Westview Press, 1994), p. 187; source and emissions data from International Energy Agency (IEA), *Key World Energy Statistics 2003* (Paris: 2003); oil share of transportation sector energy consumption from DOE, EIA, *Annual Energy Review 2003* (Washington, DC: 2004), p. 42. Box 6–1 from the following: top corporations from "Forbes 500s," *Forbes Magazine Online*, viewed 4 August 2004; revenues from *Oil and Gas Journal*, 15 September 2003, p. 46; 10 largest corporations worldwide from Sarah Anderson and John Cavanaugh, *Top 200: The Rise of Corporate Global Power* (Washington, DC: Institute for Policy Studies, 2000); automobile production and fleet data from Worldwatch Institute, "World Automobile Production, 1950–2003" and "World Passenger Car Fleet, 1950–2003, in *Signposts 2004*, CD-ROM (Washington, DC: 2004); Chinese auto data from Peter S. Goodman, "Car Culture Captivates China," *Washington Post*, 8 March 2004; air travel from Worldwatch Institute, "World Air Travel, 1950–2002," in *Signposts 2004*, op. cit. this note.

4. Culture of energy consumption from Paul Roberts, *The End of Oil: On the Edge of a Perilous New World* (Boston: Houghton Mifflin, 2004), p. 31.

5. IEA, *Analysis of the Impact of High Oil Prices on the Global Economy* (Paris: 2004); Robert D. McTeer, "The Gas in Our Tanks" (op-ed), *Wall Street Journal*, 11 May 2004.

6. DOE, op. cit. note 3, p. 9; British Petroleum

(BP), *Statistical Review of World Energy 2004* (London: 2004).

7. IEA, op. cit. note 5; Philip K. Verleger, "Energy: The Gathering Storm" (Washington, DC: Institute for International Economics, unpublished paper, 29 September 2004).

8. Stipends from Robert Baer, *Sleeping With the Devil: How Washington Sold Our Soul for Saudi Crude* (New York: Crown, 2003), p. 170; James Woolsey, presentation to the National Commission on Energy Policy conference, "Global Challenges for U.S. Energy Policy: Economic, Environmental, and Security Risks," Washington, DC, 5 March 2004.

9. Spot prices from DOE, EIA, "U.S. Petroleum Prices," at www.eia.doe.gov/oil_gas/petroleum/info_glance/prices.html; Figure 6–2 data from New York Mercantile Exchange, e-mail to Molly Aeck, Worldwatch Institute, 10 August 2004.

10. Figure 6–3 data from DOE, op. cit. note 3, pp. 303, 313, and from DOE, EIA, *International Petroleum Monthly*, September 2004.

11. DOE, EIA, *Country Analysis Brief: OPEC* (Washington, DC: 2004).

12. Carl Mortished, "OPEC Set to Raise Target Oil Price," *The Times* (London), 13 September 2004; price in 1998 from DOE, op. cit. note 3, p. 165; spare capacity from BP, op. cit. note 6; Russian production from IEA, *Oil Market Report* (Paris: 2004); market sensitivity from PFC Energy, "PFC Energy's Global Crude Oil and Natural Gas Liquids Supply Forecast," PowerPoint presentation, Washington, DC, September 2004.

13. Conventional view described in Klaus Rehaag, IEA, "Is the World Facing a 3rd Oil Shock?" presented at FVG & IBP Workshop, Rio de Janeiro, Brazil, 12 July 2004; projection from IEA, *World Energy Outlook 2004* (Paris: 2004), p. 81.

14. Colin J. Campbell and Jean H. Laherrere, "The End of Cheap Oil," *Scientific American*,

March 1998; Kjell Aleklett, Upsala University, "The Peak and Decline of World Production of Oil," PowerPoint presentation, Asia Pacific Energy Conference, Osaka, Japan, September 2004.

15. Annual discoveries from Campbell and Laherrere, op. cit. note 14; other analysts' production decline projections from A. M. Samsam Bakhtiari, "World Oil Production Capacity Model Suggests Output Peak by 2006–2007," *Oil and Gas Journal*, 26 April 2004.

16. Campbell and Laherrere, op. cit. note 14; Figure 6–4 from DOE, EIA, at www.eia.doe.gov/emeu/aer/petro.html, viewed 22 October 2004.

17. Production declines from PFC Energy, op. cit. note 12; futures prices from Verleger, op. cit. note 7.

18. Production peak estimate in PFC Energy, op. cit. note 12; Rogers cited in Jeffrey Ball, "As Prices Soar, Doomsayers Provoke Debate on Oil's Future," *Wall Street Journal*, 21 September 2004.

19. Matthew R. Simmons, "The Saudi Arabian Oil Miracle," PowerPoint presentation at the Center for Strategic and International Studies, Washington, DC, 24 February 2003; F. Jay Schempf, "Simmons Hopes He's Wrong," *Petroleum News*, 1 August 2004.

20. Iraqi production from IEA, op. cit. note 12.

21. Royal Navy from David L. Rousseau, "History of the Oil Industry," September 1998, at www.ssc.upenn.edu/polisci/psci260/OILHIST.HTM, and from Daniel Yergin, *The Prize: The Epic Quest for Oil, Money, and Power* (New York: Simon & Schuster, 1991), p. 156; Suez crisis in ibid., pp. 479–98.

22. Strategic value and Japanese and German invasions from Rousseau, op. cit. note 21; "blood of victory" from Yergin, op. cit. note 21, p. 183.

23. Baer, op. cit. note 8, pp. 81–83.

24. U.S. oil fields from Yergin, op. cit. note 21, p. 406; United States as net importer and business

relationships from Rousseau, op. cit. note 21.

25. Feis cited in Luke Mitchell, "Blood for Oil," *Harper's Magazine*, July 2004, p. 78; Truman and Eisenhower cited in ibid.

26. Contingency plans from Mitchell, op. cit. note 25; oil as a weapon from Dennis Pirages and Theresa DeGeest, *Ecological Security: An Evolutionary Perspective on Globalization* (Lanham, MD: Rowman & Littlefield, 2004); Kissinger and Carter from Michael Klare, "Bush-Cheney Energy Strategy: Procuring the Rest of the World's Oil," *Foreign Policy in Focus, PetroPolitics Special Report*, January 2004.

27. Carter Doctrine and Persian Gulf output from Klare, op. cit. note 26; Persian Gulf as source of one fifth of U.S. imports from DOE, op. cit. note 3, p. 135.

28. Production loss consequences from James Placke, Cambridge Energy Research Associates, presentation to the National Commission on Energy Policy conference, "Global Challenges for U.S. Energy Policy: Economic, Environmental, and Security Risks," Washington, DC, 5 March 2004; analysis of the strategic and business advantages to the United States and U.S. firms of the Iraq invasion from Michael Renner, "Fueling Conflict," *Foreign Policy in Focus, PetroPolitics Special Report*, January 2004; Iraq as swing producer from Independent Task Force on Strategic Energy Policy, *Strategic Energy Policy: Challenges for the 21st Century*, Baker Institute Study No. 15 (Houston, TX: James A. Baker III Institute for Public Policy of Rice University, and the Council on Foreign Relations, April 2001); control of Iraqi oil reserves from Strategic Forecasting, Inc., "War Plan: Consequences," *Stratfor Weekly*, 18 March 2003.

29. Estimate of $49 billion a year from Milton R. Copulos, *America's Achilles Heel: The Hidden Costs of Imported Oil* (Washington, DC: National Defense Council Foundation, 2003), p. 42.

30. Renner, op. cit. note 28.

31. See Michael T. Klare, *Blood for Oil* (New York: Metropolitan Books, 2004), pp. 146–79; rights abuses from various Human Rights Watch reports, at www.hrw.org, viewed 9 August 2004.

32. Natural resource curse from, for example, Renner, op. cit. note 28, and from Michael Renner, *The Anatomy of Resource Wars*, Worldwatch Paper 162 (Washington, DC: Worldwatch Institute, 2002); Norm Dixon, "Sudan: Oil Profits Behind West's Tears for Darfur," *Green Left Weekly*, 9 August 2004; United States from Robert Bryce, *Cronies: Oil, the Bushes, and the Rise of Texas, America's Superstate* (New York: Public Affairs, 2004), pp. 30–32; indigenous peoples from, for example, Kevin Koenig, "ChevronTexaco on Trial," and Ed Ayres, "The Hidden Shame of the Global Industrial Economy," both in *World Watch*, January/February 2004.

33. Oil effect on democracy from Michael L. Ross, "Does Oil Hinder Democracy?" *World Politics*, April 2001, pp. 325–61; terrorism as response to lack of rights from Alan Krueger and Jitka Maleckova, *Education, Poverty, Political Violence, and Terrorism: Is There a Causal Connection?* (Cambridge, MA: National Bureau of Economic Research, 2002); oil-dependent countries and corruption from Thomas L. Palley, "Lifting the Natural Resource Curse," *Foreign Service Journal*, December 2003, pp. 54–60; "devil's tears" from Lutz Kleveman, "The New 'Great Game'," *The Nation*, 16 February 2004.

34. Radicals from Ahmed Rashid, "Osama bin Laden: How the U.S. Helped Midwife a Terrorist," Special Report, The Center for Public Integrity (Washington, DC: 13 September 2001); Saudi charities from David E. Kaplan, "The Saudi Connection: How Billions in Oil Money Spawned a Global Terror Network," *U.S. News & World Report*, 15 December 2003; terrorist assaults from Rashid, op. cit. this note, and from "Six Charged in U.S.S. Cole Bombing," *Associated Press*, 7 July 2004.

35. Kaplan, op. cit. note 34.

36. White House report statement from U.S. Global Change Research Program, *Our Changing Planet: The U.S. Climate Change Science Program* *for Fiscal Years 2004 and 2005* (Washington, DC: 2004), p. 47; oil share of carbon emissions from IEA, op. cit. note 3.

37. Scientists' quote from "The Changing Atmosphere: Implications for Global Security," Toronto, Canada, 1988, cited in Jon Barnett, *Security and Climate Change*, Tyndall Centre for Climate Change Research, Working Paper 7 (Norwich, U.K.: 2001), p. 4; Sir David King cited in Steve Connor and Andrew Grice, "Scientist 'Gagged' by No. 10 After Warning of Global Warming Threat," *Independent* (London), 8 March 2004.

38. Threat to island nations from Robert Watson, Presentation of the Chair of the Intergovernmental Panel on Climate Change to the Sixth Conference of the Parties of the Framework Convention on Climate Change, 13 November 2000; potential cause of conflict from Federal Ministry for the Environment, Nature Conservation, and Nuclear Safety (BMU), *Climate Change and Conflict: Can Climate Change Impacts Increase Conflict Potentials? What Is the Relevance of this Issue for the International Process on Climate Change?* (Berlin: November 2002), pp. 8, 50, and from U.S. Department of Defense, Office of the Deputy Under Secretary of Defense, *Climate Change, Energy Efficiency, and Ozone Protection* (Washington, DC: undated).

39. Civilization and stable climate in Peter Schwartz and Doug Randall, *An Abrupt Climate Change Scenario and Its Implications for United States National Security* (Emeryville, CA: Global Business Network, 2003), p. 14; carbon dioxide concentration from Daniel Schrag, Harvard University, "Back to the Future: The Great Climate Experiment," speech given at State of the Planet 04: Mobilizing the Sciences to Fight Global Poverty, Columbia University, New York, spring 2004, and from Maggie Fox, "Climate Change Experts Despair Over U.S. Attitude," *Reuters*, 15 June 2004; accelerating rate of increase from Tim Whorf, Scripps Institute of Oceanography, San Diego, CA, discussion with Janet Sawin, 25 February 2004; "Supercomputer Finds Climate Likely to Heat Up Fast," *Environment News Service*, 24 June 2004.

40. BMU, op. cit. note 38, p. 8; amplification of existing threats from Intergovernmental Panel on Climate Change, Working Group II, *Climate Change 2001: Impacts, Adaptation, and Vulnerability* (Cambridge, U.K.: Cambridge University Press, 2001).

41. Thomas F. Homer-Dixon, "On the Threshold: Environmental Changes as Causes of Acute Conflict," *International Security*, fall 1991, pp. 76–116.

42. Large-country impacts from Wallace Broecker, Columbia University, interview, spring 2004, at www.earthfiles.com/news/; use of food as weapon from Peter Wallensteen, "Food Crops as a Factor in Strategic Policy and Action," in Arthur H. Westing, ed., *Global Resources and International Conflict*, pp. 151–55, cited in Homer-Dixon, op. cit. note 41; conflicts over water from Thomas H. Karas, "Global Climate Change and International Security," Advanced Concepts Group, Sandia National Laboratories, Albuquerque, NM, November 2003, p. 15.

43. Conflict contingencies in BMU, op. cit. note 38, p. 46; insurgencies and civil wars in Barnett, op. cit. note 37, pp. 5, 6; A. Rahman, "Climate Change and Violent Conflicts," in M. Suliman, ed., *Ecology, Politics and Violent Conflict* (New York: Zed Books, 1999), pp. 181–210, cited in Barnett, op. cit. note 37, p. 8; past migrations have spawned conflicts from A. Swain, *The Environmental Trap: The Ganges River Diversion, Bangladeshi Migration and Conflicts in India* (Uppsala, Sweden: Department of Peace and Conflict Research, Uppsala University Report, 1996), cited in Barnett, op. cit. note 37, p. 8.

44. Study from Alistair Doyle, "160,000 Die Yearly from Global Warming," *Reuters*, 30 September 2003, and from A. Haines and J. A. Patz, "Health Effects of Climate Change," *Journal of the American Medical Association*, 7 January 2004, pp. 99–103.

45. Shell and Stockholm Environment Institute cited in David Elliott, "A Sustainable Future? The Limits to Renewables," in Foundation for the Economics of Sustainability, *Before the Wells Run*

Dry: Ireland's Transition to Renewable Energy (Dublin, Ireland: 2002).

46. Among many such statements, see, for example, Independent Task Force on Strategic Energy Policy, op. cit. note 28, as well as Timothy E. Wirth, C. Boyden Gray, and John D. Podesta, "The Future of Energy Policy," *Foreign Affairs*, July/August 2003, and Nicholas Varchaver, "Brainstorm: Energy Security—How To Kick the Oil Habit," *Fortune*, 9 August 2004.

47. Joseph Eto, *The Past, Present, and Future of U.S. Utility Demand-Side Management Programs* (Berkeley, CA: Lawrence Berkeley National Laboratory, 1996).

48. Avoided capacity in DOE, op. cit. note 3, p. 265; decline of demand-side management programs in Thomas Prugh et al., *Natural Capital and Human Economic Survival*, 2nd ed. (Boca Raton, FL: Lewis Publishers, 1999), pp. 142–43.

49. Lovins, op. cit. note 2, pp. 9, 11.

50. Fuel economy improvement and new technologies from ibid., pp. 13–15; fleet efficiency doubling estimate from National Research Council, *Effectiveness and Impact of Corporate Average Fuel Economy (CAFE) Standards* (Washington, DC: National Academy Press, 2002).

51. Electricity savings estimate from Joseph Romm, *The Hype About Hydrogen: Fact and Fiction in the Race to Save the Climate* (Washington, DC: Island Press, 2004), p. 193; consumption and generation shares from DOE, op. cit. note 3, pp. 39–41, 224.

52. Janet L. Sawin, *Mainstreaming Renewables in the 21st Century*, Worldwatch Paper 169 (Washington, DC: Worldwatch Institute, 2004), pp. 16–17. Box 6–2 from the following: Oxburgh statement from David Adam, "'I'm Really Very Worried for the Planet,'" *The Guardian* (London), 17 June 2004; Statoil injection and 280 years' worth of storage from Koichi Sasaki, *Carbon Sequestration Technology: Current Status and Future Outlook* (Tokyo: Institute of Energy Economics–Japan, 2003), p. 3; other estimates from

Robert F. Service, "The Carbon Conundrum," *Science*, 13 August 2004, p. 962; cost estimates from Office of Fossil Energy, DOE, "Carbon Capture Research," fact sheet (Washington, DC: 2 August 2004); suitability only for point sources from Meyer Steinberg, Letter to Editor, *Science*, 5 September 2004, p. 1326; vehicle share of global carbon emissions from IEA, op. cit. note 3; monitoring studies from Service, op. cit. this note; leakage problems from Climate Action Network Australia, "Carbon Leakage and Geosequestration," press release (Eveleigh, NSW, Australia: September 2004).

53. Teresa Malyshev, IEA, "Biofuels for Transport: An International Perspective," presentation at the International Conference for Renewable Energies, Bonn, Germany, 3 June 2004.

54. See, for example, Romm, op. cit. note 51, and Matthew Wald, "Questions About a Hydrogen Economy," *Scientific American*, May 2004; nuclear power problems from Rob Edwards, "Nuclear Share of Electricity Predicted to Fall," *New Scientist*, 26 June 2004, and from Mark Gielecki and James G. Hewlett, "Commercial Nuclear Electric Power in the United States: Problems and Prospects," in DOE, EIA, *Monthly Energy Review*, August 1994.

55. Pessimists include Romm, op. cit. note 51; optimists include Amory Lovins, *Twenty Hydrogen Myths* (Snowmass, CO: Rocky Mountain Institute, 2003), and James Mason, *Electrolytic Production of Hydrogen Gas With Photovoltaic Electricity as a Replacement Fuel for Motor Gasoline in the United States: Land, Water, and Photovoltaic Resource Requirements* (Farmingdale, NY: Solar Hydrogen Education Project, 2003); fuel cells versus hybrids from "Clean Machine," *Economist*, 2 September 2004; hybrids in developing electronic controls and systems from Lovins, op. cit. this note.

56. Subsidies from Greenpeace, "The Subsidy Scandal: The European Clash Between Environmental Rhetoric and Public Spending," undated, at archive.greenpeace.org/comms/97/climate/eusub.html; Greenpeace, "Greenpeace Exposes the Dirty Face of Europe's Energy Subsidies,"

press release (Brussels: 2 April 2004); Taxpayers for Common Sense, "Fossil Fuel Subsidies: A Taxpayer Perspective," fact sheet (Washington, DC: undated); American Wind Energy Association, "Energy Subsidies: How Do Energy Subsidies Distort the Energy Market," FAQ (Washington, DC: 2000); European Environment Agency, *Energy Subsidies in the European Union: A Brief Overview* (Copenhagen: 2004); World Bank, "Massive, Inequitable Energy Subsidies Hinder Development," press release (Washington, DC: 9 June 2002).

57. Policy lessons from Sawin, op. cit. note 52, pp. 34–44.

58. Estimated value of global energy infrastructure from John Holdren, presentation at the National Commission on Energy Policy conference, "Global Challenges for U.S. Energy Policy: Economic, Environmental, and Security Risks," Washington, DC, 5 March 2004, and Roberts op. cit. note 4; required investment from Hiroyuki Kato, IEA, "World Energy Investment Outlook: Prospects and Challenges," presentation at KEEI-IEA Joint Conference on Northeast Asia Energy Cooperation, 16 March 2004.

59. China pledge from Mark Landler, "China Pledges to Use More Alternatives to Oil and Coal," *New York Times*, 5 June 2004; states from Terri Suess and Cheryl Long, "Go Solar, Be Secure," *Mother Earth News*, February-March 2002.

Nuclear Energy

1. World Nuclear Association (WNA), "World Nuclear Power Reactors 2003–04 and Uranium Requirements," 30 September 2004, at www.world-nuclear.org/info/reactors.htm, viewed 11 October 2004.

2. Charles Digges, "FSB Plants Fake Bomb at RT-2 SNF Storage Facility," *Bellona* (Bellona Foundation), 13 January 2003; Union of Concerned Scientists, *Backgrounder on Nuclear Reactor Security* (Cambridge, MA: 2002); Greenpeace, "Greenpeace Volunteers Get Into 'Top Security' Nuclear Control Centre," press release (London:

13 January 2004); Rosa Prince, "We Join Raid on Nuke Power Centre," *Daily Mirror*, 14 January 2003.

3. Greenpeace, "Calendar of Nuclear Accidents and Events," at archive.greenpeace.org/comms/nukes/chernob/rep02.html, viewed 10 September 2004; WNA, *Chernobyl Accident* (London: 2004); David R. Marples, "Chernobyl's Toll After Ten Years: 6,000 and Counting," *Bulletin of the Atomic Scientists*, May/June 1996.

4. Nuclear Information and Resource Service, "Davis-Besse Nuclear Plant Comes Close to Disaster As Lax Regulator Places Company Interests Ahead of Public Safety," press release (Washington, DC: 13 March 2002); David Lochbaum, *U.S. Nuclear Plants in the 21st Century: The Risk of a Lifetime* (Cambridge, MA: Union of Concerned Scientists, 2004).

5. Heinrich Graul, *The Global Nuclear Fuel Market, Supply and Demand 2003–2025*, Symposium Proceedings, 3–5 September 2003 (London: WNA, 2003); Organization for Security and Co-operation in Europe, "OSCE Campaign to Focus on Health Hazards of Old Kyrgyz Uranium Mines," press release (Bishkek, Kyrgyzstan: 9 January 2004); Rob Edwards, "Flooding of Soviet Uranium Mines Threatens Millions," *New Scientist*, 16 May 2002.

6. Unsecured waste from Michael T. Burr, "The Nuclear Non Starter," *Public Utilities Fortnightly*, 15 June 2003, p. 27; Ken Silverstein, "Yucca Mountain Faces Tough Hurdles Ahead," *Utilipoint International*, Issue Alert, 9 March 2004; Nuclear Information and Resource Service, Southeast Office, *Why Yucca Mt Will Leak* (Augusta, GA: undated); Suzanne Struglinski, "State Test Shows Corrosion at Yucca," *Las Vegas Sun*, 12 May 2004.

7. Nigel Hunt, "Russian Nuclear Warheads Help to Power US," *Planet Ark*, 16 March 2004.

8. Oeko-Institut, *Global Emission Model for Integrated Systems* (Darmstadt, Germany: 2002); Uwe R. Fritsche and Sui-San Lim, *Comparison of Greenhouse-Gas Emission and Abatement Cost of Nuclear and Alternative Energy Options from a*

Life-Cycle Perspective (Darmstadt, Germany: Oeko-Institut, 1997).

9. France from WNA, op. cit. note 1; countries planning to eliminate from Rob Edwards, "Nuclear Disarray as Europe Pushes East," *New Scientist*, 1 May 2004; new reactors from WNA, op. cit. note 1; Finland from idem, *Nuclear Energy in Finland*, Information and Issue Brief (London: June 2004); increase in nuclear capacity from Graul, op. cit. note 5.

Chapter 7. Disarming Postwar Societies

1. Victims since 1975 and civilian share from James Grant, Executive Director, UNICEF, and Cyrus Vance and Herbert A. Okun, Statements before U.S. Senate Appropriations Committee, Subcommittee on Foreign Operations, "Hearing on the Global Landmine Crisis," Washington, DC, 13 May 1994; International Campaign to Ban Landmines, *Landmine Monitor Report 2001: Executive Summary* (Washington, DC: 2001), pp. 3–4.

2. Debbie Hillier and Brian Wood, *Shattered Lives: The Case for Tough International Arms Control* (London and Oxford: Amnesty International and Oxfam International, 2003).

3. Frieda Berrigan, "Small Arms? Big Problem," CommonDreams.org, at www.commondreams.org/views04/0709-01.htm, viewed 9 July 2004.

4. Central America from Michael Klare and David Andersen, *A Scourge of Guns: The Diffusion of Small Arms and Light Weapons in Latin America* (Washington, DC: Arms Sales Monitoring Project, Federation of American Scientists, August 1996); Southern Africa from Jacklyn Cock, "A Sociological Account of Light Weapons Proliferation in Southern Africa," in Jasjit Singh, ed., *Light Weapons and International Security* (Delhi: Indian Pugwash Society and British American Security Information Council, December 1995); Sri Lanka from Graduate Institute of International Studies (GIIS), *Small Arms Survey 2003* (New York: Oxford University Press, 2003), p. 138, and from idem, *Small Arms Survey 2004* (New York: Oxford University Press, 2004), p. 59.

5. Latin American firearm deaths from GIIS, *Survey 2004*, op. cit. note 4, p. 53; Rio de Janeiro from idem, *Survey 2003*, op. cit. note 4, p. 147; El Salvador from Hillier and Wood, op. cit. note 2, p. 48.

6. GIIS, *Survey 2003*, op. cit. note 4, p. 140.

7. Ibid., pp. 142–43, 147; Hillier and Wood, op. cit. note 2, p. 50.

8. South Africa from GIIS, *Survey 2003*, op. cit. note 4, pp. 144–45, 149; Latin America from "The Backlash in Latin America," *The Economist*, 30 November 1996.

9. GIIS, *Small Arms Survey 2003*, op. cit. note 4, pp. 57–58; Table 7–1 compiled from GIIS, *Small Arms Survey* (New York: Oxford University Press, various annual editions).

10. Estimate of military-type weapons from GIIS, *Small Arms Survey 2002* (New York: Oxford University Press, 2002), p. 73; number of assault rifles produced from GIIS, *Small Arms Survey 2001* (New York: Oxford University Press, 2001), p. 63, and from Michael Renner, "The Cycle of Supply and Demand of Light Weapons in the North and the South, Including the Relationship Between Violent Conflict and the Ready Availability of These Weapons," prepared for Sustainable Disarmament for Sustainable Development conference, Brussels, 12–13 October 1998; Bonn International Center for Conversion (BICC), *Conversion Survey 2004* (Baden-Baden, Germany: Nomos Verlagsgesellschaft, 2004), pp. 95–96.

11. GIIS, *Survey 2004*, op. cit. note 4, pp. 7–8, 10, 27; ammunition production from idem, *Survey 2003*, op. cit. note 4, p. 13.

12. GIIS, *Survey 2003*, op. cit. note 4, pp. 9, 14; number of companies and countries from idem, *Survey 2004*, op. cit. note 4, pp. 9–10; non-state groups from Stephanie G. Neuman, "The Arms Trade, Military Assistance, and Recent Wars: Change and Continuity," *Annals of the American Academy*, September 1995, and from Christopher Smith, "Light Weapons and the International Arms Trade," in Christopher Smith, Peter

Batchelor, and Jakkie Potgieter, UNIDIR Disarmament and Conflict Resolution Project, *Small Arms Management and Peacekeeping in Southern Africa* (New York and Geneva: United Nations, 1996); illicit production from GIIS, *Survey 2003*, op. cit. note 4, pp. 26–36.

13. GIIS, *Survey 2003*, op. cit. note 4, pp. 98–102.

14. Michael Renner, "Arms Control Orphans," *Bulletin of the Atomic Scientists*, January/February 1999, p. 23.

15. BICC, op. cit. note 10, p. 130; Michael Brzoska and Herbert Wulf, "Clean Up the World's Glut of Surplus Weapons," *International Herald Tribune*, 5 June 1997.

16. BICC, op. cit. note 10, pp. 138–42; BICC, *Conversion Survey 2003* (Baden-Baden, Germany: Nomos Verlagsgesellschaft, 2003), pp. 86–87.

17. Michael Renner, *Small Arms, Big Impact: The Next Challenge of Disarmament*, Worldwatch Paper 137 (Washington, DC: Worldwatch Institute, 1997), pp. 33–34; idem, *The Anatomy of Resource Wars*, Worldwatch Paper 162 (Washington, DC: Worldwatch Institute, 2002).

18. Argentine arms from GIIS, *Survey 2003*, op. cit. note 4, p. 116, and from idem, *Survey 2001*, op. cit. note 10, p. 185; Saudi stocks lost from idem, *Survey 2004*, op. cit. note 4, p. 102; Philippines from idem, *Survey 2001*, op. cit. note 10, p. 183.

19. Somalia from Smith, op. cit. note 12; Albania from GIIS, *Survey 2004*, op. cit. note 4, pp. 54, 56; Table 7–2 from Renner, *Small Arms, Big Impact*, op. cit. note 17, pp. 39–40, from GIIS, *Small Arms Survey*, 2003 edition (p. 119), 2002 edition (p. 99), and 2001 edition (pp. 181–183), and from "U.N. Liberia Mission Investigating Weapons-Smuggling Claims," *UN Wire*, 22 July 2004.

20. Iraq from GIIS, *Survey 2004*, op. cit. note 4, pp. 44–50; Iraqi weapons in Saudi Arabia from Anonymous, *Imperial Hubris: Why the West Is Losing the War on Terror* (Washington, DC: Brassey's, Inc., 2004), p. 73.

21. Estimate of lost or stolen firearms from GIIS, *Survey 2004*, op. cit. note 4, pp. 43, 55, 65.

22. Box 7–1 from the following: "Uniendo Esfuerzos por Colombia" (Joining Efforts for Colombia), conference organized by Georgetown University, Ecole de la Paix de France, and Friedrich Ebert Foundation, Washington, DC, 24 June 2002; "La Solución Política y la Democracia son el Camino," a report by Colombian grassroots, human rights, and peace organizations, prepared for London donors' meeting, 9 June 2003; International Crisis Group, "Demobilising the Paramilitaries in Colombia: An Achievable Goal," *Latin America Report*, No. 8, 5 August 2004; Economic Commission for Latin America, *Panorama Social de América Latina 2002–2003* (Santiago, Chile: 2003); World Bank, *World Development Indicators 2002* (Washington, DC: 2002), p. 296; Juan Gabriel Tokatlián, *Globalización, Narcotráfico y Violencia* (Bogotá: Norma, 2000); U.N. Office on Drugs and Crime, "The Role of Alternative Development in Drug Control and Development Cooperation," International Conference, Feldafing, Germany, 7–12 January 2002.

23. Organization for Security and Co-operation in Europe and Wassenaar data sharing from GIIS, *Survey 2004*, op. cit. note 4, p. 115 (the members of the Wassenaar Arrangement are primarily East and West European nations, plus the United States, Canada, Australia, New Zealand, Japan, South Korea, and Argentina); reluctance to make information public from Edward Laurance, "Reflections on the 'Decade of Disarmament'—1986–1999," presented at Promoting Security: But How and For Whom?, BICC 10-Year Anniversary Conference, Bonn, Germany, 1–2 April 2004, p. 9.

24. BICC, op. cit. note 10, pp. 85, 131–36; GIIS, *Survey 2004*, op. cit. note 4, pp. 112–14, 125; Hillier and Wood, op. cit. note 2, pp. 71, 78–79; Organization of American States, at www.oas.org/juridico/english/treaties/a-63.html; South Eastern Europe Clearinghouse for the Control of Small Arms and Light Weapons, at www.seesac.org; "Protocol Curbing Illegal Gun Trade Signed by 11 African Nations," *UN Wire*, 22 April 2004.

25. Weaknesses of regional efforts from Hillier and Wood, op. cit. note 2, p. 71.

26. Britain from GIIS, *Survey 2004*, op. cit. note 4, p. 99; Hillier and Wood, op. cit. note 2, pp. 40–41.

27. GIIS, *Survey 2004*, op. cit. note 4, pp. 249–58.

28. BICC, op. cit. note 10, p. 98; idem, op. cit. note 16, p. 83; GIIS, *Survey 2004*, op. cit. note 4, pp. 259–60; Mireille Widmer and Atef Odibat, "Focus on the Middle East and North Africa," *Small Arms and Human Security Bulletin*, February 2004, pp. 1–3.

29. United Nations Office on Drugs and Crime, at www.unodc.org/unodc/en/crime_cicp_signa tures_firearms.html, viewed 9 September 2004.

30. GIIS, *Survey 2004*, op. cit. note 4, pp. 68–71; Larry Rohter, "Brazil Adopts Strict Gun Controls to Try to Curb Murders," *New York Times*, 21 January 2004; Brazilian victims from Hillier and Wood, op. cit. note 2, p. 14; U.S. ban on assault weapons from Sheryl Gay Stolberg, "Effort to Renew Weapons Ban Falters on Hill," *New York Times*, 9 September 2004, and from Fox Butterfield, "As Expiration Looms, Gun Ban's Effect is Debated," *New York Times*, 10 September 2004.

31. European Union embargoes from Stockholm International Peace Research Institute, "International Arms Embargoes," at projects.sipri .se/armstrade/embargoes2003.html, viewed 4 September 2004; Table 7–3 derived from ibid. and from Ken Epps, *International Arms Embargoes*, Ploughshares Working Paper 02-4 (Waterloo, ON, Canada: Project Ploughshares, 2002).

32. Seventeen operations since 1990 from GIIS, *Survey 2003*, op. cit. note 4, p. 293.

33. Experiences from different operations are analyzed in detail in several volumes in the series *Managing Arms in Peace Processes*, UNIDIR Disarmament and Conflict Resolution Project (New York and Geneva: United Nations, 1995 and 1996), and are addressed in GIIS, *Survey 2004*, op.

cit. note 4, pp. 57–58, and in idem, *Survey 2003*, op. cit. note 4, pp. 298–99.

34. GIIS, *Survey 2003*, op. cit. note 4, pp. 277, 282–96, 302–04; *Managing Arms in Peace Processes*, op. cit. note 33.

35. Renner, *Small Arms, Big Impact*, op. cit. note 17; Sami Faltas and Wolf-Christian Paes, *Exchanging Guns for Tools*, BICC Brief 29 (Bonn, Germany: BICC, 2004).

36. Edward Laurance, *The New Field of Micro-Disarmament: Addressing the Proliferation and Buildup of Small Arms and Light Weapons*, BICC Brief 7 (Bonn, Germany: BICC, September 1996); Sverre Lodgaard, "Demobilization and Disarmament: Experiences to Date," *UNIDIR Newsletter*, no. 32, 1996; Peter Batchelor, "Disarmament, Small Arms, and Intra-State Conflict: The Case of Southern Africa," in Smith, Batchelor, and Potgieter, op. cit. note 12.

37. UNDP from GIIS, *Survey 2003*, op. cit. note 4, p. 154; Albania and Cambodia from BICC, op. cit. note 16, pp. 84–85; cross-sectoral approach from GIIS, *Survey 2003*, op. cit. note 4, p. 155.

38. GIIS, *Survey 2003*, op. cit. note 4, pp. 60, 155, 255; Viva Rio from "Holding Up Development: The Effects of Small Arms and Light Weapons in Developing Countries," *id21 Media*, at www.id21.org/id21-media/arms.html, viewed 7 August 2004. Table 7–4 adapted from the following: GIIS, *Survey 2003*, op. cit. note 4, pp. 279–81 (which lists 40 "major" programs); Thailand from idem, *Survey 2004*, op. cit. note 4, p. 71; Mozambique from Faltas and Paes, op. cit. note 35, p. 20.

39. Total of 8 million destroyed since 1990 from GIIS, *Survey 2004*, op. cit. note 4, p. 58; Russia, Ukraine, and Bulgaria from ibid., p. 60, and from BICC, op. cit. note 10, pp. 139–40; Table 7–5 is adapted from GIIS, *Survey 2004*, op. cit. note 4, p. 58, and from idem, *Survey 2002*, op. cit. note 10, p. 74.

40. Kees Kingma and Vanessa Sayers, *Demobilization in the Horn of Africa*, BICC Brief 4 (Bonn,

Germany: BICC, June 1995); Batchelor, op. cit. note 36; BICC, *Conversion Survey 1996* (New York: Oxford University Press, 1996); BICC, op. cit. note 16, pp. 65–66, 116.

41. BICC, op. cit. note 40; Kingma and Sayers, op. cit. note 40.

42. Box 7–2 from the following: Movimento Popular para a Libertação da Angola (MPLA) and União Nacional para a Independencia Total da Angola (UNITA) from Inge Tvedten, ed., "Introduction," in *Angola 2001/2002: Key Development Issues and Aid in a Situation of Peace* (Bergen, Norway: Christian Michelsen Institute, 2002); resource income, corruption, and lack of transparency from Renner, *Anatomy of Resource Wars,* op. cit. note 17, and from Justin Pearce, *War, Peace and Diamonds in Angola: Popular Perceptions of the Diamond Industry in the Lundas* (Pretoria, South Africa: Institute for Security Studies, 25 June 2004); misappropriation of state oil revenues from Global Witness, *All the President's Men* (London: 2002), from Human Rights Watch, *Some Transparency, No Accountability* (New York: March 2004), from The Economist Intelligence Unit, *Angola: Country Profile 2003* (London: 2003), and from Mabel González Bustelo, "Angola: Una Monarquía Apoyada por las Petroleras," and David Sogge, "Otra Adquisición Petrolera de EE UU," both in Manuela Mesa and Mabel González Bustelo, eds., *Escenarios de Conflicto: Irak y el Desorden Mundial*, CIP Yearbook 2004 (Barcelona, Spain: Icaria-CIP, 2004); disarmament, demobilization, and reintegration data from *Angola's Future,* Report from the British-Angola Forum Conference, 13–14 November 2003 (London: Royal Institute for International Affairs, 2004); Cabinda from The Open Society Initiative for Southern Africa, *Angola: Human Development Opportunities and Threats: A Programme of Action* (Johannesburg, South Africa: 2003), p. 26; lack of social service provision and class, ethnic, and regional divides from author's interviews with Angolan civil society organizations in September 2003, April 2004, and July-August 2004.

43. BICC, op. cit. note 16, pp. 104–05, 107; United Nations, General Assembly and Security Council, "The Situation in Afghanistan and its Implications for International Peace and Security," New York, 12 August 2004.

44. BICC, op. cit. note 16, pp. 71–72. Table 7–6 compiled from the following: El Salvador from Larry Rohter, "In U.S. Deportation Policy, a Pandora's Box," *New York Times,* 10 August 1997, from *Dialogo Centroamericano,* January 1997 (Centro para la Paz, Arias Foundation for Peace and Human Progress, San José, Costa Rica), from Klare and Andersen, op. cit. note 4, and from Laurance, op. cit. note 36; Nicaragua from GIIS, *Survey 2003,* op. cit. note 4, p. 282; Mozambique from Batchelor, op. cit. note 36; Angola from BICC, op. cit. note 10, pp. 74–76, and from idem, op. cit. note 16, pp. 65–66; Sierra Leone from GIIS, *Survey 2003,* op. cit. note 4, pp. 307–08, from BICC, op. cit. note 10, p. 77, and from Somini Sengupta, "Warriors in West Africa Need Jobs as Well as Peace Treaties," *New York Times,* 23 May 2004; Liberia from "U.N. Liberia Mission Investigating Weapons-Smuggling Claims," *UN Wire,* 22 July 2004, from "Lack of Funding Leads to Slow Progress in Liberia's Reconstruction, UN Says," *UN News Service,* 2 July 2004, and from "UN Agencies Train Former Liberian Soldiers to Rebuild Their War-Ravaged Country," *UN News Service,* 6 August 2004; Afghanistan from BICC, op. cit. note 16, pp. 65, 112–14; Sri Lanka from ibid., p. 70.

45. BICC, op. cit. note 10, p. 83; idem, op. cit. note 16, pp. 111, 117; GIIS, *Survey 2003,* op. cit. note 4, p. 152.

46. United Nations Foundation, "Disarming Children: Quick Facts" (Washington, DC: undated); Coalition to Stop the Use of Child Soldiers, at www.child-soldiers.org.

47. UNICEF from BICC, op. cit. note 10, p. 80; Optional Protocol from Coalition to Stop the Use of Child Soldiers, op. cit. note 46.

48. GIIS, *Survey 2003,* op. cit. note 4, pp. 306–07; BICC, op. cit. note 16, p. 71.

49. GIIS, *Survey 2003,* op. cit. note 4, pp. 310–11.

Nuclear Proliferation

1. Number of weapons and of states from Carnegie Endowment for International Peace, "Nuclear Weapon Status 2004," at www.carnegie endowment.org/images/npp/nuke.jpg.

2. Larry Rohter, "Brazilian's Remark About Nuclear Weapons Creates Alarm," *International Herald Tribune*, 10 January 2003; "Japan's Opposition Leader Raps Remarks Against Non-Nuclear Policy," *Agence France Presse*, 1 June 2002; David E. Sanger and William J. Broad, "South Korea Says Secret Program Refined Uranium," *New York Times*, 3 September 2004.

3. Improvements over 20 years from George Perkovich et al., *Universal Compliance: A Strategy for Nuclear Security* (Washington, DC: Carnegie Endowment for International Peace, 2004), p. 10; "Treaty Between the United States of America and the Russian Federation on Strategic Offensive Reductions" (Moscow Treaty) at www.state.gov/t/ac/trt/18016.htm#1.

4. Paul Kerr, "Libya Vows to Dismantle WMD Program," *Arms Control Today*, January/February 2004, pp. 29–30; South Africa from Joseph Cirincione et al., *Deadly Arsenals: Tracking Weapons of Mass Destruction* (Washington, DC: Carnegie Endowment for International Peace, 2002).

5. Perkovich et al., op. cit. note 3, p. 11; for more information, see European Community, *A Secure Europe in a Better World: European Security Strategy* (Brussels: 2003).

6. These obligations and their implications are discussed in detail in Perkovich et al., op. cit. note 3.

Chemical Weapons

1. U.S. Department of State, *Saddam's Chemical Weapons Campaign: Halabja, March 16, 1988* (Washington, DC: Bureau of Public Affairs, 14 March 2003); Christine M. Gosden, *Chemical and Biological Weapons Threats to America: Are We Prepared?* Testimony before the Senate Judiciary Subcommittee on Technology, Terrorism, and Government, and the Senate Select Committee on Intelligence, Washington DC, 22 April 1998; Human Rights Watch, *The Anfal Campaign Against the Kurds* (New York: 1993); Aum Shinrikyo attack from Gilmore Commission, *First Annual Report of the Advisory Panel to the President and the Congress to Assess Domestic Response Capabilities for Terrorism Involving Weapons of Mass Destruction* (Santa Monica, CA: RAND Corporation, 1999), and from National Police Agency, *1996 Police White Paper*, translated by Emiko Amaki and Robert Mauksch, Center for Nonproliferation Studies (Monterey, CA: Monterey Institute of International Studies).

2. The United States declared some 31,500 tons of chemical weapons at nine stockpile sites; Russia, some 40,000 tons at seven stockpile sites. Exact quantities for the other four are not publicly available, but it is thought they represent a combined total of about 1,000 tons. See official statements of States Parties and the Director-General of the Organization for the Prohibition of Chemical Weapons (OPCW) at www.opcw.org.

3. In addition to the four originally declared stockpile countries, Albania declared a small chemical weapons stockpile in 2003 and Libya declared its stockpile in 2004. See OPCW, *Note by the Technical Secretariat: Status of Participation in the Chemical Weapons Convention as at 24 September 2004* (The Hague: 30 September 2004), for the latest list of member states and a fuller explanation of membership status.

4. For Russian figures, see the Russian Munitions Agency, at www.munition.gov.ru; the only destroyed weapons to date have been at the Gorny stockpile in the Saratov Oblast. For U.S. figures, see the U.S. Army Chemical Materials Agency, at www.cma.army.mil. For Libya, see three recent OPCW press releases, *Destruction of Chemical Weapons in Libya Commences on 27 February 2004*, *Libya Completes the First Phase of Chemical Weapons Destruction*, and *Libya Submits Initial Chemical Weapons Declaration* (The Hague: 26 February, 4 March, and 5 March 2004). Libya has declared 23 tons of mustard agent. See also OPCW, *The Chemical Weapons Ban: Facts and Figures* (The Hague:

8 October 2004), where limited information is available on Albania and South Korea.

5. For a discussion of community opposition, see Paul F. Walker, *The Demilitarization of Weapons of Mass Destruction in Russia: The Case of Chemical Weapons—From Architecture to Implementation*, Discussion Paper for the Forum for Chemical Weapons Destruction, Geneva, Switzerland, 26–27 June 2003, and "Russian Chemical Weapons Destruction: Successes and Challenges," in R. J. Einhorn and M. A. Flournoy, eds., *Protecting Against the Spread of Nuclear, Biological, and Chemical Weapons* (Washington, DC: Center for Strategic and International Studies, January 2003).

6. For opposition regarding the Delaware River, see, for example, Jeff Montgomery, "N.J. Lawmaker Presses Army on VX," *The News Journal*, 7 October 2004, and Lawrence Hajna, "Material is Scheduled for Disposal in Salem County," *Courier Post Online*, 7 October 2004; for a discussion of Russian issues, see Green Cross Russia, *Fifth National Dialogue Forum* (Moscow: Agency Rackurs Production, 2004).

7. See, for example, the 22 July 2004 press statement of the Chemical Weapons Working Group, *Utah Regulators Deny Public Access to Records on Nerve Agent Alarm at Weapons Burner*, at www.cwwg.org/pr_07.22.04udeqnix.html.

8. Viktor Kholstov, Deputy Chief of the Federal Agency for Industry (formerly Russian Munitions Agency), in "Russian Agency Reports Progress in Chemical Weapons Disposal," *BBC Monitoring/Interfax Military News Agency*, 11 October 2004; Alexander Kharichev, Secretary of the Russian State Committee for Chemical Disarmament, in "Russia to Double Spending on Disposal of Chemical Weapons Stocks," *BBC Monitoring/ITAR-TASS News Agency*, 8 October 2004; S. I. Kislyak, Head of the Russian Delegation to the Eighth Conference of the States Parties, OPCW, The Hague, 20–24 October 2003; "Destruction of the U.S. Chemical Weapons Stockpile—Program and Status," before the Subcommittee on Terrorism, Unconventional Threats, and Capabilities of the Armed Services Committee, U.S. House of Representatives, 1 April 2004; "Russian

and U.S. Chemical Weapons Destruction: Global Security and Nonproliferation," *Global Green USA Legacy Forum 2004* (Washington DC: 17 September 2004); G-8 Global Partnership activities and pledges at www.sgpproject.org.

Chapter 8. Building Peace Through Environmental Cooperation

1. Raul P. Saba, "The Arizona-Sonora and Ecuador-Peru Borderlands: Common Interests and Shared Goals in Diverse Settings." *Journal of the Southwest*, winter 2003, pp. 633–48; Carlos F. Ponce and Fernando Ghersi, "Cordillera del Condor (Peru-Ecuador)," prepared for workshop on Transboundary Protected Areas in the Governance Stream of the 5th World Parks Congress, Durban, South Africa, 12–13 September 2003.

2. Alexander Carius and Wiebke Roessig, *Battle Fields and Peace Parks: Nature Conservation and Conflict in the Cordillera des Condor* (Berlin: Adelphi Research, in press).

3. Alan Nicol, "The Dynamics of River Basin Cooperation: The Nile and Okavango Basins," in Anthony Turton, Peter Ashton, and Eugene Cloete, eds., *Transboundary Rivers, Sovereignty and Development: Hydropolitical Drivers in the Okavango River Basin* (Pretoria, South Africa: African Water Issues Research Unit and Green Cross International, 2003), pp. 167–86.

4. Isidro Pinherio, Gabaake Gabaake, and Piet Heyns, "Cooperation in the Okavango River Basin: The OKACOM Perspective," in Turton, Ashton, and Cloete, op. cit. note 3, pp. 105–18.

5. Ibid.

6. Anton Earle and Ariel Méndez, "An Oasis in the Desert: Navigating Peace in the Okavango River Basin," *PECS News* (Woodrow Wilson International Center for Scholars), spring 2004, pp. 1, 13–14.

7. Michael T. Klare, *Resource Wars: The New Landscape of Global Conflict* (New York: Henry Holt and Company, 2001); Thomas F. Homer-Dixon, "Environmental Scarcities and Violent

Conflict: Evidence from Cases," *International Security*, summer 1994, pp. 5–40; idem, *Environment, Scarcity, and Violence* (Princeton, NJ: Princeton University Press, 1999); Günther Baechler et al., *Kriegsursache Umweltzertörung: Ökologische Konflikte in der Dritten Welt und Wege ihrer friedlichen Bearbeitung* (Zurich: Rüegger, 1996); Günther Baechler and Kurt R. Spillmann, *Environmental Degradation as a Cause of War: Regional and Country Studies of Research Fellows*, and Günther Baechler and Kurt R. Spillmann, *Environmental Degradation as a Cause of War: Regional and Country Studies of External Experts* (Zurich: Rüegger, 1996). See also Günther Baechler, *Violence Through Environmental Discrimination: Cause, Rwanda Arena, and Conflict Model* (Dordrecht, Netherlands: Kluwer Academic Publishers, 1999), and Colin H. Kahl, "Population Growth, Environmental Degradation, and State Sponsored Violence: The Case of Kenya, 1991–1993," *International Security*, fall 1998, pp. 80–119. For an overview of the debate surrounding research on environment and conflict, see Ken Conca and Geoffrey D. Dabelko, eds., *Environmental Peacemaking* (Washington, DC, and Baltimore, MD: Woodrow Wilson Center Press and Johns Hopkins University Press, 2002). Table 8–1 from the following: Donella Meadows et al., *The Limits to Growth* (New York: Universe Books, 1972); U.S. Council on Environmental Quality and Department of State, *The Global 2000 Report to the President. Entering the Twenty-first Century,* vol. 1 (Charlottesville, VA: Blue Angel, 1981); Independent Commission on Disarmament and Security Issues, *Common Security* (New York: Simon & Schuster, 1982); World Commission on Environment and Development, *Our Common Future* (Oxford: Oxford University Press, 1987); Sverre Lodgaard et al., *Environmental Security: A Report Contributing to the Concept of Comprehensive International Security* (Oslo: PRIO/U.N. Environment Programme (UNEP), Programme on Military Activities and the Human Environment, 1989); Arthur H. Westing, ed., *Cultural Norms, War and the Environment* (Oxford: Oxford University Press, 1988); idem, ed., *Comprehensive Security for the Baltics* (London: Sage, 1989); Eduard Shevardnadze, "Ekologiya i Diplomatiya," in *Literaturnaya Gazeta*, 22 November 1989 (published in English as "Ecology and Diplo-

macy," *Environmental Policy and Law*, vol. 20, no. 1–2 (1990), pp. 20–24); Johan Jørgen Holst, "Security and the Environment: A Preliminary Exploration," *Bulletin of Peace Proposals*, vol. 20, no. 2 (1989), pp. 123–28; U.N. Development Programme, *Human Development Report 1994* (New York: Oxford University Press, 1994); Alexander Carius and Kurt M. Lietzmann, eds., *Environmental Change and Security. A European Perspective* (Berlin: Springer, 1999); World Conservation Union–IUCN, "State-of-the-Art Review on Environment, Security, and Development Cooperation," draft report for the Working Party on Environment, OECD Development Assistance Committee (Gland, Switzerland: 1999); Kurt M. Lietzmann and Gary D. Vest, eds., *Environment & Security in an International Context* (Washington, DC: U.S. Department of Defense, for NATO Committee on the Challenges of Modern Society, 1999); European Commission, "Communication from the Commission on Conflict Prevention" (Brussels: 4 November 2001); German Federal Foreign Office, "Civilian Crisis Prevention, Conflict Resolution and Post-Conflict Peace-Building Summary" (Berlin: 12 May 2004).

8. On the meaning of environmental security, see Geoffrey D. Dabelko and P. J. Simmons, "Environment and Conflict: Core Ideas and U.S. Government Initiatives," *The SAIS Review*, winter/spring 1997, pp. 127–46; Jon Barnett, *The Meaning of Environmental Security: Ecological Politics and Policy in the New Security Era* (London: Zed Books, 2001); United Nations, *Interim Report of the Secretary General on the Prevention of Armed Conflict*, Report of the Secretary-General on the Work of the Organization, 12 September 2003.

9. European Union, "EU Non-paper on the Preparation for the Review of Agenda 21 and the Programme for the Further Implementation of Agenda 21" (Brussels: unpublished, 2000); Dabelko and Simmons, op. cit. note 8.

10. Nils Petter Gleditsch, "Armed Conflict and the Environment: A Critique of the Literature," *Journal of Peace Research*, May 1998, pp. 381–400; Marc A. Levy, "Is the Environment a Security Issue?" *International Security*, fall 1995, pp.

35–62; Daniel Deudney, "The Case Against Linking Environmental Degradation and National Security," *Millennium: Journal of International Studies*, winter 1990, pp. 461–76.

11. John Deutsch, speech to the Los Angeles World Affairs Council, 1996.

12. Marian A. L. Miller, *The Third World in Global Environmental Politics* (Boulder, CO: Lynne Rienner, 1995); Somaya Saad, "For Whose Benefit? Redefining Security," in Ken Conca and Geoffrey D. Dabelko, eds., *Green Planet Blues: Environmental Politics From Stockholm to Kyoto* (Boulder, CO: Westview Press, 1998); Adil Najam, "An Environmental Negotiation Strategy for the South," *International Environmental Affairs*, summer 1995, pp. 249–87; Tanvi Nagpal and Camilla Foltz, eds., *Choosing Our Future: Visions of a Sustainable World* (Washington, DC: World Resources Institute, 1995); Thomaz Guedes da Costa, "Brazil's SIVAM: As it Monitors the Amazon, Will It Fulfill its Human Security Promise?" *Environmental Change and Security Project Report*, summer 2001, pp. 47–58.

13. Emanuel Adler, "Imagined (Security) Communities: Cognitive Regions in International Relations," *Millennium: Journal of International Studies*, June 1997, pp. 249–77; Emanuel Adler and Michael Barnett, eds., *Security Communities* (Cambridge, U.K.: Cambridge University Press, 1998); Michael N. Nagler, "What Is Peace Culture?" in Ho-Won Jeong, ed., *The New Agenda for Peace Research* (Aldershot, U.K.: Ashgate, 1999); Dejan Panovski, "Fallbeispiel Lake Ohrid (The Case of Lake Ohrid)," paper presented at the Conference Naturschutz–(Aus)Loeser von Konflikten (Nature Conservation as a Trigger for Conflict or Cooperation), Berlin, 26–27 November 2002.

14. See Ken Conca, Fengshi Wu, and Joanne Neukirchen, *Is There a Global Rivers Regime? The Principled Content of International River Agreements*, research report of the Harrison Program on the Future Global Agenda (College Park, MD: University of Maryland, September 2003).

15. For a review of literature on interdependence as a force for peace, see Ken Conca, "Environmental Cooperation and International Peace," in Paul F. Diehl and Nils Petter Gleditsch, eds., *Environmental Conflict* (Boulder, CO: Westview Press, 2000); Robin Broad and John Cavanagh, "The Death of the Washington Consensus?" *World Policy Journal*, fall 1999, pp. 79–88; John Markoff, "Globalization and the Future of Democracy," *Journal of World-Systems Research*, summer 1999, pp. 277–309.

16. UNEP, *Understanding Environment, Conflict, and Cooperation* (Nairobi: 2004).

17. UNESCO, Moscow Office, at www.unesco .ru/eng/articles/2004/Admin1415.php, UNEP, Organization for Security and Co-operation in Europe (OSCE), and U.N. Development Programme (UNDP), *Environment and Security: Transforming Risks into Cooperation. The Case of the Southern Caucasus* (Geneva, Bratislava, and Vienna: 2004).

18. Saleem H. Ali, "The K-2-Siachen Peace Park: Moving from Concept to Reality," in UNEP, op. cit. note 16, pp. 34–36; R. Prasannan, "K2 The Peak of Potential Dispute," *The Week* (Kerala, India), 12 September 2004.

19. Pinherio, Gabaake, and Heyns, op. cit. note 4.

20. Aaron T. Wolf, "Water, Conflict, and Cooperation," in Ruth S. Meinzen-Dick and Mark W. Rosegrant, eds., *Overcoming Water Scarcity and Quality Constraints*, 2020 Vision Focus 9 (Washington, DC: International Food Policy Research Institute, 2001); N. Manoharan, "India-Pakistan Composite Dialogue 2004: A Status Report," article (New Delhi: Institute of Peace and Conflict Studies, 22 September 2004).

21. Pamela Doughman, "Water Cooperation in the U.S.-Mexico Border Region," in Conca and Dabelko, op. cit. note 7, pp. 190–219.

22. Larry Swatuk, "Environmental Cooperation for Regional Peace and Security in Southern Africa," in Conca and Dabelko, op. cit. note 7, pp. 120–60.

23. Erika Weinthal, *State Making and Environmental Cooperation: Linking Domestic and International Politics in Central Asia* (Boston: The MIT Press, 2001).

24. Erica Weinthal, "The Promises and Pitfalls of Environmental Peacemaking in the Aral Sea Basin," in Conca and Dabelko, op. cit. note 7, pp. 86–119.

25. Ibid.

26. "An Environment Agenda for Security and Cooperation in South Eastern Europe and Central Asia," at www.envsec.org, viewed 23 September 2004.

27. Box 8–1 is based on UNEP, OSCE, and UNDP, op. cit. note 17. For further information and the full report on the Southern Caucasus countries, including a color map that includes further details of security concerns, see www.envsec.org.

28. On the environment and globalization, see Hilary French, *Vanishing Borders* (New York: W.W. Norton & Company, 2000); Timothy M. Shaw and Sandra J. MacLean, "The Emergence of Regional Civil Society: Contributions to a New Human Security Agenda," in Jeong, op. cit. note 13; J. Lewis Rasmussen, "Peacemaking in the Twenty-First Century: New Rules, New Roles, New Actors," in I. William Zartman and J. Lewis Rasmussen, eds., *Peacemaking in International Conflict: Methods and Techniques* (Washington, DC: United States Institute of Peace, 1997).

Environmental Impacts of War

1. U.N. Environment Programme (UNEP) and U.N. Human Settlements Programme (UNCHS), *The Kosovo Conflict, Consequences for the Environment and Human Settlements* (Nairobi: 1999); UNEP, *Post-Conflict Environmental Assessments* as follows: *FYR of Macedonia* (2000), *Albania* (2000), *Depleted Uranium in Kosovo* (2001), *Depleted Uranium in Serbia and Montenegro* (2002), *Depleted Uranium in Bosnia and Herzegovina* (2003).

2. UNEP *Assessments*, op. cit. note 1.

3. UNEP, *Afghanistan, Post-Conflict Environmental Assessment* (Nairobi: 2003).

4. Ibid.

5. UNEP, *Desk Study on the Environment in Iraq* (Nairobi: 2003); UNEP, *Environment in Iraq: UNEP Progress Report* (Nairobi: 2003).

6. UNEP, Desk Study, op. cit. note 5; UNEP, Environment in Iraq, op. cit. note 5.

7. UNEP, Desk Study, op. cit. note 5; UNEP, Environment in Iraq, op. cit. note 5.

8. UNEP, Desk Study, op. cit. note 5; UNEP, Environment in Iraq, op. cit. note 5.

9. UNEP, Desk Study, op. cit. note 5; UNEP, Environment in Iraq, op. cit. note 5.

10. Marsh discussions from internal UNEP information.

11. "Convention on the Prohibition of Military or Any Other Hostile Use of Environmental Modification Techniques," at www.unog.ch/disarm/distreat/environ.pdf; UNEP, "If There Must Be War, There Must Be Environmental Law," press release (Nairobi: 6 November 2003).

Chapter 9. Laying the Foundations for Peace

1. "Statement by the President in His Address to the Nation," Office of the Press Secretary, The White House, Washington, DC, 11 September 2001.

2. Rise of civil conflicts from Richard P. Cincotta, Robert Engelman, and Daniele Anastasion, *The Security Demographic: Population and Civil Conflict After the Cold War* (Washington, DC: Population Action International, 2003), p. 22; new security challenges from Carla Koppell with Anita Sharma, *Preventing the Next Wave of Conflict* (Washington, DC: Woodrow Wilson International Center for Scholars, 2003).

3. "International Campaign Against Terror Grows," remarks by President George W. Bush

(Washington, DC: 25 September 2001); broader hopes for international unity after 9/11 from Hilary French, "Reshaping Global Governance," in Worldwatch Institute, *State of the World 2002* (New York: W.W. Norton & Company, 2002), p. 197; U.N. creation from Stephen C. Schlesinger, *Act of Creation: The Founding of the United Nations* (Boulder, CO: Westview Press, 2003), p. 37; United Nations, "Charter of the United Nations and Statute of the International Court of Justice," New York.

4. Role of citizens' groups in U.N. creation from Dorothy B. Robins, *Experiment in Democracy: The Story of U.S. Citizen Organizations in Forging the Charter of the United Nations* (New York: The Parkside Press, 1971), and from Schlesinger, op. cit. note 3, p. 67; Lester M. Salamon, S. Wojciech Sokolowski, and Regina List, *Global Civil Society: An Overview* (Baltimore, MD: Johns Hopkins Center for Civil Society Studies, 2003).

5. United Nations, "Secretary-General's Address to the General Assembly," New York, 23 September 2003.

6. United Nations, op. cit. note 3.

7. Relative shortage of cross-border annexations from Schlesinger, op. cit. note 3, p. 287; rise of civil conflicts from Cincotta, Engelman, and Anastasion, op. cit. note 2, p. 22 ; 170 peace settlements from "The United Nations: A Snapshot of Accomplishments," fact sheet (Washington, DC: Council for a Livable World, undated); U.N. peacekeeping missions from ibid. and from U.N. Department of Public Information (UNDPI), Peacekeeping Operations, at www.un.org/Depts/dpko/dpko/bnote.htm.

8. Office of Public Information, op. cit. note 3; belief in relationship between economic issues and war from Schlesinger, op. cit. note 3, p. 27, and from Bruce Rich, *Mortgaging the Earth* (Boston: Beacon Press, 1994), p. 52.

9. Historical context from H. W. Singer, "An Historical Perspective," in Mahbub ul Haq et al., *The UN and the Bretton Woods Institutions: New*

Challenges for the Twenty-First Century (New York: St. Martin's Press, 1995), p. 17; Erskine Childers with Brian Urquhart, *Renewing the United Nations System* (Uppsala, Sweden: Dag Hammarskjöld Foundation, 1994), pp. 77–79; quote from Rich, op. cit. note 8, p. 70; Austrian Federal Chamber of Labour, "Minimum Labour Standards in the WTO," position paper (Brussels: 1 August 2003); U.N. Environment Programme (UNEP) from WWF-UK and WWF International, "5th WTO Ministerial Conference, Cancún WWF Briefing Series: Observer Status," policy brief, undated.

10. World Health Organization (WHO), "Smallpox Eradication—A Global First," "Disease Eradication or Elimination," and "Disease Prevention and Control," fact sheets (Geneva: 1998).

11. United Nations, op. cit. note 3; basic information on United Nations Population Fund, UNEP, and the Global Environment Facility from their Web sites.

12. Edward C. Luck, "Global Terrorism and the United Nations: A Challenge in Search of a Policy," prepared for the United Nations and Global Security Project, United Nations Foundation, Washington, DC, undated; United Nations, "Secretary-General Addressing General Assembly on Terrorism, Calls for 'Immediate, Far-Reaching Changes,' in UN Response to Terror," press release (New York: 1 October 2001).

13. "International Conference on Population and Development, 5–13 September 1994," at www.unfpa.org/icpd/icpd.htm.

14. United Nations, "World Leaders Adopt 'United Nations Millennium Declaration' at Conclusion of Extraordinary Three-Day Summit," press release (New York: 8 September 2000); Box 9–1 adapted from United Nations, Millennium Development Goals (MDGs), available at www.un.org/millenniumgoals; Box 9–2 from UNDPI, "The Road from Johannesburg: What Was Achieved and the Way Forward" (New York: January 2003); U.N. role in MDG implementation efforts from United Nations, MDGs, op. cit. this note.

15. Truman speech from Schlesinger, op. cit. note 3, pp. 289–94; United Nations, op. cit. note 5; United Nations Foundation, "Secretary-General's High-Level Panel," fact sheet (Washington, DC: undated).

16. Debates on veto from Schlesinger, op. cit. note 3, pp. 193–94, 284; anachronistic nature of Security Council from "The United Nations: A Winning Recipe for Reform?" *The Economist*, 24 July 2004.

17. "The United Nations," op. cit. note 16; "Four Nations Launch UN Seat Bid," *BBC News*, 22 September 2004.

18. On an enhanced role for the Security Council, see Susan E. Rice, "The Social and Economic Foundations of Peace and Security: Implications for Developed Countries," prepared for the United Nations and Global Security Project, United Nations Foundation, Washington, DC, winter 2004; other proposals from Jens Martens, *The Future of Multilateralism after Monterrey and Johannesburg*, Dialogue on Globalization Occasional Papers (Berlin: Friedrich Ebert Stiftung, 2003), from Stanley Foundation and United Nations Foundation, "Issues Before the UN's High-Level Panel—Development, Poverty, and Security," 10–11 May 2004, and from Johanna Mendelson Forman, "Don't Leave it to the Military Alone," *Open Democracy*, 24 June 2004.

19. Reform proposals from Hilary F. French, *After the Earth Summit: The Future of Environmental Governance*, Worldwatch Paper 107 (Washington, DC: Worldwatch Institute, 1992), p. 36, from United Nations Foundation and Woodrow Wilson International Center for Scholars, "Environment and Security—The Role of the United Nations," Roundtable Conference, 2 June 2004, and from Daniel C. Esty, "The Case for a Global Environmental Organization," in Peter B. Kenen, ed., *Managing the World Economy: Fifty Years After Bretton Woods* (Washington, DC: Institute for International Economics, 1994); French initiative from President Jacques Chirac, "Address to the U.N. General Assembly," New York, 25 September 2003, and from "UN Organization for the Environment," daily press briefing (Paris: Ministry

of Foreign Affairs, 23 September 2004).

20. Bank Information Center et al., "60 Years of the World Bank and the International Monetary Fund," Civil Society Strategy Meeting, Summary Report, Washington, DC, January 2004; Martens, op. cit. note 18, p. 37; Global Policy Forum, "Global Governance and the Three Sisters," at www.globalpolicy.org/socecon/bwi-wto/indsisters.htm; criticism of World Bank and International Monetary Fund from www.50years.org/institutions/index.html.

21. Aldo Caliari and Frank Schroeder, "Reform Proposals for the Governance Structures of the International Financial Institutions," A New Rules for Global Finance Briefing Paper (Washington, DC: undated), pp. 7–8; panel discussion on Threats, Challenges and Change: The United Nations in the 21st Century, Woodrow Wilson International Center for Scholars, Washington, DC, 20 September 2004.

22. French, op. cit. note 3, pp. 194–96; United Nations, "Report of the Secretary-General on the Implementation of the Report of the Panel of Eminent Persons on United Nations–Civil Society Relations," prepared for the 59th session of the U.N. General Assembly, fall 2004.

23. International Institute for Strategic Studies (IISS), *The Military Balance 2004–2005* (London: 2004); Koppell with Sharma, op. cit. note 2.

24. "World Leaders Adopt," op. cit. note 14; UNDPI, op. cit. note 14.

25. World Economic Forum, *Global Governance Initiative Annual Report 2004* (Washington, DC: 2004).

26. World Bank, *Global Monitoring Report 2004* (Washington, DC: 2004), pp. 41–45; World Economic Forum, op. cit. note 25. Table 9–1 from the following: hunger from U.N. Food and Agriculture Organization, "Proportion of Population Below Minimum Level of Dietary Energy Consumption," in *The State of Food Insecurity in the World* (Rome: annual); water from WHO and UNICEF, "Proportion of Population with Sustainable Access to an Improved Water Source, Urban and Rural," *Global Water Supply and Sanitation Assessment, 2000 Report* (Geneva: 2000). All data available in Millennium Indicators Database, U.N. Statistics Division, at unstats.un.org/unsd/mi/mi_goals.asp. Table 9–2 from U.N. Development Programme (UNDP), *Human Development Report 2004* (New York: Oxford University Press, 2004), p. 133.

27. UNDP, *Thailand's Response to HIV/AIDS: Progress and Challenges* (Bangkok, Thailand: 2004).

28. World Bank, *World Development Report 2004* (New York: Oxford University Press, 2003), pp. 30–31; Government of Mexico, "Oportunidades Cumple La Meta Del Presidente Fox; Atiende Ya A 5 Millones de Familias," press release (Mexico: 24 August 2004); 2004 GDP from World Bank, "Mexico at a Glance," *Country at a Glance Tables* (Washington, DC: 16 September 2004), and from "Mexico's Gil Diaz Expects Economy to Grow More Than 4% in 2004," *Bloomberg.com*, 14 October 2004.

29. World Bank, *World Development Report*, op. cit. note 28, pp. 44–45.

30. World Bank, "Participatory Budgeting in Brazil," project commissioned by the World Bank Poverty Reduction Group, at poverty.worldbank.org/files/14657_Partic-Budg-Brazil-web.pdf.

31. Lester Brown, "The Aral Sea: Going, Going..." *World Watch*, January/February 1991, pp. 20–27.

32. Current size of Aral Sea from UNEP, *Afghanistan: Post-Conflict Environmental Assessment* (Geneva: 2003), p. 60; fish production from Brown, op. cit. note 31; impacts from UNDP, *State of Environment of the Aral Sea Basin* (Kazakhstan: 2000).

33. Mangroves from Ryuichi Tabuchi, "The Rehabilitation of Mangroves in Southeast Asia," Proceedings from the International Symposium on Alternative Approaches to Enhancing Small-Scale Livelihoods and Natural Resources Management

in Marginal Areas: Experience in Monsoon Asia, United Nations University, Tokyo, 29–30 October 2003, and from N. J. Stevenson, "Disused Shrimp Ponds: Options for Redevelopment of Mangrove," *Coastal Management*, September 1997, pp. 423–25; National Space Research Institute of Brazil, *Monitoring of the Brazilian Amazon Forest by Satellite 2000-2001* (São José dos Campos, Brazil: 2002); "Annual 'Dead Zone' Spreads Across Gulf of Mexico," *Environmental News Network*, 4 August 2004.

34. Santosh Mehrotra and Enrique Delamonica, "Public Spending For Children: An Empirical Note," *Journal of International Development*, November 2002, pp. 1105–16; for the impact of the consumer society, see Worldwatch Institute, *State of the World 2004* (New York: W.W. Norton & Company, 2004).

35. Mahmood Hasan Khan, "Rural Poverty in Developing Countries," *Finance and Development*, December 2000; Ma Shenghong, "The Brightness and Township Electrification Program in China," presentation at the Renewables 2004 Conference, Bonn, Germany, June 2004; National Renewable Energy Laboratory, *Renewable Energy in China: Township Electrification Program* (Golden, CO: 2004).

36. "China Plans to Set Up Green GDP System in 3–5 Years," *China Daily*, 12 March 2004; field-testing from "China's 'Green GDP' Index Facing Technology Problem, Local Protectionism," *People's Daily*, 3 April 2004; 1.2 percent from "Blind Pursuit of GDP To Be Abandoned," *China Daily*, 5 March 2004.

37. Brazil from Alex Bellos, "Brazil's New Leader Shelves Warplanes to Feed Hungry," *The Guardian* (London), 4 January 2003; Costa Rica from UNDP, *Human Development Report 2003* (New York: Oxford University Press, 2003), p. 98, and from idem, op. cit. note 26, pp. 139–42; developing countries compiled from IISS, op. cit. note 23 ("developing countries" includes those categorized by the World Bank as low and lower-middle income).

38. WHO cited in Mari Pangestu and Jeffrey

Sachs, *Interim Report of the Millennium Project Task Force on Poverty and Economic Development* (New York: U.N. Millennium Project, 2004), p. 34.

39. Development Assistance Committee, "ODA Statistics for 2003 and ODA Outlook," Background Paper (Paris: Organisation for Economic Co-operation and Develoment, 14 April 2004) ("donor countries" refers to countries that are members of the Development Assistance Committee; GNI is a similar measure to GDP but adds income earned abroad by domestic entities while subtracting income earned by foreign entities in the referenced country); Johannesburg from Pangestu and Sachs, op. cit. note 38, p. 54; $50 billion from UNDP, op. cit. note 37, p. 146, and from Pangestu and Sachs, op. cit. note 38, pp. 56–57 (various sources suggest achieving the MDGs will require from $40–60 billion in aid; $50 billion represents a conservative estimate); Belgium and Ireland from UNDP, op. cit. note 37, p. 147.

40. Tied aid from UNDP, *Development Effectiveness Report 2003: Partnerships for Results* (Washington, DC: 2003), p. 41; aid for services from IBON Foundation, *The Reality of Aid 2004: Focus on Governance and Human Rights in International Cooperation* (Manila: 2004), pp. 181–94, and from Santosh Mehrotra, *The Rhetoric of International Development Targets and the Reality of Official Development Assistance* (Florence: UNICEF Innocenti Research Centre, 2001).

41. UNDP, op. cit. note 37, pp. 152–53; World Bank, *Global Development Finance* (Washington, DC: 2004).

42. Agricultural subsidies from UNDP, op. cit. note 37, p. 155; 200 million people from William R. Cline, *Trade Policy and Global Poverty* (Washington, DC: Center for Global Development and Institute for International Economics, 2004), pp. 128–31.

43. Figure 9–1 from Development Assistance Committee, op. cit. note 39, and from IISS, op. cit. note 23; Marcus Corbin and Miriam Pemberton, *A Unified Security Budget for the United States* (Washington, DC: Center for Defense Infor-

mation and Foreign Policy In Focus, 2004).

44. *Shared Responsibility: Sweden's Policy for Global Development*, Government Bill 2002/03:122, Stockholm, Sweden; John Zanchi, Ministry of Foreign Affairs Press Secretary, Government of Sweden, discussion with Erik Assadourian, 12 October 2004.

45. United Nations Statistics Division, *Progress Towards the Millennium Development Goals, 1990–2003*, unofficial working paper (New York: 23 March 2004).

46. Box 9–3 from Salamon, Sokolowski, and List, op. cit. note 4, p. 2; growth of international organizations from Helmut K. Anheier, "What Kind of Third Sector, What Kind of Society?" presentation to Citizen's Forum, Council of Europe, Strasbourg, October 2002; second superpower from Patrick E. Tyler, "Threats and Responses: News Analysis; A New Power in the Streets," *New York Times*, 17 February 2003.

47. U.S. defeat in Security Council from David Cortright, "Civil Society: The 'Other Superpower'," *Disarmament Diplomacy*, March/April 2004; Fernando Henrique Cardoso, "Civil Society and Global Governance," contextual paper, Panel of Eminent Persons on UN–Civil Society Relations, New York, undated; Jonathan Schell, "The Will of the World," *The Nation*, 10 March 2003.

48. Bill Dobbs, media coordinator, United for Peace and Justice, discussion with Gary Gardner, 29 September 2004.

49. Chris Nineham, "The European Social Forum in Florence: Lessons of Success," *ZNet*, 17 December 2002; "World Social Forum Says NO to War, Pan American Free Trade," *Agence France Presse*, 28 January 2003.

50. Mario Pianta and Federico Silva, "Parallel Summits of Global Civil Society," in Centre for the Study of Global Governance, *Global Civil Society 2003* (Oxford: Oxford University Press, 2003), pp. 387–94.

51. February 15 demonstrations tally from Phyllis Bennis, *The Roller Coaster of Relevance: The Security Council, Europe and the U.S. War in Iraq* (Washington, DC: Institute for Policy Studies, 2004); March 2004 demonstrations from "Around the World, Repudiation of Iraq Invasion," *Inter-Press Service*, 21 March 2004.

52. Attendance in 2001 from Naomi Klein, "A Fete for the End of History," *The Nation*, 19 March 2001; 2004 attendance from "March Closes Bombay Social Forum," *BBC*, 21 January 2004; Arundhati Roy, "Public Power in the Age of Empire," *The Socialist Worker*, 3 September 2004.

53. Dobbs, op. cit. note 48.

54. Signatories from International Campaign to Ban Landmines, at www.icbl.org, viewed 12 October 2004.

55. Daniel Feakes, "Global Civil Society and Biological and Chemical Weapons," in Centre for the Study of Global Governance, op. cit. note 50, p. 100.

56. Julia Scheeres, "Pics Worth a Thousand Words," *Wired*, 17 October 2003.

57. Mexico from Ann Hornaday, "A Lens on the World," *Washington Post*, 21 November 2002; Natalie Bauer, "In Their Eyes," National Underground Railroad Freedom Center, undated.

58. Table 9–3 from Thorsten Benner et al., *Shaping Globalization: The Role of Global Public Policy Networks* (Berlin and Geneva: Global Public Policy Institute, 2002), and from network Web sites.

59. Kimberley Process Certification Scheme, at www.kimberleyprocess.com, viewed 12 October 2004.

60. "Broken Vows," report on a survey of U.S. diamond retailers (London: Global Witness, 2004).

61. UNDPI, *Partnerships for Sustainable Development* (New York: 2003).

62. Norimitsu Onishi, "Nongovernmental Organizations Show Their Growing Power," *New York Times*, 22 March 2002.

63. Access to Security Council from "Security Council Relations with Civil Society," Note, Panel on Civil Society, United Nations, New York, undated; proposed reforms from Panel of Eminent Persons on UN–Civil Society Relations, *We the Peoples: Civil Society, the United Nations and Global Governance*, Final Report (New York: United Nations, 2004).

64. Andrew Kohut and Melissa Rogers, "Americans Hearing About Iraq from the Pulpit, But Religious Faith Not Defining Opinions," Survey Report (Washington, DC: Pew Research Center for the People and the Press, March 2003).

65. Gandhi from R. C. Zaehner, *Hinduism* (Oxford: Oxford University Press, 1966); civil rights from David Chappell, *A Stone of Hope: Prophetic Religion and the Death of Jim Crow* (Chapel Hill: University of North Carolina Press, 2004); baby formula from Robin Broad and John Cavanagh, *The Corporate Accountability Movement: Lessons and Opportunities*, a study for the World Wildlife Fund's Project on International Financial Flows and the Environment (Washington, DC: 1998), pp. 12, 30; Lawrence S. Wittner, *Toward Nuclear Abolition: A History of the World Nuclear Disarmament Movement, 1971 to the Present* (Stanford, CA: Stanford University Press, 2003); debt from Martin Wroe, "An Irresistible Force," *Sojourners*, May-June 2000; Pandit Madampagama Assagi Nayaka Thero, "Interreligious Co-existence," at www.wcc-coe.org.

Index

WORLDWATCH BOOKS AND ELECTRONIC PRODUCTS

State of the World 2004

Worldwatch's flagship annual is used by government officials, corporate planners, journalists, development specialists, professors, students, and concerned citizens in over 120 countries. Published in more than 20 different languages, it is one of the most widely used resources for analysis. In the 2004 edition of *State of the World*, the Institute's award-winning research team examines how we consume, why we consume, and what impact our consumption choices have on the planet and other people. With chapters on food, water, energy, the psychology of consumption, and redefining the good life, *State of the World 2004* asks whether a less-consumptive society is possible, and then argues that it is essential.

Worldwatch Papers

Worldwatch Papers are written by the same award-winning team that produces State of the World. Each 50–70 page *Paper* provides cutting-edge analysis of an environmental topic that is making—or is about to make—headlines worldwide. Selected available *Papers* appear in the Price List.

Worldwatch Book Series

In *Eat Here*, learn why eating local food is one of the most significant choices you can make for the planet and yourself. Discover why local food products are better for your health, farmers, and the environment. Find out why long-distance food can be dangerous. Get practical advice on finding homegrown pleasures in an anonymous food chain.

Vital Signs 2003

This book provides comprehensive, user-friendly information on key trends and includes tables and graphs that help readers assess the developments that are changing their lives for better or for worse. In *Vital Signs 2003*, the authors find that despite progress toward a more equitable distribution of resources and opportunities, the twin goals of protecting Earth's fragile ecosystems and improving the prospects of billions of people will not be achieved as long as humanity remains divided into the extremes of rich and poor.

Signposts 2004

More comprehensive than ever! Includes 238 datasets of global trends (100 brand new and 138 updated datasets). Each dataset is accompanied by PowerPoint slides of charts and graphs, Excel files, and HTML pages—ready for your use in classroom and boardroom presentations. Also includes full text of *State of the World 2001, 2002, 2003*, and *2004* and *Vital Signs 2001, 2002*, and *2003*, with more than 50 years of environmental, economic, and social indicators, plus an electronic timeline with links to 449 web resources. Contains a powerful, full-text search engine! *Signposts 2004* is Windows, Macintosh, Linux, and Unix compatible.

Check
www.worldwatch.org
for our newest products.

To order these and other Worldwatch publications, call us at 888-544-2303 or 570-320-2076, fax us at 570-320-2079, e-mail us at wwpub@worldwatch.org, or visit our website at www.worldwatch.org.

PRICE LIST

WORLDWATCH BOOKS AND ELECTRONIC PRODUCTS

ANNUALS: .... *Worldwide*
State of the World 2004 .... $16.95
Vital Signs 2003 .... $14.95

Multiple copy discounts:
State of the World 2004
5+ copies $13.95/ea

Vital Signs 2003
5+ copies $11.95/ea

Signposts 2004
Individual $100.00
Non-profit Institution $125.00
Corporation $225.00

Other Worldwatch Products
State of the World 2003,
Christopher Flavin et al., 2003 .... $16.95
Vital Signs 2002 .... $14.95
Eat Here: Defending Homegrown Pleasures
in a Global Supermarket, *Brian Halweil , 2004* $13.95
Vanishing Borders: Protecting the Planet in the
Age of Globalization, *Hilary French, 2000* $13.95
Pillar of Sand: Can the Irrigation Miracle Last?,
Sandra Postel, 1999 .... $13.95
Beyond Malthus: Nineteen Dimensions of the
Population Challenge, *Lester Brown et al., 1999* ... $13.00
Life Out of Bounds: Bioinvasion in a Borderless
World, *Chris Bright, 1998* $13.00
The Natural Wealth of Nations: Harnessing
the Market for the Environment,
David Malin Roodman, 1998 .... $13.00
Last Oasis: Facing Water Scarcity,
Sandra Postel, 1997 .... $10.95
Fighting for Survival: Environmental Decline,
Social Conflict, and the New Age of Insecurity,
Michael Renner, 1996 .... $11.00
Tough Choices: Facing the Challenge of Food
Scarcity, *Lester Brown, 1996* $11.00
Environmental Milestones Timeline, *2004* $10.00

WORLDWATCH PAPERS

On Climate Change, Energy, and Materials
169: Mainstreaming Renewable Energy in
the 21st Century, *2004* $5.00
160: Reading the Weathervane: Climate Policy
from Rio to Johannesburg, *2002* $5.00
157: Hydrogen Futures: Toward a Sustainable
Energy System, *2001* $5.00
151: Micropower: The Next Electrical Era, *2000* $5.00

149: Paper Cuts: Recovering the Paper
Landscape, *1999* $5.00
144: Mind Over Matter: Recasting the Role of
Materials in Our Lives, *1998* $5.00
138: Rising Sun, Gathering Winds: Policies To Stabilize
the Climate and Strengthen Economies, *1997* $5.00

On Ecological and Human Health
165: Winged Messengers: The Decline of Birds, *2003* . $5.00
153: Why Poison Ourselves? A Precautionary
Approach to Synthetic Chemicals, *2000* $5.00
148: Nature's Cornucopia: Our Stakes in Plant
Diversity, *1999* $5.00
145: Safeguarding the Health of Oceans, *1999* $5.00
142: Rocking the Boat: Conserving Fisheries
and Protecting Jobs, *1998* $5.00
141: Losing Strands in the Web of Life:
Vertebrate Declines and the Conservation of
Biological Diversity, *1998* $5.00
140: Taking a Stand: Cultivating a New
Relationship with the World's Forests, *1998* $5.00

On Economics, Institutions, and Security
168: Venture Capitalism for a Tropical Forest:
Cocoa in the Mata Atlântica, *2003* $5.00
167: Sustainable Development for the Second
World: Ukraine and the Nations in Transition, *2003* . $5.00
166: Purchasing Power: Harnessing Institutional
Procurement for People and the Planet, *2003* $5.00
164: Invoking the Spirit: Religion and Spirituality
in the Quest for a Sustainable World, *2002* $5.00
162: The Anatomy of Resource Wars, *2002* $5.00
159: Traveling Light: New Paths for International
Tourism, *2001* $5.00

On Food, Water, Population, and Urbanization
163: Home Grown: The Case for Local Food in a
Global Market, *2002* $5.00
161: Correcting Gender Myopia: Gender Equity,
Women's Welfare, and the Environment, *2002* $5.00
156: City Limits: Putting the Brakes on Sprawl, *2001* .. $5.00
154: Deep Trouble: The Hidden Threat of
Groundwater Pollution, *2000* $5.00
150: Underfed and Overfed: The Global Epidemic
of Malnutrition, *2000* $5.00
147: Reinventing Cities for People and the Planet, *1999* . $5.00
136: The Agricultural Link: How Environmental
Deterioration Could Disrupt Economic Progress, *1997* . $5.00

*Multiple copy discounts are available
on any combination of Worldwatch Papers*
Single copy $5.00
5–10 copies $4.50/ea
11–20 copies $4.00/ea
21+ copies $3.50/ea